THE ARCHITECTURE OF HOME IN CAIRO

To my late parents Gamal Abdelmonem & Samia AlMatbouly.
To my wife Gehan, and kids, Nadeen & Hisham.
To the wonderful people of Cairo.

The Architecture of Home in Cairo

Socio-Spatial Practice of the Hawari's Everyday Life

Mohamed Gamal Abdelmonem
Queen's University Belfast, UK

First published 2015 by Ashgate Publishing

2 Park Square, Milton Park, Abingdon, Oxfordshire OX14 4RN
711 Third Avenue, New York, NY 10017

Routledge is an imprint of the Taylor & Francis Group, an informa business

First issued in paperback 2017

British Library Cataloguing in Publication Data
A catalogue record for this book is available from the British Library.

Library of Congress Cataloging-in-Publication Data
Abdelmonem, M. Gamal.
  The architecture of home in Cairo / By Mohamed Gamal Abdelmonem.
    pages cm
  Includes bibliographical references and index.
   ISBN 978-1-4094-4537-1 (hardback) -- ISBN 978-1-4094-4538-8 (ebook) -- ISBN 978-1-4724-0614-9 (epub) 1. Architecture and society--Egypt--Cairo. 2. Domestic space--Egypt--Cairo. 3. Alleys--Social aspects--Egypt--Cairo. 4. Cairo (Egypt)--Buildings, structures, etc. 5. Cairo (Egypt)--Social conditions. I. Title.
  NA2543.S6A23 2014
  728.0962'16--dc23

                                                                            2014030043

ISBN 13: 978-1-4094-4537-1 (hbk)
ISBN 13: 978-1-138-56732-0 (pbk)

Contents

List of Figures

Acknowledgements

Cairo's everlasting story has always been told as a tale of two contrasting paths of history, the genuine past and the declining present. This sense of time-space division is rarely relevant in the everyday life and fabric of the city. A short walk in Cairo's downtown, the Bourgeoisie masterpiece of the European Ismaili quarter of the late nineteenth century, is enough to experience a millennia of uninterrupted living history of Cairo's urban fabric and culture through its built environment; interconnected and overlaid fragments that with time make a complex socio-cultural landscape. From the heart of Parisian Cairo's Opera Square, Abdeen's Palace, through to Garden City and ending at Al-Azhar mosque in al-Gammaliyyah quarter, the story of the city could be told by its architecture, streets, and landscape. Intriguingly enough, each visitor, tourist, or resident can develop their own story and images about Cairo. However, it is the stories of everyday Cairenes that most engage with the reality of the city's living history. These stories are rarely written.

As an architect, I was overwhelmed by one question: How could a city with such notable history of architecture have given up its wealth to borrow 'alien forms', as described by Hassan Fathy, that have neither fitted into its fabric nor into the lives of its residents? To me, there was a gap in the history of the city; the gap between the end of its homogeneous medieval fabric and the contemporary condition of disrepair. What went wrong for the city's architects and architecture over the past two centuries? How far could we get in recognising today's architecture in the city as Cairene? Yet, it became increasingly clear that there is one reading of the city that may answer my questions; the Cairene's everyday life. Regardless of the fancy image of the classical forms of precious domes, arches and thick city walls of the medieval times, I learned to understand that architecture stems its meaning from everyday reality of living and socio-spatial interactions between the walls. It was never the walls that mattered to the Cairenes; rather, the city is about little stories of micro-scale events. This project was an enlightening journey in architecture as a profession, a practice and a context for living, in which I owe a great deal of appreciation to many parties.

This book would have never been possible without the help, support, and contribution of many people. First and above all, I have to acknowledge the kind generosity of the residents of local communities and houses in Cairo, who welcomed me to their streets, houses and invited me to their events and ceremonies and into their eventful social lives. The magnificent and generous families of al-Darb al-Asfar, in particular, deserve special acknowledgement for inviting me to meals and allowing me access to their homes and families. It is in those communities where the real story of Cairo, with its glamour and suffering, generosity and poverty, is a living reality. Casual conversations, observation and attending local events and meals have revealed layers of social processes that otherwise would have gone unnoticed. Equally important were my continuous discussions and debates with colleagues and scholars during the course of my work. In particular, debates about the role of the architect in contemporary society were central to theory discussions with colleagues at University of Sheffield, Queen's University Belfast, and Cairo University and were helpful in shaping up much of the theoretical foundation of this manuscript. In particular, I need to acknowledge several discussions, and critiques with Renata Tyszczuk, Peter Blundell Jones, Tatiana Schneider, Rosie Parnell, Magda Sibely, Cristina Currelli, Jeremy Till and Ruth Morrow. I wish also to thank my publishing editor Valerie Rose, and the editorial team of Ashgate Publishing, Caroline Spender and Charlotte Edwards for their enormous help and support throughout the development and production of this book.

I owe my understanding of the context of the old city and its historical evolution to my lengthy discussions and interviews with several Egyptian intellectuals, researchers and architects. Among those, Nezar AlSayyad, Ayman Fuad Sayyid, Salah Zaki, Nelly Hannah, and Dina Shihayyeb are most notable. I have to thank the late architect and designer Asa'ad Nadim for providing drawings and personal accounts of the restoration and development of al-Darb al-Asfar buildings that he managed for a long time. Those researchers and professionals, as I witnessed, are an integral part of the fabric of the city of Cairo and its history. I wish also to acknowledge that some exerts of this text were used in a series of articles published in academic journals. Among those are the *Journal of Architecture*, *Traditional Dwellings and Settlements Review*, *Journal of Civil Engineering and Architecture*, and *Hospitality and Society*. The text benefited from various reviews and comments from anonymous peer-reviewers for those Journals.

Due to the nature and content of this book, I had to rely on the assistance of many specialists and archivists in seeking, reading and analysing archival and historical materials. I must thank the director and staff of the National Centre for Archival Documents in Cairo, who greatly facilitated my access to historical documents and maps. I am indebted to Yasmin Abd-Allah, Ola Seif, and Conchita Anorve-Tschirgi, the curators at the Rare Books and Special Collections Library of the American University in Cairo for granting me access and comprehensive guidance with great passion to original drawings and photographs of Cairo's nineteenth- and twentieth-century architecture and urban life. Without their help, several parts and materials of this volume would not have been possible. Most notably the special collections of Hassan Fathy and Ramses Wissa Wasif were instrumental to developing a comprehensive understanding of architectural practice in Cairo during the first half of the twentieth Century. Thanks are also due to the library staff and to the President of the Egyptian Society for Geography, Professor Safi El-Din Abu El-Ezz, for allowing me the access to rare historical maps of the city. An equal appreciation is a must to several librarians at the University of Sheffield, the British Library, and The Egyptian National Archives for their enormous help with tracking down complex records of historical material, archival documents and books. My active research assistant, Basma Abu-Elfadl, played a significant role in handling the female side of the Cairene hawari, a very sensitive context, by conducting interviews and engaging with the local women in their houses, to which I could not have access.

On the personal level, I have much appreciation to the relentless and timely support of my writing companion, Rachel Grantham, who has proofread my manuscripts and with time she became an expert in Cairo and in architecture and to Mohamed Othman, for his help with very complex drawings of floor plans and building sections with high accuracy. To Mohammad Selim, my father-in-law, I am truly grateful for his extensive help and support on several fronts. Through his vast contacts, I was able to get access to several archival libraries, material and to arrange interviews with several officials and intellectual figures. The help and support of my wife and colleague Gehan Selim has been integral to the development of this book, spending months in reading and commenting over drafts and engaging tirelessly in endless conversations, critiques and even walks in Cairo, a city we both admire.

Above everyone else, my late parents are the invisible and implicit power behind the production of this book. For my parents Gamal and Samia, I have embarked on this project and to them I dedicate it. I learned from them to seek knowledge from the simplest reality in life, the original ordinary. To my Kids, Nadeen and Hisham, who reluctantly accepted my absence on many trips and research work, thanks for your continuous love and unlimited support.

Notes on Copyright and Interviewee Coding System

All pictures, drawings and diagrams included are the author's work unless otherwise specified.

I am inclined to use Arabic terminology wherever possible and tend to use its popular and common dialect in everyday life.

The standard code used for interviewees in this book is [Rx.n.yy], where Rx refers to the resident code; n refers to the number of the interview; and yy refers to the year when the interview took place. Intellectuals inclusive of architects, scholars and officials have similar system, but R is replaced with I (intellectuals). It was difficult to distinguish between architects, officials and scholars as most of them are of architectural background.

For example: [R5.1.10] means Resident no. 5, first interview, in the year 2010.

Glossary of Arabic Terms

a'mma; general Public (Arabic: عامة)

Afrangi; foreign (Arabic: أفرنجي)

Al-Askar; capital of Egypt before Cairo, 750–868AD, (Arabic: العسكر)

Al-Bab al-A'ali Court; the provincial court in medieval Egypt (Arabic: الباب العالي)

al-Bayt; Home (Arabic: البيت)

Al-Darb al-Ahmar; local quarter in Old Cairo (Arabic: الدرب الأحمر)

Al-Darb al-Asfar; local community in Old Cairo (Arabic: الدرب الأصفر)

Al-Fustat; capital of Egypt before Cairo, 641–750AD (Arabic: الفسطاط)

al-Mandharah; male reception and living space (Arabic: المندرة)

al-Mansaj; medieval sewing frame (Arabic: المنسج)

al-mashrabiyya; wood-lattice window typology in medieval houses (Arabic: المشربية)

Al-Qahira; Cairo, capital of Egypt (Arabic: القاهرة)

Al-Qata'I; capital of Egypt before Cairo, 868–969AD (Arabic: القطائع)

al-Rab'; medieval form of work-live complex with focus on commerce and trade
  (Arabic: الربع)

Al-Salihyah al-Najmiyyah; court in medieval Egypt (Arabic: الصالحية النجمية)

aman; safe-safety (Arabic: أمان)

atffa; narrow and twisted alley (Arabic: عطفة)

baladi; commoner, popular of lower order (Arabic: بلدي)

bawwab; doorman (Arabic: بوّاب)

bee'a; low tast, from lower class (Arabic: بيئة)

darb; local alley (Arabic: درب)

durqa'a; space in the living quarter (Arabic: دورقاعة)

Eid al-Adhha; Al-Adhha religious feast (Arabic: عيد الأضحى)

Eid al-Fitr; Al-Fitr religious feast (Arabic: عيد الفطر)

el-harafish; local groups of street gangs (Arabic: الحرافيش)

farah; wedding ceremony (Arabic: فرح)

futuwwa; tough character in medieval city, a guardian or defender of the community
  (Arabic: فتوة)

hajj; pilgrimage (Arabic: الحج)

hammam; public bath (Arabic: حمّام)

harah (pl. hawari); local alleyway, a community (Arabic: حارة ، حواري)

harem; women's quarter (Arabic: حريم)

hidjab; veil (Arabic: حجاب)

howsh; courtyard (Arabic: حوش)

ibn al-balad; the villagers, the son of the country (Arabic: إبن البلد)

Khassa; elite or private groups (Arabic: خاصة)

Khitta; plan, quarter of land (Arabic: خطة)

kursi raha; toilet facility (Arabic: كرسي راحة)

lijan sha'biyyah; public patrols (Arabic: لجان شعبية)

majlis; living space for men guests (Arabic: مجلس)

malqaf; ventilation tower in medieval houses (Arabic: ملقف)

manzel; house (Arabic: منزل)

maskan (pl. masakin); residence (Arabic: مسكن ، مساكن)

masharee'i al-Eskan; housing projects (Arabic: مشاريع الإسكان)

mastabah; stone seat (Arabic: مصطبة)

mawa'id Rahman; charity funded meals during the month of Ramadhan (Arabic: موائد الرحمن)

mawlid; commemoration ceremony of the birth of religious figures (Arabic: مولد)

meqa'ad; area for collective male gathering in the house (Arabic: مقعد)

mubaya'a; selling contract, mainly for properties (Arabic: مبايعة)

muhtaseb; tax collector/official in medieval Egypt (Arabic: محتسب)

muqarnas; ceiling dome ornaments and structure (Arabic: مقرنس)

nokkout; money gifts during community occasions (Arabic: نقوط)

ouqda; central space in medieval alleways that acts as socia space (Arabic: عقدة)

ruwwaq; hall (Arabic: رواق)

saha; relatively large space for social gathering within urban area (Arabic: ساحة)

sakka; water Carrier (Arabic: سقا)

sha'bi; popular amongst low middle class (Arabic: شعبي)

share'i; street (Arabic: شارع)

Shari'ah; Islamic law (Arabic: شريعة)

shaykh; scholar/ religious leader (Arabic: شيخ)

sheikh mi'mar; master builder (medieval Arabic: شيخ معمار)

sheisha; smoking waterpipe (Arabic: شيشة)

sikkah; narrow alley (classical Arabic: سكة)

sobou'i; popular celebration of 7 days after birth (Arabic: سبوع)

takhtabush; outdoor sheltered sitting area in the courtyard house (Arabic: تختابوش)

tanzim; organisation or institution (Arabic: تنظيم)

tuhoor; boy's admission to adulthood (Arabic: طهور)

tumn (pl. atman); one-eighth (Arabic: ثمن، أثمان)

ulama; scholars (Arabic: علماء)

urf; tradition (Arabic: عُرف)

usta; chief worker, or skilled worker (Arabic: أسطى)

waqfiyyah; charity endowement (Arabic: وقفية)

wekalat; commercial and trade place or building (Arabic: وكالة)

zikr; religious and vocal prayers (Arabic: ذكر)

zuqaq; narrow and dead-end alley (Arabic: زقاق)

Introduction

I.1 Contemporary everyday scene of the Palace Walk in Old Cairo

The *hawari* (sing. *harah*) of Cairo are narrow non-straight alleyways that represent basic urban units that have formed the medieval city since its foundation back in 969AD, and continued to be until early in the twentieth century.[1] Covering an area of of 3.87 sq. Km. with a population of 310,500 people in 1986[2] (Figure I.2), they are defensible territories and inclusive powerful institutions of peculiar social structure and value system. The 1988 Nobel Laureate for Literature, Naguib Mahfouz, based his Trilogy of *Al-Sukkariyyah*, *Palace Walk* and *Desire Palace*, on everyday life drama of a family in three adjacent hawari in Old Cairo in the early twentieth century. In *Palace Walk*, Mahfouz displayed the socio-cultural systems of the harah of 1910, represented through the tyranny of the family's master, exclusive nature of daily activities (male and female activities), and the powerful institution of community. The contrasting image of the liberated domains of men and the imprisoned women was an intellectual protest against the hawari's culture, spatial distinction between public and private, and the oppressive social constraints. In the above scene, Mahfouz was sensitive to the idea that the public life of the hawari was not separated from the private life of imprisoned wives. Rather, he realised how the alleys' public life was engraved in the memory of women, even though it was only seen from tiny openings of harem windows. This image of the harah, however, was not entirely accurate as it reflected only the life of upper-middle-class families, which were influential but not dominant. Rather, *Palace Walk*, like other hawari, has always included members of all social levels.

> She seemed to be in a hurry as she wrapped her veil about her and headed for the door to the balcony. Opening it, she entered the closed cage formed by the wooden latticework and stood there, turning her face right and left while she peeked out through the tiny, round openings of the latticework panels that protected her from being seen from the street. The balcony overlooked the ancient building housing a cistern downstairs and a school upstairs which was situated in the middle of Palace Walk … There was nothing to attract the eye except the minarets of the ancient seminaries of Qala'un and Barquq, which loomed up like ghostly giants enjoying a night out by the light of the gleaming stars. It was a view that had grown on her over a quarter of a century.[3]
>
> Naguib Mahfouz, Palace Walk

Contemporary hawari, in contrast, are dominated by lower class population, and increasingly, by commercial and industrial activity in their tight, non-straight lanes and alleyways. The medieval urban maze, of extremely short, broken, zigzag streets with innumerable dead ends,[4] has been transforming into a modern neighbourhood, busy with commercial traffic and apartment buildings. The exclusive and predominantly residential environment of the hawari has apparently changed under the pressure of modern economical needs. The harah today becomes a site of paradox, a contest between the medievality of its domestic environment and the modernity of its busy and noisy commercial and industrial life. The spatial order of each harah (buildings and spaces) is a socially and culturally distinct entity that is informed by intrinsic but implicit mechanisms of everyday needs and activities. To understand that architecture of home, one needs to realise how such a home is structured and how it developed historically. Each harah has its distinct story, meaning and architecture, so that any disruption to this organisation could cause effects which are counter to the desired change. Living in harah in Old Cairo is living a distinctive world of interaction, social systems and architecture, of which, few people has some knowledge. Moving away from the generic typological structures, through which historians and architects prefer to percieve the hawari of Old Cairo, this book looks to get deeper insights of what makes a harah and how does it become a home, socially, spatially and culturally? Hence, we need to gain knowledge and learn from the local context in order to be able to undrstand its practices, and processes that developed its architecture and built fabric.

Contemporary architects, on the other hand, deal with Old Cairo's fabric from their intellectual perspectives with little knowledge about the effects any intervention may have on the lives of many people. In its October 2009 issue, the *Journal of Architecture Education* featured an article by Reese Campbell and Demetrios Comodromos called 'Urban Morphology + The Social Vernacular: A Speculative Skyscraper for Islamic Medieval Cairo'. The architects proposed a futuristic vision for the hawari of medieval Cairo that is largely removed from the complexity of its context or perculiarity it social-cultural processes. To propose a *skyscraper* that simply *"verticalizes the complex interrelationship of informal social networks and urban/civic form"*.[5] Their proposal was a spatial and morphological abstraction of the community in a layered pile of services and land uses, ascending in social significance to the mosque at the top.[6] The proposal was built upon a misconception image of Old Cairo, as prototypical communities with specific arrangement of services and structures. Campell and Comodromos might have benefited from more reliable and in-depth information and resources about this specific context, which could have taken their proposal in an entirely different direction. There is no doubt that sociological, cultural and economical information and knowledge will give rise to architectural forms and create more possibilities with meaningful implications. They give the architect the privilege of being involved, understanding and becoming part of the context within which he/she works.

This proposal provides a strong ground for the urgency to study the Cairene harah and to closely investigate its socio-spatial associations in everyday life. There is a need to know why the socially informed spatial organisation and architecture of Old Cairo worked for such an extended period of time, while many modern buildings struggle to survive for few decades. We need to understand why the hawari took their irregular urban form, why houses were built as small interconnected pieces (horizontally and vertically) and how spaces and activities were and still are associated and interact on a daily basis. Comprehensive accounts of Edward Lane detailed everyday life, habits and practice of the early nineteenth century, to which contemporary living patterns have obvious similarities. Such consistency of time-practice would surely imply consistency and continuity of local practices and processes of production. Contemporary architectural production in the hawari, however, is left frozen in time, lacking comprehensive strategies required to deal with inherent processes of consumption and production of space. Instead of developing their proposals within the confines of deep understanding of processes, architects align themselves with the creative reproduction of the image.

I. 2   Central
Cairo 2008

Note: Shaded area
represents zones where
the hawari are still the
dominant pattern today.

500m 1Km   2Km

What remains absent in our knowledge about the hawari today, is the spatial and social transformation of its built fabric. How do local people make use of their domestic spaces, what is the relationship between indoor and outdoor spaces from one side and the pattern of everyday life from another? In the centuries-old hawari, there is no single typology of houses or behaviours, as suggested by Stefano Bianca.[7] Rather, there is a multiplicity of typologies that developed over a long period and what we see today is a natural development of previous structures and behavioural patterns. It is insufficient to look at the hawari and their houses in the present day without investigating their historical development. Each harah has developed its own typologies of houses and behavioural patterns that evolved from the immediate need at each historical stage. Each house today, was part of or built on the foundation of another house in the past. Residents have their own historical routes of lineage, traditions, habits and social networks: all of which are at work in their everyday life. As such, homes of today are not a contemporary production, rather, they are a historical construct of people-building-behaviour association in everyday life, or what is called throughout this book, a socio-spatial practice of home.

This book, while targeting architects, planners, and urbanists especially those connected to historical contexts such as old towns, does not exclude sociologists, anthropologists and historians: who might find it useful to consolidate a cross-disciplinary approach to architecture and the use of space. Moreover, it targets those interested in understanding material culture and lifestyle of ordinary people in Middle Eastern historic cores. Amongst a vast list of titles on the modernist cities in the Arab States, inspired by the recently emerging new urban centres and their global dynamics in the Arabian Gulf, it becomes necessary to look at what makes the Arab city and its fabric unique. I contend in this book that it is not in the grand scheme of city images or mega structures where the idenity of the city is to be explored or interrogated, rather it is in the detilas of everyday life, where material culture, value systems and social dynamics come into play to shape the identity of society. No more testimony on this than the way the Arab Spring surprised the world in the year 2011, with new production of old values and social practices, as I shall argue in part of this book, that were in fact part of the local communities's inherited socio-spatial systems. By virtue of their existence, old buildings and spaces materialise social memories and histories to inform our everyday social practice and develop meaningful perception of the urban experience. I hope students in architecture, urban studies, material culture, social studies, and anthropology would be particularly interested in the methodological structure and order of documentation, spatial and qualitative analysis of buildings and spaces of old and traditional quarters.

While the built environment in Old Cairo has been investigated in different volumes, it leaned mostly on either the characterestics and features of its historical buildings, or on the socio-economic processes of the present. It has never been studied through a rational analytical approach that tackles the association between social activities and spatial organisation of space. Previous analysis of houses in medieval Cairo have been overwhelmingly grounded in the theory that every building and house was a spatial demonstration of religious morals and laws, of which Besim Hakim's *Arabic-Islamic Cities* is a good representatives.[8] This appropriated the organisation of spaces in these houses to the universal rules of Islam: which denied Cairene domestic environments the potential of being studied according to their relevance to everyday life and patterns of activities. While Nelly Hanna's social analysis of seventeenth- and eighteenth-century houses[9] and the historical accounts of Edward William Lane of domestic environments in early nineteenth-century Cairo exemplified exceptions to this line of thought,[10] Mohamed El-Sioufi's study, *A Fatimid Harah: Its Physical, Social and Economic Structure*, represented the first serious attempt to link the architecture of the hawari to their socio-economic conditions.[11] However, El-Sioufi's effort was limited to a survey of the existing situation, with limited focus on spatial organisation or development. Architectural studies have yet to provide a body of research that correlates the study of domestic spaces with actual everyday life dynamics in Old Cairo.

While Islamic morals and code of behaviour continue to be prevalent in everyday activities of Muslim Cairenes, their influence appears as implicit references rather than direct rules for architectural production.[12] Instead of approaching Old Cairo houses as examples of Islamic architecture, I study Cairene houses and hawari spaces as a product of Cairenes' everyday life. Hence, houses of Cairo are liberated from loaded religious references that used to consider architecture as a constructed spirituality (as a pure, religious quest).[13] Islamic and architectural historians, such as Robert Hillenbrand and Mohamed Arkoun, used to project the values and morals of religious spirituality and meanings on the organisation and form of spaces and places of medieval Islamic societies.[14] Even though this book does not deny the religious influence in everyday life, it does not bind itself within the religious domain as the ultimate source of architecture. Rather, it transcends the threshold between the religious and moral from one side and the social and cultural from the other. With this in mind, less reference to religious resources, covered by many other studies,[15] has been made. Instead, in adopting the rationale of everyday life as the central analytical approach, this study will make use of literature from diverse disciplines, including sociology, social anthropology and psychology, planning, urban studies, in addition to architecture.

There are two main streams of studying urban communities and their built fabric. First, the sociologists' approach of understanding communities as symbolic constructions with built-in meanings and structural self-identity, in which the built fabric, including houses, represents the idea of local identity, or what Anthony Giddens called: *locale*.[16] Its principal argument is that all social interactions are situated interactions: situated in space and time. Their routinised occurrences, constituted within different areas of time-space, collectively represent institutionalised features of social systems, through tradition, custom or habits.[17] The second is the anthropological and geographical-based approach: which rejects the notion of exclusive identity, tradition and local culture and sees communities as integrated structures in a globalised world with more interlinked cultural influences and economic interdependence.[18] Much of the sociological and community research into Cairo (e.g. Diane Singerman's *Avenues of Participation*, Evelyn Early's *Baladi Women of Cairo*; Farha Ghannam's *Remaking the Modern*, Hanna Hoodfar's *Women in Cairo's Invisible Economy*, and Heba ElKholy's *Defiance and Compliance*) followed one of these two approaches. While most of them complement Michel De Certeau's notion of *Everyday Life,* they do so without considering the spatial experiences and practices in the investigated contexts.

This book takes integrative account of both approaches. While the Cairene harah is a type of community that has developed as a historic construct over centuries, with distinct spatial settings and social relevance, it cannot be analysed without considering the forces of globalisation and relevant changes in the contemporary situation. Cutting through the extended period of two centuries brings the above approaches into their contemporary situations. During the nineteenth century, a local sense of identity was essential to the survival of the community, while forces of modernity are obviously prevalent in the contemporary harah. This study, being interdesciplinary in nature, reads information and data from different disciplines: trying to explore the interconnections and mutual influences. Deploying anthropologial, social, and architectural information to investigate social conditions is not, however, a new trend. Several studies have demonstrated the use of archaeological evidence, ornaments and domestic spaces to build social models of communities.[19] Anthropological analysis of domestic interaction and behaviour and the way spaces were used and developed has been found to be crucial to a fundamental understanding of social practices within architectural space.[20]

Such an interdisciplinary approach suggests an alternative model for studying the organisation of domestic spaces in the hawari from the radical, spatial-only analysis of house structures and physical characteristics. Hence, space organisation and architectural characteristics are analysed within the larger domains of social spheres: which are viewed as social domains of everyday activities that bring together events, rituals and social interaction within particular spatial settings and their particular moment in time. While the book is basically architectural in its structure and methodology, sociological, anthropological and geographical information is used, where relevant, to clarify the architecture of home within the hawari of the old city.

This introduction is tailored to, briefly, display principal intentions, methodology and structure. Thus, it is essential to introduce, tentatively, the principal terms and their meanings from my perspective. This part is designed to situate the contextual and theoretical stand of the book, while each term will be comprehensively discussed in the relevant places/chapters in the text. These terms are: the hawari (singular: harah), the architecture of home, everyday life, spatial practice, everyday architecture and socio-spatial practice.

By the hawari of Cairo, I refer to the group of individual harah(s) that form the urban context and patterns of Old Cairo, the plot surrounded by old city walls. While the hawari environment extends beyond the walls, those inside the walls remained distinctive for their spatial organisation and social structures. The hawari are basically residential communities with diverse professional and commercial activities. By harah, I refer to a single community in Old Cairo.[21] Even though my focus is on the residential harah, the harah could be a residential, commercial or trade/profession-based harah. Therefore, wherever the term harah is mentioned, throughout the text, it refers to a residential community unless otherwise specified. Multiples of harah are either harat, as commonly used by writers or authors, or hawari, the popular term used by ordinary Cairenes in their everyday language. As such, I preferred, for contextual adjustment, to use hawari to put the study in its relevant context of everyday langauge. The harah is usually structured of a group of attached houses, whose sizes and components vary from one to another. Historically, each extended family occupied a single house, called a bayt. Today, those houses have largely been replaced by apartment buildings, which host single families of different origins. The harah's main public space is the dead-end area, reached via non-straight bystreets, lanes and alleyways (ranging from 3–5 metres wide, with the height of adjacent buildings varying from 2–6 storeys).

Through the architecture of home, this book liberates architecture from its physical determinants, characteristics and aesthetics, while working on the process of planning: enhancing and playing with the objectives and characteristics of space to accommodate changing subjective needs. While architecture is seen to focus on the process and product of planning, designing and constructing buildings, it encompasses aspects of realising spatial and aesthetic qualities of created spaces. It is a comprehensive action and process of building.[22] However, by the architecture of home, I refer to the process and product associated with creating activity spaces that attend to the social and cultural needs of home environments. It is a discipline of architecture that considers the planning and creation of coherent domestic socio-spatial environments that transcend the physical boundaries of residential spaces and reflect the social construction and activities of their occupiers. By home, here, I don't mean the physical characteristics of the house. Rather, I mean the flexible and dynamic domestic environment of everyday life that keeps changing and developing on a daily basis.

This leads to the question of what I mean by everyday life. By everyday life, activity, architecture, I mean ordinary behavior of people in their daily routines.[23] In the simplest terms, it is what people do on daily basis. Hence, everyday activity determines style of life, routine habits of people in a certain context. Everyday could be investigated through people's daily stories and narratives that traverse and organise places in a meaningful order of activities.[24] The notion of everyday, then, connects and links spaces and places together through spatial trajectories. In this book, everyday patterns of activities, events and rituals are used to investigate the organisation of home as a set of social domains against the spatial order of spaces. That association between space and activity is referred to, throughout this manuscript, as the spatial practice of everyday life in the hawari of Old Cairo. Michel De Certeau intelligibly described the spatial practice in his study, *The Practice of Everyday Life*:

> *Narrative structures [of stories] have the status of spatial syntaxes. By means of a whole panoply of codes, ordered ways of proceeding and constraints, they regulate changes in space (or moves from one place to another) made by stories in the form of places put in linear or interlaced series from here, one goes there; this place includes another … every story is a travel story – a spatial practice.*[25]

In this context, places such as the hawari of Cairo cannot be analysed in terms of their operational suitability for serving local people's needs without investigating these people's everyday lives: which turn these dead places (determined through objects and physical characteristics) into used and active architectural venues (determined through activities by living human beings). Spaces are, as per De Certeau, places where practice occurs.[26]

This project while established a theoretical basis for the architecture of home as a discpline of practice and building socio-spatial experiences, it used Old Cairo as its primary case of investigation. My approach was built on the coherence of study with its principles. Through generalising this theory over a broader range of case studies, I could run the risk of losing the credibility of its details, and building on shallow appearances rather than deep understanding of the phenomenon. However, my argument, here, has developed on the Cairene context, but is open to be tested elsewhere, constituting a discipline of research. When Jane Jacobs argued long ago that street spaces and environment are integrated into the intimate and safe experience of the American cities in the 1960s, it was a moment to direct our thoughts to the notion of homes as fundamental urban constructs. Yet, since then, the work of sociologists and anthropologists, rather than architects, have called for a new agenda of the home that considers the spatial systems of domestic space. Despite several studies on housing designs and policies, the home remained an alien territory in architecture literature, and it was only possible for such inter-disciplinary approach to try to bridge such a gap.

Architecture literature on old urban quarters, on the other hand, tends to focus on the image and ignores ongoing active life and local processes of adaptation to the changing needs of modern life. The harah has been extensively studied on two analytical levels: as an exemplar for urban features of medieval Islamic urbanism,[27] and at the level of individual buildings in terms of documenting and studying elements and aesthetics of medieval structures.[28] Both focus on quality of production and relevant contextual and cultural issues at the time of construction and planning. The lack of understanding regarding how people live, communicate and relate to each other within their houses over long duration, leaves architecture with several dimensions and meanings that are yet to be uncovered. The contemporary harah, for example, is considered a medieval construct that is irrelevant to contemporary architectural discourse. Therefore, it is a historical object, and its mechanisms and patterns of social and spatial life were not discussed in any volume on the contemporary architecture of the city. Such a lack of knowledge hinders and in some cases misleads the contribution of contemporary architects, with potentially damaging effects on the localities of their projects. It is essential, today, for Old Cairo to re-establish the link between architecture past and present and through the understanding of its architecture of home.

Architecture should be dealt with, this book argues, as an effective socio-spatial phenomenon and practice that is informed by everyday activity and life. However, as a collective home, there is no single way to analyze the complexity and interconnectedness of the components and elements of the Cairene harah. Thus, layering its narratives to render visible its life story and inherent dynamics appears to be the most effective approach to achieve this target. Hence, it is the aim of this study to layer its narratives of architectural and social developments as a domestic environment over the past two hundred years, with an attempt to explore the in-depth social meaning and performance of spaces, both private and public. Hence, this book is constructed around a central question: *By what processes/strategies, the architecture of home, over the past two centuries, has developed in the hawari of Old Cairo an organisation of space sufficiently flexible to respond to the changing needs of everyday life?* The selected period spans from the period denoted as the end of medieval Cairo during the French occupation (1798–1801) and the national revolt of 1800 to the present day. This resembles a period of transformation, both social and spatial, of local communities and the city in general.

Undertaking an investigation that linked the present to its past, and crossed the boundaries between different areas of study, rendered it difficult to find a suitable and previously tested method that could be confidently used to achieve desired objectives. Architectural research has traditionally distinguished between historical and theoretical investigation, between research and practice, and, ultimately, between design process and people's perceptions of that design.[29] This was apparent in the structure of Linda Groat and David Wang's book, *Architectural Research Methods*, in which they listed seven principal strategies connected to architectural research.[30] Amongst these, interpretive-historical, qualitative, correlational and logical argumentation were clearly distinguishable from each other.[31] While some architects combined design with socio-cultural investigations (examples include Roderic Lawrence, and Amos Rapoport), their attempts were limited to experimental investigations, and did not constitute a consistent approach to architectural enquiry.[32]

As a method, qualitative research was selected as the principal domain that allows for the study of interrelated wholes in the form of naturally occurring data rather than as divided, discrete, predetermined variables.[33] Its combined analysis is appropriate to study human activities that utilises analysis of physical evidence and records to reach the symbolism beyond their explicit exposure as well as to draw conclusions and prove interrelationships between evidence and its providers.[34] The role of the researcher, accordingly, is to gain a holistic overview of the context under study: its logic, arrangements, explicit and implicit rules. This is essential to the investigation of the architecture of home that relies on analysing non-determined, non-measurable but expressive and meaningful interactions within spaces, while managing a process of discovery and interpretation of diverse materials and data.[35] In line with my investigation, qualitative methods are inductive and relate to the generation, not the testing, of hypotheses.[36] Within this context, Michel De Certeau's *The Practice of Everyday Life* is considered a prolific example in investigating everyday dynamics and spatial practices within a defined and specific context.

While De Certeau's work investigates immediate situations by means of direct observation and analysis, there was a need to develop his strategy to link social and anthropological information, records and literature more closely to their spatial order and architectural context throughout a journey across time in the Cairene harah. Initially, there was a need to take an early pilot investigation, using the narratives drawn from Edward William Lane, Clot Bey and AbdulRahman al-Jabarity's accounts of Cairenes' lives during the first quarter of the nineteenth century to explore the ways in which spaces of that period were used, organised and ordered.[37] These primary resources were supported by another range of accounts, such as Sophia Lane Poole's *The Englishwoman in Egypt: Letters from Cairo* and Gerard De Nerval's *Voyage en Orient* accounts. These socio-cultural and anthropologically-rich narratives were applied to the organisation of space of surviving houses in Old Cairo, in general, and in certain instances, in haret al-Darb al-Asfar, as well as archival records of another range of contemporary houses.[38] In context of these evidence and accounts, homes emerged to be of dominant existence in the hawari of Old Cairo, which remained in action today. Buildings and spaces have much justifications on social grounds than they do on spatial-sequence only. Architecture of home, I argue, was the force that led to the production of these domestic structures and buildings in association with the local alley, and not to be isolated from it, as many manuscripts on Islamic Architecture suggest.

This field work of this project was undertaken mostly between 2006–11, through several local visits, interviews and documentation of several events, venues and buildings. The analysis of Old Cairo, as context of home, extended over three historical periods, each of which lies in a different historical era with relevant resources and materials. For 1800, resources, as explained above, had to be drawn from reliable historical accounts. While Edward William Lane, Clot Bey and Gerard De Nerval's accounts are considered to express the orientalist's view of nineteenth-century Cairenes,[39] I relied in most cases on descriptive, factual information about certain occasions, events and activities within spaces. I distanced myself from any interpreted information/images or impressions displayed or influenced by the authors. In addition, those accounts were verified by other local resources/authorities such as Abdul-Rahman al-Jabarti, which covered the same period. Archival records supported certain information about space organisation and order as well as increasing our knowledge about the leading characters and senior personalities of that time. They also helped me to relate many characters and their jobs/affiliations within the social structure of community.

Around the turn of the twentieth century, similar strategy was adopted, with two additional but principal resources. The first consisted of interviews with a few senior members (six members) who had lived in the harah since the 1930s and 1940s. The second resource was novels and books written by contemporary writers who lived at the time. The *Trilogy of Naguib Mahfouz*, describing the life and culture of the hawari in 1910 and beyond, and Qasim Amin's book, *Liberating Woman* (published 1899), were of particular importance for feeding into the socio-cultural patterns of everyday life and activities of the hawari. Archival records of houses of the 1900s remained an essential source of verification of social-spatial information derived from other resources.[40]

The present day, on the other hand, was investigated by myself and a female assistant, Ms. Bassma Reda Abu ElFadl, who managed communications and interviews with local female members at their homes.[41] Domestic spaces were surveyed, while outdoor social, commercial and cultural activities were observed periodically over the span of three years (July 2006–September 2011). Field investigation included extended interviews with some residents (35 residents: 12 female, and 23 male) as well as with the younger generation and workers. I was invited to participate in local events and attended occasional parties, shared a Ramadhan breakfast meal with several residents and workers in the middle of the alleyway.[42] Building plans and façades were drafted, while only selected examples were presented in this book. The investigation of the contemporary situation utilised extensive social and anthropological research of modernity and modern practice in Cairo's popular quarters.

The analysis of the three periods are used to determine socio-spatial model and transformation of the practice of home. These were followed by thematic discussions and intellectual debate on the layered experiences and forces of homes in Old Cairo. The architecture of home, hence, emerged as a field of interactive and integrative practice in which the fluidity of the home allows human activities (public and private) to expand, shrink and flow between spaces, free of rigid physical determination of each unit and according to available capacity of spaces (indoor and outdoor).

In persuading coherent presentation of its argument, this book is divided into three main parts and 14 chapters, in addition to an Introduction and Afterwards. Every section is designed to provide a coherent structure of a particular area of the argument.

Part I: On Home and Architecture, lays the contextual and theoretical foundation by analysing the notions, concepts and methodological strategy of the architecture of home. Chapter 1: Homes: A History of a Human Practice, recognises the evolution of home as a human practice that develops according to the complexity of the social structure of the habitual unit in the pre-modern culture and addresses the fundamental principles of what homes as social units of habitation and shelter mean. This chapter follows the progress and complexity of spatial manifestation of home environments. Chapter 2: The Idea of Home between Social Reality and Cultural Production, on the other hand, interrogates the notion of home, its sociological and anthropological constructs and the way this notion develops into different spatial systems in response to human pattern of activities and needs in everyday life. It defines the home and differentiates it from the house and the household, while building an in-depth view of the process of producing homes as a set of practices and rituals in different cultures. This is reviewed through an interdisciplinary approach that bridges the gap between architecture as a discipline and studies in anthropology and sociology.

Chapter 3: Architecting Homes: A Practice in Question, offers a critique to the contemporary professional context, with the authoritative position of the architect and conventional processes of design being debated using examples of practice throughout the twentieth century. Reference is made to the studies of Amos Rapoport, Roderick Lawrence, Christopher Alexander, and Peter Hübner, reaching to the most contemporary work of the Spatial Agency. Chapter 4: Making Homes in Cairo: Constructing the Socio-spatial Architecture of Home, builds the rational analytical approach of the book, in which the particular idea and concept of home is analysed and grounded in clear methodological framework. The notion of social spheres is used to make case for architecture as a socio-spatial practice that responds to everyday dynamics.

Part II: Homes of Old Cairo over Two Centuries, is concerned with the detailed investigation of the ḥawari of Old Cairo and their specific context and architectural and social development as homes over the period of the last two centuries. This includes six chapters from Chapter 5 to Chapter 10. Chapter 5: Cairo: Structural and Contextual Consideration, introduces the reader to Old Cairo from its foundation and its urban development. It follows the evolution of the harah as part of its urban development, highlighting the city's structural development. I draw interconnections between the home, the ḥarah, and Cairo, as different levels of manifestation of the practice of home. In Chapter 6: The Cairene Harah and the Embodiment of Home in Cairo, I discuss how the harah emerged as an urban unit of home and complex political and social structure as an urban home to a group of people. It, then, interrogates the connection between the social and spatial characteristics of the Cairene harah, with specific focus on the architecture of the harah and its meaningful practices.

Chapters 7 to 10 deal with empirical investigation of the processes and everyday practices of home in the hawari of Old Cairo during three particular periods since 1800. Chapter 7: Medieval Homes of Cairo in 1800, investigates everyday homes of Cairo by 1800 with detailed investigation of individual families, social structures, spatial organisation of houses, and the way individuals and groups communicated within space. It develops a rational analysis of the houses of each class in association with their respective lifestyles and social interaction. Chapter 8: Changing City 1880s–1930s, follows with analysis of the change of space organisation in a transitional period around the turn of the twentieth century under the profound cultural reform of the time. The central argument here is the influence of the changing position of women, changing social structure and economic conditions on spatial order and practices in the old city. In Chapter 9: Contested Territories of Modernity, I investigate the influence of modernity on Old Cairo, as reflected in everyday practice in public and private spheres. This chapter gives particular attention to the experience of modernity in Old Cairo and the emergence of commerce and industry as powers in the local public sphere that contest its domestic norms on a daily basis. Chapter 10: Narratives of Spatial Transformation, hence, compiles empirical analysis and review of change and development of the socio-spatial transformation throughout these two centuries to help explain contemporary spatial practices in Old Cairo and their roots. It develops a theory of socio-spatial development that works out how architecture responds to the changing needs, culture and social structure of everyday life.

Part III: Modernity and the Architecture of Home, wraps up the theory and practice of architecture of home through thematic discussions on the realities of everyday practice in Old Cairo. This part includes Chapters 11, 12, 13 and 14. In Chapter 11: Why Architects Fail in Cairo, I bring the perspectives of residents, architects and intellectuals and officials, on what architecture of home means to them. It introduces a critique of the profession, institutional structure and system from one side, and the architects' perceptions and involvement on another. This involves analysis of formal decrees, codes, and statutory documents as well as interviews with officials, and architects. Chapter 12: Architecting Homes: From Theory to Practice, introduces a case for the architecture of home in practice through locally developed socio-spatial architecture. It reflects on case studies in which architects ground the making of homes in the everyday lives of local inhabitants.

Chapter 13: Architecture and the Construction of Memory, explores the relationship between architecture, memory and contemporary home through determining the way architecture moderates experiences and communicates narratives among generations in Cairo. By the virtue of its durability and ability to survive, architecture brings events and traditions of the past alive into the present through the spatial transformation, social practice and the value of the historical fabric. Chapter 14: Closing the Loop: Gender, Education and Sustainable Homes in Cairo, reviews the discourse of modernity and the making of homes in Old Cairo, through the lens of gender, position of women, curricula of architectural education, as well as discussing the harah community as a potential model of sustainable living in Cairo. Finally, Afterwards: Towards a Progressive Practice, summarises the findings and develops a pathway for research and practice of the architecture of home in Cairo. It furthermore, puts forward a set of principles, assets, and factors as requisites for a progressive and socially-informed practice of architecture.

NOTES

1   Ali Pasha Mubarak had listed 188 harahs (plural: hawari) in 1888AD, which had represented the majority of urban areas in Cairo at that time. See, Mubarak, *Al-Khitat al-Tawfiqiyyah*.

2   Latest available statistics for the old city were compiled during 1986 via the *Rehabilitation of Historic Cairo Project* funded by United Nations Development Programme (UNDP) published in December 1997.

3   Naguib Mahfouz describes, in *Palace Walk*, a night scene of Old Cairo in the early twentieth century through the eyes of a wife awaiting her husband's return at midnight. Mahfouz, *Palace Walk*, p. 2.

4   Jomard, as quoted in Abu-Loughd, *Cairo*, p. 65.

5   Campbell and Comodromos, *Urban Morphology*, p. 6.

6   These included a mechanical substation, burial chambers, and hospitals below ground; madrassa, green bridge, housing and mosques above.

7   Stefano Bianca, *Urban Form in the Arab World, Past and Present*, p. 12.

8   To this point, we can understand why Islamic architecture resembles the broadest and most diverse style in the history of Architecture. As such, it is inclusive of architectural production in the area between Indonesia (east) and Spain (west), and from the boundaries of Russia (north) to the south of India (south), for periods of up to 12 centuries. See the numerous volumes about Islamic architecture; for specific examples, refer to Behrens-Abuseif, *Islamic Architecture in Cairo*.

9   Hanna, *Habiter au Caire aux XVIIe et XVIIIe siécles*.

10   Lane, *The Manners and Customs of the Modern Egyptians*.

11   El-Sioufi, *A Fatimid Harah: Its Physical, Social and Economic Structure*.

12   While the houses of Old Cairo today (despite Islamic behavioural codes still being dominant), could not, by any means, be described as Islamic architecture, there is no reason to believe the situation was different at any single moment in history. The only factor is the dominant culture, everyday patterns of activities, and available technology and stylistic craftsmanship.

13   Arkoun, *Spirituality and Architecture*, p. 4.

14   Arkoun, ibid.; Hillenbrand, *Islamic Architecture*; Creswell, *The Muslim Architecture of Egypt*.

15   See for example: Hakim, *Arabic-Islamic Cities*. Refer also to Creswell, ibid.

16   Giddens, *The Constitution of Society*; Cohen, *Symbolic Construction of Community*; Jenkins, *Social Identity*.

17   Giddens, *The Constitution of Society: Outline of the Theory of Structuration*, p. 86.

18   Massey, *London Inside-out*; Olwig and Hastrup, *Siting Culture: The Shifting Anthropological Object*.

19   Gazda, *Roman Art in the Private Sphere*, p. 3.

20   Blundell Jones, *Social Construction of Space*. See also the work of Mary Douglas, *The Idea of Home*, and Victor Turner, *The Ritual Process*.

21   Later in the book we shall learn that this term is interpreted differently according to different periods and its official and its popular use. However, mainly it is perceived as referring to a distinct alleyway with its local community/residents.

22   *Oxford English Dictionary*. Online edition (last accessed on 3rd March 2010).

23   For the most comprehensive study of everyday life, refer to Michel De Certeau's *The Practice of Everyday Life* (2 volumes), especially, *Volume Two: Living and Cooking*.

24   De Certeau, *The Practice of Everyday Life*, vol. 1, p. 115.

25   De Certeau, ibid., p. 115.

26   Ibid., p. 117.

27   See, al-Sayyad, *Medieval Modernity*; Hakim, *Arabic Islamic Cities*; Akbar, *Crisis in the Built Environment*.

28   Behrens-Abuseif, *Islamic Architecture in Cairo*.

29   Rendell, *Architectural Research and Disciplinarity*, p. 145. Jane Rendell differentiated theoretical research by identifying historical, philosophical, qualitative and practice-led research; disciplines varying according to the desired outcomes.

30   Groat and Wang, *Architectural Research Methods*.

31   Ibid., p. 118.

32   Instead, their research used to be connected with non-architectural research disciplines, such as environmental psychology in the case of Rapoport, or systems design in the case of Alexander, and computer simulation and modelling in the case of Lawrence. Recent topics that question the social dimensions of architectural practice such as Alternative Praxis, Spatial Agency, and Everyday Architecture remained at their embryonic stage as theoretical debate. They are yet to be investigated in real situations and actual case studies.

33   McKinlay, *Social Work and Sustainable Development*, p. 16.

34   Richardson, *Handbook of Qualitative Research Methods*, p. 4.

35   Qualitative methods provide the research with wide flexibility to study selected issues in depth, and overcome others and handle them in a pilot manner: as they identify and attempt to understand particular information that comes out of the gathered data. See Silverman, *Interpreting Qualitative Data*, p. 26.

36   Scheurman, *Research and Evaluation in the Human Services*, p. 107; Silverman, ibid., p. 38.

37   Lane, *An Account of the Manners and Customs of the Modern Egyptians*; Clot-Bey, *Apercu general sur L'Egypte*; Al-Jabarti, *A'ajib al-A'thar*. This pilot investigation took place in 2006 in Old Cairo, where I toured relevant historical buildings and houses of 1800s period.

38   Archival records are those relating to the sale or exchange of houses (kept under the state's control) that describe the property in question. The writer of the document used to describe the surveyed property as he walked through it. The description was not always clear, but it displayed the hierarchy of spaces and their sequential order. The records do not show survey data (measurements/areas), but display rooms and names, and in many cases, the jobs and addresses of involved people and witnesses.

39   Campo, *Orientalist Representations of Muslim Domestic Space in Egypt*, pp. 35–6; see also: Raymond, *Islamic City, Arab City: Orientalist Myths and Recent Views*, pp. 3–5.

40   Abdelmonem, *The Cairene Harah in 1800: Reading Domestic Space as a Spectrum of Social Spheres*.

41   Basma Abo El-Fadl, an Egyptian architect and research associate who works in Cairo.

42   This is a part of using the alleyway for private activities and reflects the habit of having cross-family meals as well as inviting personal guests to a *meal in the street*.

PART I
On Home and Architecture

1

Homes: A History of a Human Practice

HOMES ARE NOT PRIVATE SPHERES. THEY HAVE NEVER BEEN

We have inherited this subtle notion of the home as a haven and refuge that transcends different conditions and relations of living within the house. Despite its enduring emotional and memory references, we incline to the perception of its personal or private nature. Homes exist in physical space that is socially constructed through various forms, uses and meanings.[1] Whereas houses are built to accommodate activities of private life, homes are constructed out of continuities of social-cultural, rituals and patterns of living. We seem to encounter a contradiction in the attributes of the home. How it could be private, while it associates its form, organisation and largely meaning with the commonality of a group, culture or society.[2] In architectural terms, homes have always been constructed in groups as a collective effort and inherited methods and techniques.

A social-spatial system of living, the home stood as central to cross-family social life. It is a domain of overlapped public and private life with obligation and trust beyond the family to the expansive formation of the city and society at large. In the second half of the twentieth century, there was a rediscovery of the home as a multidimensional concept[3] a multidisciplinary understanding of its formation and evolution as human practice. Indeed, homes are political arenas in which ideologies, positions and dissidence emerge, develop and gather momentum. *"The home is a major political background – for feminists, who see it in the crucible of gender domination; for liberals, who identify it with personal autonomy and a challenge to state power; for socialists, who approach it as a challenge to collective life and the ideal of a planned and egalitarian social order."*[4] They were are remain central to the society's interest, ideological reform and social movement in human history.[5] But, how has the institution and human practice of home developed to shape our society as we see it today?

Since the mid-1980s, home has been subject a comprehensive investigation from various points of view and research areas: sociologists, environmental psychologists, anthropologists, architects, and historians. However, the abundance of titles on architecture of dwellings, houses, and homes ironically did very little to interrogate the spatial systems of everyday living and as a human practice. The practice of being at home helps clarify the ambiguity behind the socio-spatial systems of domestic environments. In this part of the book, I look at the evolution of homes as socio-spatial systems of the human practice. I tend to confront the perception of homes as housing typologies that are developed out of geographic or temporal condition in isolation from the narratives of accumulating human knowledge. I argue that homes, while being informed by local traditions and social meanings, are a continuum in a long process of a human practice in progress. This chapter tries to unravel the home as a learning process, in which humans are developing forms and spatial systems to respond to fundamental needs. Through this journey, I will look at six key stages of development of the home as human practice; *"Home as a shelter"*; *"From unit to cluster"*; *"Medieval homes of production"*; *"Living downtown"*; *"Homes of modernity"*; and *"Reversal of the contemporary"*. In no case are these categories conclusive. Rather, they are elected to mark an enquiry in the narratives of living that operates beyond the stereotypes of housing forms and function.

MAKING SENSE OF THE HOME

> *Architecture is the thoughtful making of spaces. The continual renewal of architecture comes from changing concepts of space.*
>
> *Louis Kahn*

Homes have been one of the most explorative and experimental missions of humans in their negotiation with nature. It was a process of discovery not only of the capability of nature, but of oneself. Hence, most philosophers have related the process of building a shelter, dwelling or a home to the state of being in the world. For Norberg-Schults *"architectural space may be understood as a concretization of environmental schemata or images, which form a necessary part of man's general orientation or 'being in the world'."*[6] From Nietzsche to Martin Heidegger and Hannah Arendt, the act of building and inhabiting space is a process of mediation with nature and contextual condition.[7] The change of this mediation forces a change in form, spatial configuration and social processes. Man has always created spaces as a medium of understanding of the world.

In the introduction to his illustrative volume, *Dwellings*, Paul Oliver offered a critique to the shortcomings of architects' vision of the tradition of building that ranged from the contextual ignorance of early Renzo Piano work to the axioms of "form follows function" and other modernists' categorical view of the world.[8] For Oliver, dwellings are products of the past in forms that satisfied societies in earlier times, yet they continue to mediate changing ways of living: *"Dwellings may outlast lineages and in some cultures may be re-occupied or adapted, their survival to the present being a testimony to their responsiveness to changing life-ways"*.[9] Aligning the three notions of "vernacular", "dwellings", "past" is explicit in the way Oliver looks at the act of building as a tradition that contrasts somehow with the professional attitudes of the present. Dwellings are conceptually seen to emphasise the dichotomy of different worlds, the past versus the present, the vernacular versus the professional, and the peripheral rural versus the progressive urban; a typical 1980s post-modernist appreciation of traditional practices. Yet, Oliver's focus on the rationale behind building processes, materials, structure and the instinctive response to environmental challenges is one of the most comprehensive and constructive accounts to date.[10]

Aside from the dichotomy of the past and present, there will always remain something about the home that is unmeasured, but dictates judgement, perception, and inhabitation of the home. Everyone accords different attributes to the environment and space he/she calls home, which could be in the most awkward of places but appear as tangible and credible experience of living. From caves and huts, to informal apartment towers that besiege Mumbai and Cairo and informal shantytowns that in Tijuana, Jakarta or Mexico City, homes could exist in vastly contrasting forms of living and socio-spatial systems. To make sense of these homes, one must engage with the patterns of everyday living and needs of their residents that enable, according to Teddy Cruz, the creative building of the built environment.[11]

HOME AS A SHELTER

Shelter was the first experiment in transforming natural materials into construction systems for the sake of protection and accommodation of social engagement. Three early unitary shelters: the tent, igloo and cave, were simple processes of carving a volumetric space within liveable contextual settings. However, they could not operate on their own or in isolation from spatial and natural surroundings. Rather, they were part of a larger system of networks: supplies and primitive infrastructure (water, food, crops, animals … etc.) and a social and ecological system of which animals and plants were essential components. These systems of unitary shelters were more complex than what their appearance suggests. Aboriginal Australian tribes, for example, draw their social territories much wider than the space of the tents. They undertook basic daily activities including cooking and sleeping outside the tents. Tents, hence, were arranged at measurable distances to allow for a sense of privacy and freedom of social practice and communication without incursion by others.[12] Similarly, the existence of the Bedouin Arabs is partly, as Ibn Khaldoun recognised, "*the negation of building*".[13] Their life depended on the mobility of the nomads, suspicious of cities and settlements. Many Bedouins tribes in Sinai, the Grand Sahara and the Arabian Peninsula refuse to settle in houses as it negates the notion of limitless boundaries that allow them to inhabit the desert as a way of life.

The shift from the tent to the house was a step in associating the individual unit with structures of permanent settlements. Settlement suggests the permanence of living patterns, alongside water resources for farming and transport, leading homes to become spatially and physically defined. Early settlement homes encompassed a series of chambers that were used for the storage of food, protection of animals as well as accommodating spaces for inhabitants. Models of ancient Egyptian houses found in ancient tombs display the basic association and need for two parts of the house, the enclosed shelter for living and the outdoor court for cooking, livestock and domestic work (Figure 1.1). When spaces were not available, the house had to be vertically organised without private courtyards. This system was later developed in the early Arab city of Al-Fustat, which at the peak of its evolution had a dense landscape of high rise residential buildings that according to the contemporary travellers featured prominent heights above city walls, in a similar manner to those we see today in the desert Yemeni town of Shibam.[14] Houses were a series of compact apartment buildings, vertically organised for the workers occupants. Large courtyard houses were mainly built by wealthy elites and merchants.[15] Extended families resided in more combined arrangements of houses with different wings of variable privacy, forming early versions of the medieval urban landscape.

1.1   Models of Ancient Egyptian houses dating back to the first century BC

Large houses of the Ancient Egyptians and the Romans followed similar arrangements. Organised around two courtyard spaces, one was for family living and activities like cooking, eating and socialising, while the other was for stables and livestock. Both offered internal yet outdoor environments to provide ventilation and fresh air in the warm Mediterranean weather; always protected by relatively high walls and organised over two storeys, with most private spaces, like bedrooms and associated services, placed on the higher level. These homes enjoyed a complex social hierarchy within the home, especiallt with the presence of servants, whose accessibility privileges are different to other occupants.

FROM UNIT TO CLUSTER: SPATIAL SYSTEMS OF HABITAT

The emergence of the community has been a subject of debate, with Jamil Akbar wondering whether the social forces derived the physical development of settlements like alleyways, lanes or communities, or whether the opposite was the case, with physical settings leading social integration and cohesion.[16] In Jane Jacobs' notions of community, the feeling of proximity within the street is what accords the attributes of home to issues of security, intimacy and direct social interaction in the locality. The home lies in-between the houses and not in the physical characteristics of the house itself.[17] This is what transforms the static and rigid physical space into dynamic socially-cohesive and culturally intense venues. In this sense, shared spaces of local communities shaped the culture of home through inter-family interaction.[18] With the idea of clustering a series of houses to make a "community of parts" dominating the early formation of the town, we find some of the old quarters of the urban metropolis remain faithful to this system today.

Clustering takes different arrangements with overlapping of territorial boundaries between individual units making the case for a habitat in which domains of interactions are shared internally but concealed from external intrusions. This relied on a spine of spaces connected through individual branches and units, as we see in the extended family or any other cohabitation such as tribal grouping. The power of the locality and shared socio-spatial systems within tribal structure, communities, or urban neighbourhoods are one form of living at home. The value of home appeared strongly in those territories that are not owned by the individuals.

In fact, the layout of tribal compounds reflects social custom, family structure and tribal hierarchy, labour and leadership. From the tribal cluster of tents, huts, or caves to the informal communities on the peripheries of large metropolitan cities, the routes of urban settlements have largely followed the structure of homogeneous groups living in adjacent quarters. The notion of clusters was largely dependent on the group's need for shared access to resources and effective participation in the management of everyday living; from protection, security, social and economic support and shared trade, craft, or production . Those early forms of clusters followed patterns of army camps of the time, where the arrangement of the tents transformed into plots of houses, resulting in informal and organic settlements. These are found in the old quarters of cities such as Rome, Cairo, Jerusalem, Athens, and others.

HOMES OF PRODUCTION: FROM MEDIEVAL TO INDUSTRIAL MODELS

Central to the medieval town and pre-industrial urban structure were trade, economic activities and the rise of communities as units of production. Homes had to be part of communal trade and production that transcended the boundaries between living and working. This forced the boundaries between these two worlds to exist inside the house. In medieval Europe, individuals and families were members of communities whose everyday lives involved a combination of productive and domestic activities, "undifferentiated and indistinguishable".[19] Either in farming or manual crafts such as blacksmiths, houses were in direct connection with the process of making. The spatial organisation of the house mainly followed different routes in the country and the town. Peasant and country houses shared simpler forms of two main parts, human living and livestock space.[20] On the other hand, urban communities were involved in the production of food, cloth, and other products with the house having to incorporate storage for these products and be in the proximity of the market.

Work and live were not separate patterns in the construction of the productive households of medieval societies as they were combined in everyday socio-spatial exchange. The majority of domestic activities took place inside or around buildings. The dwelling was the physical form of the way to survive, and the form was decided by practical and ecological conditions.[21] Time might apparently be a simple concept, but it was always used to coordinate these two different processes of activities and engagement, each with peculiar requirements. Processes involved manual production were largely integrated in time and space within the domestic arena. This had to incorporate separate venues for raw materials, storage, processing and production that were different from those for living, socialising and sleeping. The contingent domestic environment led to a sophisticated spatial system whereby production was central to the survival of the home, and family living withdrew to smaller zones for use at night.

The invention of the factory and the development of the City as a financial institution initiated the division between "work" and 'live' patterns in the industrial age. The home is hence reduced to Le Corbusier's infamous connotation to industrial production: a "machine to live in".[22] Work has become a profession with standard and systematic techniques that cannot be performed at home. The contingent home is no longer a space for industrial production. The demand of an exact and chronological order in production and distribution changed the concept of time from circular, as in the agrarian society, to linear, as in the modern industrial process.[23] Dwellings acquired sleeping rooms, bathrooms and living rooms as the ultimate private arena. The need for structuring and classifying private spaces became important and living, as a pattern, became more complex. This new reality resulted in the division of roles. Domestic space became primarily feminine, for women to occupy and fill with domestic work, while men moved elsewhere for day work. Overlapping domains or work and live were no longer essential and soon after house building became a standardised product itself, with distinct typologies such as terraced houses with standard spans, and typical facades and roofs. However, with work leaving the domestic space, it offered new opportunities for other public activities, namely socialising and entertainment with neighbours, guests and friends. The home hence emerged to become a domain of social representation; a venue for placing oneself in the social hierarchy of the group.[24] But with this new role, home started to lose the sense of permanency of its existence, an essential part of its meaning, in search for possible upgrade of social status. Homes required social mobility and dissociation with permanent site, location, or context.

LIVING DOWNTOWN: THE PARISIAN DREAM AND THE RISE OF TENANCY SOCIETY

"The dwelling's value as a social institution that organized acquaintanceship and Certified public character emerged in contradiction to the perceived limitation and dangers of alternative housing forms. In contrast to the carefully regulated social traffic of home hospitality, boarding and tenant houses appeared social promiscuous, nonselective, and immediately vulnerable to market determinations of personal worth. Tenant housing relations, in contrast, were perceived as imposing no social accountability for moral transgressions".[25]

In *Housing and Dwelling*, Barbara Miller Lane made the case for the transition of living from locales and communities to the downtown of large cities as a shift of social and spatial significance.[26] Apartment buildings in New York, Paris and London as well as those emerging later in Cairo and Istanbul allowed families to move out of their productive communities with long inherited socio-cultural systems. The move towards prestigious downtowns with wide boulevards and apartment buildings was inspired by Europeans who had been living in apartment buildings for generations, and by the second half of the nineteenth century, Parisian apartment buildings were the fashionable model to imitate in big cities (Figure 1.2).[27] The new tenancy system was based on the idea of being independent and was accompanied by new lifestyles of mobility and connectivity in the new city quarters. No self-sufficiency, no intimate social coherence within the locality. Marketed at the time in New York as "Parisian dwellings", "Parisian Buildings" or "French Flats",[28] while late nineteenth-century Cairo was ascribed the name of "Paris on the Nile", the then European inspiration exemplifying how far the French model of living was a subscription for enlightened and elitist social class and an abandonment of the traditional and old pre-industrial community living, with freedom to move easily and frequently "both in location and in social status; anxieties of families to assert their social status in shockingly new circumstances".[29]

It was argued that this transition was a by-product of early nineteenth-century middle-class efforts in planning reform, developing household technologies, and the changing aspiration of working class people.[30] The rise of the bourgeois classes and their delicate living culture and lifestyle owe much to the rise of those European housing models associated with liberal values that flourished until the late 1920s and 1930s, when new forces of modernity became apparent. The highly picturesque and highly decorated nineteenth-century apartment buildings in Cairo, Alexandria or in Albany and Broadway in New York appeared massive in terms of size and height within their local contextual settings. Their multiple apartments per floor, occupied by tradesmen, merchants and foreign businessmen, extended to more than the size of four houses of the time. Perhaps, the wide wrought-iron balconies were the most unique element of what represented a substantial shift in the perception of domestic living in the nineteenth century. For the Cairenes, for example, such explicit outward exposure to stranger-onlookers from as deep as the heart of the private home, was totally rejected in the old quarters that relied chiefly on inner, visually protected courtyards for services and storage of supply and livestock, hence providing freedom for women to use and move through them without being seen. Moreover, to live adjacent to stranger tenants in the same building and sometimes on the same floor meant a shift in the socio-spatial boundaries of privacy from the medieval community edge and/or gates to the very door of the apartment and only a few metres away from the bedroom.

1.2   Apartment: buildings in the Boulevards of Haussmann's Paris

1.3   Tenement: apartments in Manhattan, New York

Kitchens had inner light wells and were shared among apartments, which could now accommodate bachelors, a practice totally prohibited in the old city.[31] In addition, there was no space for the traditional large team of servants in the new smaller apartments. The connection with everyday activities in the street was lost in favour of more mobility and opportunities for engagement without the locality. This led to an early transition towards modernity and disjuncture with the notion of locality and social coherence. Instead, modern society with the centrality of the self became the criterion of the new life in the newly shaped urban living.

As tenancy housing proliferated in big capital cities from the late nineteenth century until the mid-twentieth century, old quarters and neighbourhoods witnessed migrant countrymen lower class residents filling in spaces and buildings abandoned by the departing wealthy merchants and professionals. While there was generous provision of services and infrastructure to the new areas, old quarters' quality of living and support was shattered. Previous courtyard houses were divided into smaller units of one-room or two room apartments for shared and boarding accommodations for new arrivals and the old homogeneous communities were polarised between old resident groups and the new settlers. On many occasions, migrant workers had to be housed in old warehouses or former middle class apartment buildings, which were divided into smaller units, not a healthy arrangement as many of the rooms were without proper ventilation. For example, in 1900, Park Avenue and 109th Street in New York City were filled with country migrants, with many of the typical two bedroomed Dumbbell floor plans turned into units of single room accommodation per family. Smaller rooms were dark, without proper ventilation. Shared courtyards were filled with services, creating an unpleasant appearance, all of which contributed to the passing of The American Tenement House Acts of 1867 and 1879 that prohibited cellar apartments unless certain health conditions were met. These forms of desperate and informal living, however, did not vanish but moved to the peripheries of the cities.

HOMES OF MODERNITY

The underlying meaning of modernity suggests a temporal condition that is largely time-flexible, thereby situating the understanding of modern homes peculiar to their contemporary relevance. Medium density terraced houses of eighteenth-century Paris and London represented the invention of house typologies that are typically geometrical and were initially occupied by wealthy households, before becoming places for low income workers a century later. This was the fashionable living style that followed the property line of the early modern European city to form lengthy rows of built fabric that shaped the industrial era. However, with the modern movement flourishing in the 1930s, all ideas of design and forms of living were questioned by affiliates of the Bauhaus. House forms and designs of the 1930s were in stark contrast to the ideals of former Georgian and Victorian terraces. Modern houses of the 1930s and 1940s were ornament-free, geometrical in form, with internal walls removed in the search for economic, industrial and efficient space that could be used for different purposes. Developed to follow the modern movement ideal of "less is more"[32] and its minimalism approach, new modes of domestic living were driven by intellectual vision and art movements represented by architects, artists and designers forums like CIAM (*Congrès internationaux d'architecture modern*, 1928–58).

1.4   Cherif: Pasha Street with overlooking apartment houses, Alexandria at the turn of the twentieth century

The New School ideals challenged the organisation of home and its design to introduce a new social system in which home was no longer a refuge, rather it became more engaged with the outside world, more flexible and permeable. The revolutionary demonstration of modern movement principles of "architecture as a social art" appeared in their envisioned models of the "functional city" in 1933, proposing the distribution of large populations into tall apartment blocks at widely spaced intervals; a model that Le Corbusier revealed nine years later and applied in his vision of Chandigarh. Combined with economic hardship, world wars and demand for fast, industrial solutions for mass housing, Modernists' models of minimalist, efficient and prefabricated apartment blocks dominated the scene of domestic living throughout much of the twentieth century, on a global scale. Open kitchens, light and movable partitions, and transparent glass façades that eliminated the traditional in-out threshold of privacy, became characteristics of modern living and modern neighbourhoods.

More recently, practices of modernity at home have been recognised as human and cultural-centric, rather than space and image-centric. People can be modern in their culture and actions, regardless of the age or style of the building in which they live.[33] The virtual and technology-driven lifestyle is becoming the main reference for modern living that can appear in old traditional quarters to the same extent as it does in modern neighbourhoods. The use and management of domestic space to accommodate modern and technologically-led lifestyles have become symptoms of many informal settlements on the peripheries of many cities, with what appeals to occupants as modern living. Architects like Teddy Cruz were able to decode inventive practices of creative adaptability of the second-hand houses that are transported to Mexico for cheap, yet modern living options.[34] Reshaping stock of old houses was an active act of applying modern attributes to the notion of home and living.

REVERSAL OF THE CONTEMPORARY: THE MEDIEVAL THAT FOLLOWS THE MODERN

Following heavy waves of formal modernity and post-modern construction of the home, two forms of urban living have emerged to shape the practice of living in the twenty-first century, ultra-rich enclaves on one side and informal settlements, whether unregulated concrete forests of tower blocks or shantytowns of adapted houses, on the other. Both forms have enjoyed widespread popularity in large cities such as Mumbai, Cairo, Mexico City and others for respective classes in recent decades. The ultra-rich enclaves are deliberately foreign to their local context and home culture. They represent a contemporary reproduction of the classical models of the early twentieth-century Garden City, but protected by guards, gates and high fences. The stylish sub-urban compounds are sites for neo-liberal economies and global forces of capital establishments that have no interest in local forms of living or associated socio-spatial systems. Rather, they tend to replicate the departure from local constraints and social systems that replicates the earlier departure of bourgeois classes from the medieval homes in the mid-nineteenth century in search of new social networks and alliances in self-sufficient environments, with their own outlets and luxurious facilities. Despite their physical enclosure, these enclaves could not match the social coherence and strong networks of the other end of the spectrum, the informal settlements.

Unregulated informal settlements, on the other hand, helped to accommodate the majority of urban dwellers who could not afford the cost of living in planned neighbourhoods, yet needed access to job markets. Faithful to informal urban living that is medieval in its spatial system, modern in practices, homes of the lower middle class had to reinvent ways to develop convenient housing models. Late twentieth-century architects such as Peter Hübner and Teddy Cruz searched for the qualities these informal settlements offer to their inhabitants, in a quest to develop socially-engaging design methods. Their work emerged to recognise the power of sharing services, social support and local amenities as an attempt at collective revival in the metropolitan context. In the scarcity of space, there is much reliance on blurred boundaries between individual units and easy development of shared locality with human traffic in and around houses.

The stark scenes of high-rise concrete forests of apartment blocks in Mumbai and Cairo, with units as small as 60–100 square metres and with a sole façade overlooking lengthy and unserved local streets of 4–6-metre-wide, present a new form of living that is medieval in process and practices, yet modern in its instruments and facilities. These giant concrete forests emerged as a result of illegal real estate trading in small chunks of abandoned farm land. They offer marginalised classes an affordable place of living, whose attraction lies not in the quality of the house, but in the manner with which it is connected to transportation, business hubs and services, infrastructure and job opportunities in the markets of the metropolis. This form of unregulated building has one façade where narrow rooms are located. Kitchens and bathrooms are mostly ventilated by light wells. They remain the only quarters where grocers, barbers and medical services can exist at a local level.

NOTES

1   Blackmar, 'Manhattan for Rent'.

2   Abdelmonem, Understanding Everyday Homes of Urban Communities.

3   Mallett, Understanding Home.

4   Saunders and Williams, The Constitution of the Home, p. 91.

5   Ibid.

6   Norberg-Schultz, Existence, Space & Architecture.

7   Abdelmonem, *The Practice of Home in Old Cairo.*

8   Oliver, *Dwellings: The Vernacular House Worldwide.*

9   Oliver, ibid., p. 9.

10   Nepomechie, *Dwellings.*

11   Cruz, *Tijuana Case Study*, and see his TED Talk, on TED website: http://www.ted.com/talks/
teddy_cruz_how_architectural_innovations_migrate_across_borders (last accessed in June
2014).

12   Barton, *Sustainable Communities.*

13   Ibn Khaldun, *The Muqaddimah.*

14   Al Sayyad, *Virtual Cairo.*

15   See the detailed description of the courtyard houses and the excavation of Ali Bahgat in
Archaeological sites of Al-Fustat.

16   Akbar, *Crisis of the Built Environment.*

17   Jacob, *The Death and Life of Great American Cities.*

18   Abdelmonem, *Understanding Everyday Homes of Urban Communities*, ibid.

19   *Holliss, Space, Buildings and the Life Worlds of Home-based Workers.*

20   Blundell Jones, *Social Construction of Space*; and Bourdieu, *Towards a Theory of Practice.*

21   Junestrand and Tollmar, *The Dwelling as a Place for Work.*

22   Holliss, ibid.

23   Kern, *The Culture of Time and Space.*

24   Bourdieu, *Distinction.*

25   Blackmar, *Manhattan for Rent.*

26   Lane, *Housing and Dwelling.*

27   Cromley, *Alone Together.*

28   Cromley, ibid., as quoted in Lane, B.M. (ed.), 2007, *Housing and Dwelling*, p. 106.

29   Ibid., p. 107.

30   Blackmar, *The Social Meanings of Housing*, p. 108.

31   Edward William Lane stated that as a single man he was denied any accommodation in a
local Cairene harah during the first half of the nineteenth century.

32   The infamous quote of the pioneer modern architect, Mies Van Der Rohe.

33   Ghannam, *Remaking the Modern*, and Abdelmonem, *The Practice of Home in Old Cairo.*

34   Cruz, *Tijuana Case Study.*

2

The Idea of Home between Social Reality and Cultural Production

THE IDEA OF HOME: IN SEARCH OF MEANING

The real meaning of beauty, the idea of houses as places which express one's life, directly and simply, the connection between vitality of the people and the shape of their houses, the connection between the force of social movements and the beauty and vigour of the places where people live – this all forgotten, vaguely remembered as the elements of some imaginary golden age.[1]

The idea of home is by no means a new or innovative field of investigation. The feeling of home has been associated with the personal perception of safety, security, comfort and passion, which have been studied in different fields. The home has mostly been approached as a subjective and cognitive idea rather than a spatial phenomenon that could be objectively studied apart from the perceived situation. For psychologists, it is a comfortable environment that is associated with intimate and relaxing behaviour. It is free from physical parameters. For sociologists, it is determined in the size and type of interaction it promotes, between people and people, people and things, things and things. Martin Heidegger, the influential German philosopher, sees the home and dwelling as a refuge from the *"object-character of technological domination"*. It *"codifies the sense of longing for shelter amidst the destructive and self-destroying properties of capitalism"*.[2] For contemporary architects, in contrast, the design of the home responds to basic physical needs and economic functions and, more recently, to concern over the environment.

The home has been recognised as an interwoven union of physical space and moral ideas of mind and soul.[3] It has its connotations to peace, faith and purity, in many religions and faiths. In Islam, the mosque is God's home on earth, and good Muslims are those who keep their homes as pure and peaceful as mosques. Moving away from the universal rules, a home should be relevant and unique to its occupiers, especially families. Christopher Alexander recognises two key issues that mark a successful home; these are the uniqueness of every family and every person, who must be able to express such uniqueness, and that home should connect them with other people and the society at large. In that respect, Alexander sees the failure of modern housing projects in expressing isolation, lack of friendship and failure to create human bonds in which people feel themselves part of fabric which connects them to their fellow men.[4]

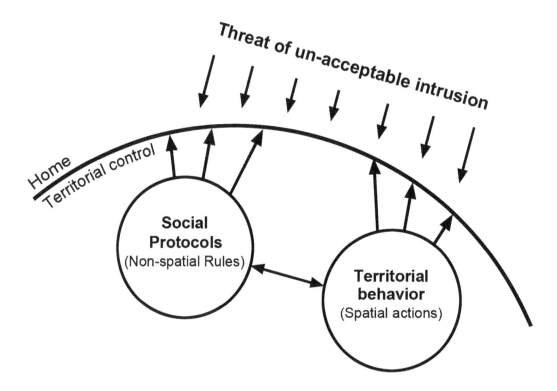

2.1   Model
of territorial
behaviour at home
After Sidney N. Bower
(1980: 183).

Home emerges as a stronghold territory, in which certain security measures, control and defence are continuously at work. The number of people who can simply walk into someone's house uninvited is limited strictly to close family: if strangers try to enter it is a great affront. Home is an intermediate territory that lies between the personal spaces, in which people maintain certain distance from others, and between public spaces, in which individual requirements are superseded by the superior shared interest of society. Homes fit the definition of human territoriality as "*the relationship between an individual/group of people and particular physical settings that is characterized by a feeling of possessiveness, attempts to control the appearance and the use of space*".[5] Homes, thus, could be realised through a network of spaces (indoor and outdoor) that a person uses routinely as a haven with high degree of comfort and intimacy. They envisage specific codes of human behaviour learned through cultural training and that require adjustment every time people cross their boundaries. Attempts to interact with people at their home territories without following the settled code of action, such as knocking on the door, is considered an unjustifiable intrusion.

Spatial configurations of homes are not limited to the physical boundaries; rather, people construct their home domain beyond what appear to be physical boundaries. The Immediate front of a house is part of home territory. Visual and audio access are essential intrusions that require responses (curtains or barriers; closed windows). Nomadic tribal homes and those of the Australian aborigines, as discussed earlier, are not exceptions. Their home territories are defined by certain configurations and spatial arrangement. Aboriginal people treat their outdoor areas as an integral and active part of their homes' living spaces.[6] They need space to sleep, cook and rest outside, without being close to neighbours doing the same, which defines the home territorial boundaries and control as per local social and cultural codes (Figure 2.1).

The power of home lies in on-going experiences and practices, in everyday rituals, in social interactions and conflicts, in material culture and the way they are implanted in certain contexts to retain the personal and cultural identity. This chapter investigates the home as an idea transformed from the individual's experience and cognition into a meaningful production of space. It looks at the production of home as a set of practices and rituals that shape its spatial organisation.

THE HOME, THE HOUSE, THE HOUSEHOLD: BETWEEN TERMINOLOGY AND ETYMOLOGY

The anthropological habit of investigating the meaning of things and practices in primitive societies is apparently due to the fact that contemporary societies are overwhelmingly dominated by political institutions that organise behavioural moods and responses, which hinders any intelligible understanding of complex phenomena such as home. The overwhelming political and economic structures in today's urban settings impose certain formality of human practice on people's daily lives. Home as an idea and practice, hence, seems diminished in today's architectural discourse, while the house and household have become more relevant. Today, the home is associated with such questions as where do you live, rather than how do you live?[7] Economical and political issues seem powerful enough to dominate architecture rather than to question the complex nature of domestic environments. So we need first to shed some light on the meaning of home as an everyday practice and investigate it in both western as well as Arabic terminology and literature. For better clarity, the term was compared with other dominant terms such as house and household.

The Home

Home is an ambiguous term that retains different meaning within different contexts. Its inclusive meaning reflects the physical parameters of a residential space (house, dwelling), place (neighbourhood, town); environment (domestic life), as well as the social determination of a particular group (community). It could, depending on the context under discussion, represent political institutions such as the *Home Office*, or make passionate reference to a country, *home country*.

The reference to home was developed from *Oikos*, which represented the basic unit of the society in the Ancient Greek culture, into the Roman *domus* with its two basic structural divisions: private/public (gender) and grand/humble (quality). Its meaning changed slowly from social[8] to a spatial determination of habitable environment, capturing the spectrum of social distinction between the "grand-public" such as the atrium, and the "humble-private" such as slaves' quarters.[9] While modern scholars define the home as a private sphere dominated by females as opposed to the public male-dominated polis, Lisa Nevett argued that such gender-based classification is entirely a modernist view of the household, which ignored the evidence that men occupied a particular zone in the home as well.[10]

The contemporary term *home* appeared during the tenth century to be very broad, ranging from the notion of a village or a collection of dwellings to the intimacy of a single household. *Home* is defined by *Oxford English Dictionary* as:

> *A dwelling-place, house, abode; the fixed residence of a family or household; the seat of domestic life and interests; the dwelling in which one habitually lives, or which one regards as one's proper abode. Sometimes including the members of a family collectively; the home-circle or household.*[11]

The intersection of several terms with distinct meanings causes this ambiguity. Dwelling, shelter, house, home, family, and residence share similar references but different meanings about domestic environments, individually or collectively. However, the contemporary literature of domestic architecture alienates itself by use of three basic terminologies: home, house and household. This emphasis shows that home is used for the analysis of the environment, house for the characteristics of physical space and its economic aspects of production. The household refers to the economic unit that occupies the house (not essentially the family), its affordability and ecology.[12]

In comparison with home, to house is to encase and to provide a physical enclosure: "To dwell or take shelter in a house; To cover or protect with a roof; To place in a secure or unexposed position; to give shelter to; to provide with a house to dwell in".[13] The household refers to occupants of this physical space (house), whether humans, goods, objects and furniture. They are 'the "holding" or maintaining of a house or family; housekeeping; domestic economy; the contents or appurtenances of a house collectively'.[14] Neither term considers the social, emotional or cultural situations, or the patterns of activities and daily rituals. The interrelationship between the house and the household, the mechanism of daily practices, power relations and material cultures seem to be recognised only in the more inclusive term home.

In the above definition of home, emphasis on the experience is obvious, as in "seat of *domestic life* and *interests*", "one's *own house*", the dwelling in which "*one habitually lives*", or which one regards as "one's *proper abode*". The underlined terms suggest the subjectivity of the experience and such categorisation raises more questions than it answers. For example: what constitutes a domestic life, what are the qualities of the house which make it my own, habitual, or proper? What are the limits of the home? And how can we define its boundaries? The answers to these questions lie in the contextual settings and the perceptions and codes of local cultures. To place this discussion into context, analysis of Arabic terminologies of home and house is a must. However, at this stage it is necessary to settle on an initial definition of the home. Home, as I recognise it, allowing for contextual adjustment, is the spatial settings that accommodate the personal and family socio-cultural and psychological needs. This includes indoor and outdoor spaces, private and public environment, individual and collective ownership. Its boundaries are mentally constructed based on individual needs and context. A local alley, a harah, could be a home to the same extent as a single room for a bachelor or a studio for an artist.

Home, in Arabic, is *bayt*, while house is *maskan* (traditionally *dar*), and residence is *manzel*. The bayt's proper meaning is a covered shelter where one may spend the night.[15] Etymologically it is derived from the verb *bata* (بات), meaning to spend the night.[16] Later, bayt became the place/venue where one would spend the night, without certain spatial determinants. Another old term is dar, from *dara* (to surround), which is a space surrounded by walls, buildings, or nomadic tents placed approximately in a circle.[17] The latter term was a meaningful description of the historically dominant courtyard houses in the Middle East, including Cairo.[18] This model represents several socio-cultural and religious metaphors. For example, it covers privacy (visual privacy), interdependence of the occupiers/neighbours and emphasis on batin (hidden) versus zahir (exposed) in neglected and simplified terms There is no factual or evident connection between the term dar and its manifestations in the form of houses. However, one can see the relevance of both the tent of Arabian tribes and the courtyard houses as spaces surrounded by a physical skin; textile/fabric in the former and a series of rooms for the latter (Figures 2.2 and 2.3). Interestingly, this connotation could reflect on houses before the Arab legacy in Egypt. Ancient houses in Egypt, before the arrival of the Arabs, were not so different from that space-skin relationship. See the models of pre-Islamic houses in Egypt, displayed in Chapter 1.

Al-bayt, on the other hand, was the historical as well as the popular term to identify the residence of a particular family. *Bayt al-Qadi* (house of the judge), for example, was the house of family of the previous principal (first) judge of Cairo in the first half of the twentieth century. Its name (bayt al-Qadi) was and still is the identification mark for the family, their descendants, and sometimes the whole harah. Other examples include Bayt al-Razzaz, Bayt Mustafa Ja'afar, Bayt Sit Waseela in Old Cairo, all of which were identified as the places of well-known families, who lived in them for an extended period of time. Families, names and buildings have become inseparable from each other. More importantly, they provide evidences on families' habits, morals and socio-cultural patterns within homes that were representatives of their contemporary periods. While Bayt al-Razzaz was a representative of the Mamluk's liberal culture that was more tolerant of male-female interactions, Bayt Mustafa Ja'afar represented the conservative ideology of the Ottomans, with harsh isolation between male and female activities. *Al-bayt*, hence, is a term associated with the people who live in it (holding the family name), essentially a family, and their culture.

Maskan, in contrast, had become the physical place and space identified by the address and spatial characteristics. The meaning of Maskan as a physical space could be grounded in its original meaning from the Arabic verb *sakan* (سكن), which is explained by Ibn Manzur[19] as static and not in motion. The physical object/space, hence, seems to be the justification for the housing unit itself. It is a relatively new term that is the equivalent socio-economical unit of the modern house in western literature. It is the planning and standard unit in modern housing projects (Masharee'l al-Eskan) in Egypt. It appeared as a common term during the twentieth century along with the modern movement and its mass production of houses. Shared accommodation, in Cairo, on the other hand, does not constitute a home, nor is it considered a house. It is a temporary setting where people attend to their needs of sleeping, studying or eating.

Different analytical approaches tend to deal with the house as a unit of accommodation with a number of objective physical attributes, such as size, amenities, state of repair and form.[20] Some reduce the house to its materialistic nature, that is, 'the bricks and mortar or other building materials that comprise the constructions within which people live'.[21] The house is used as a unit of planning and target of the governmental policies (by setting design standards, requirements, and structures). As a physical product, it is determined; it could be counted, improved and measured in terms of housing stock and future requirements according to population growth.

2.2 Arabian tribal tents: tribal dar in Arabia in the pre-settlement eras

It could be flexible, rigid or adaptable.[22] In this sense, the house is similar to any other physical and technical object or product: the car, the plane, the computer, etc. Housing in the most extreme view is becoming a consumable product, bought, used and disposed of based on the changing needs and conditions of its occupants.[23] As such, it is a kind of service, whose lifetime lasts only as long as its usability.[24] According to the well-known view of Le Corbusier, the house/dwelling is "a machine for living".[25]

The terms *house* and *housing* limit the domestic experience and associated practices to a set of standard dimensions, physical and spatial arrangement represented easily in an abstracted two-dimensional image, whatever the plans, elevations, sections or perspectives. At the basic level, "*housing provides the space which frames and often defines many activities that constitute primary relationships in the* home".[26] This led to dissatisfaction with the focus on housing research and studies began to call for the home to come to the fore, as Peter Saunders and Peter Williams[27] suggested, as the principal territory of investigation, when the target is the quality of life and semantic development of the domestic environment.

In contrast with the house, as mentioned above, the home represents the flexibility of experiences that change with time and socio-cultural and economic development. The same apartment plan, which looks in Cairo the same as it looks in London, will be experienced differently in either cultural context. While the home is our concern here, we should focus on the household and the social processes that are associated with it. The house represents the consumption, while the home represents the provision.[28]

The home starts from its locality. It is here or there. "*It is a localizable idea. Home is located in space but it is not necessarily a fixed space and does not need bricks and mortar. It can be a wagon, a caravan, a boat, or a tent*".[29] On the other hand, Karsten Harries[30] argued that in the modern world, in the light of transportation and communication development, there is no such place that can be called home. When everything becomes close to the man, then there is no *there* that contrasts with *here*. The concept of distance vanishes and so does the idea of home. Man feels at home on earth, which is brought to his home through the TV, or he takes it with him through material objects.[31]

Saunders and Williams have their own take on the notion of home, adopting Giddens's concept of the "locale". They see home as "a crucial 'locale' in the sense that it is the setting through which basic forms of social relations and social institutions are constituted and reproduced". They approach the home as a domain within which "gender and age relations are centrally structured" and "class differentiation, ethnic inequality, the status order and even distinctive regional and national cultures and identities are reproduced".[32] The home is a dynamic arena that responds to the contextual socio-cultural changes. For Martin Heidegger, the home, as a dwelling, "is the capacity to achieve a spiritual union between humans and things".[33] It is not bound by particular functions. The emergence of the home as workplace in the twenty-first century is one of the most recent examples. Susanne Tietze and Gill Musson[34] discuss the experiences of home-workers and their families when distinct and traditionally separate functions come to meet within the confines of the home. Home then becomes the place where the process of industrial production meets with the process of household production, which encompasses conflicting and competing demands and values (ibid.), and where the traditional spatial and temporal boundaries between home and work are contested.[35]

In this book, I confine myself to the interpretation of the home as a set of experiences that are represented through the spatial and temporal organisation of domestic environment. The home represents an institution of collective social processes associated with domestic environment. In this sense, I am aligned with the interpretation of Saunders and Williams of the constitution of the home. Therefore, by home I mean *the socio-cultural settings that elaborate a set of practices during which the actors (the household) work and interact within the spatial configuration (the house) to provide a certain living context*. Under this understanding the home is a *process*, the house is a *product* and the household is the *producer*.

Through the process (home), rituals turn everyday practice into spatial requirements. The home, thus, becomes a dynamic arena that takes its form and shape from actual practices taking place within its confines. It is not bound by physical boundaries of the house, rather it creates its own boundaries that might or might not match with those of the house. In such a context, the players, the households, and their own socio-cultural norms, play an essential part in forming the spatial requirements.[36] We look at home, architecturally, as a domestic environment that is a space of social and "*cultural inscription structured by the collective and symbolic organization of its residents*", and whose physical layout provides the metaphor of its socio-cultural context.[37]

WHAT CONSTITUTES A HOME?

The constitution of home stems from two basic forms; the social organisation of the home, the way it is structured, and socio-cultural diversity of the contextual model. The home is ordered by customs, habits and classification of the home's inhabitant: the *Homer*.[38] The categorisation of rooms and furniture usually conforms to a consistent set of rules in a specific society, representing the code of socio-cultural norms and traditions. The dwelling in that sense is a *"warehouse of material culture and social customs handed down from previous generations"*.[39] The idea of home appears clearer within subcultures and local communities than hybrid urban housing communities determined by class, income group or sector. Traditional quarters of metropolitan cities emerge as interesting case-studies to that model of thought, where houses were planned in collective settings and grew in congested settings that affiliated the house with the neighbourhood to become a home. For example, traditional organisation of the home in a city like Cairo complements domestic tasks and gives them much attention and space (such as the kitchen) until the present day, while western conception of space tends to promote the quality of the collective image of the house as scenic spaces producing a blurred sequence of privacy, giving less attention to individual spaces and internal boundaries, as we see in modern open-plan houses.[40]

In his comparative analysis of the British and Australian domestic typologies and space organisation, Roderick Lawrence[41] realised the importance of everyday routine in shaping the image and spaces of the house relevant to its specific culture. The regularity of everyday activities, often referred to as rituals, defines the essential meaning of the home to its occupants.[42] The practice of sleeping is defined by the system of activities which are the procedures and the situation of sleeping, socially and spatially. The space needs to promote a relevant sense of activity and includes the essential material objects (comfortable bed, dark walls, heavy curtains, indirect lighting) which are culturally-specific and associated with the act (of sleeping, for example). This perspective considers the significance of household structure and routine. How many members of the household, how and where do they eat, and sleep? The decline of cooking and sharing meals among some working-class British families, for example, reduced the importance of the kitchen as a space for social gatherings, giving way to other activities.[43] According to that perspective, architecture is a product of the home, and not about producing home as a physical instrument.

But, it is perhaps naïve to differentiate between the student of culture and the study of the home. The study of cultural production of homes has led to extensive debate; many researchers have found rituals to be heavily loaded with sacred meaning and inherent religious and moral emphasis on architectural production which extends from the places of worship to the home, denying the home the opportunity to be analysed rationally.[44] On the other hand, sociologists look at rituals as actions and beliefs in the symbolic order which do not necessarily involve religious commitment on the part of the actor (Douglas 1996). In this sense, rituals are not fixed, but flexible, part of people's social lives expressed at home. When a discrepancy appears between the ritual situation and the form of expression, then the latter has to change. This explains the way a house may be modified by new occupants to suit the newcomers' ritual practices.

The *human body*, according to Mary Douglas,[45] is represented by the *social body* through the medium of *expression*, which is the image of another body (the home).

The association of the social and the personal is to be represented in the organisation of home. To that extent, the English working-class home, mentioned above, attempts to provide privacy in a very tight layout. The spatial organisation of bodily functions with respect to privacy corresponds to the distinction between social and private occasions. The back of the house is for cooking, washing and disposal of body waste. The front parlour, on the other hand, is distinguished from the living room-kitchen, as a special space for public, social representation. It is the front of the house, in terms of composition, and the public face of the body.[46] By breaking the barrier between zones, the private and the public, the family demonstrates their belonging to the public unstructured organisation by expressing their unstructured personal system of control. In such a situation, the family adopts a unique spatial and cultural order that disassociates them from the dominant model. The space organisation here, as a form of expression, has been adapted to suit the required ritual situations and emerging new pattern of practices.

Therefore, there has never been a single dominant model of the home in any culture, society or even city. A home is determined according to the particular cultural context within which it is constructed and lived. Alienation and integration with the structured organisation (socially, spatially, culturally) could be seen as different uses of symbolic codes derived from different social systems. This could, moreover, explain the open-plan arrangement of modern Cairene homes in westernised-urban zones in Cairo, which contrasts with the structured, relatively complex plans of traditional quarters. The former expresses disintegration of the dominant model, brought about by rebellion against the traditional ritual situation represented in the spatial organisation, and personalisation of a particular non-aligned form that belongs to the desired culture and ideological destination.

A close look at such differences between models enables us to define the home within interconnected spatial settings such as the hawari: in which the dominant cultural model unifies indoor spaces of individual units (houses) with the shared outdoor space (local alleys). Furthermore, it justifies why people in Old Cairo consider the hawari as ultimate homes that extend beyond the physical limitations of the house to the neighbouring buildings, the alleyways and shared public spaces. It constitutes a sophisticated social identity that gathers neighbours, partners and local businesses under a particular identity of social and cultural significance. It is an experience rather than a representation. In this context, males take meals (breakfast, lunch) in collective settings in the public space, while females eat separately at home.[47] Females, on their side, attend to the needs of each other by, for instance, taking turns to go shopping, serving larger groups of housewives in one trip.[48]

On the other hand, the homes of people from higher social orders in Garden City and El-Agouza districts of Cairo, for example, represent a different cultural ideology: "the progressive westernised" identity, associated with the liberal images of open plan, European decoration, and stylish dress. The central cultural image is one that is at odds with the traditional context. Therefore, in contrast with the strongly linked community of the hawari, the main feature of the modern quarters is their isolated environments and self-orientated small family units. This stems from the variety of members' cultural backgrounds and lack of the authenticity required to promote a distinct cultural model.

2.3   The spatial organisation of open plan houses in Cairo
This structural analysis follows a similar model developed by Roderick Lawrence, 1987.

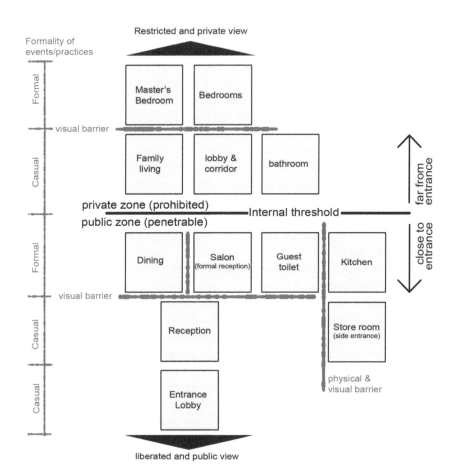

However, the dominant model aspires to emulate an alien culture; and is a model that is reproduced, for example, in the strict formality of dining rooms and large, highly decorated and rarely used reception spaces. However, the community, as a collective setting, has little to demonstrate in terms of social network or ritual practice. Homes end where the physical boundaries of the houses end, at the front door. Rituals are therefore centred within the house, with the image reflecting the uniqueness of the self, not the group.

HOME AS AN EVERYDAY PRACTICE: EPISTEMOLOGICAL AND THEORETICAL PERSPECTIVES

The turn of the twentieth century saw the emergence of everyday practice in theoretical and philosophical discourse along with the beginnings of the structuralist movement led by Swiss linguist Ferdinand De Saussure.[49] For theorists, structure was the only independent and definable thing that the mind could comprehend.[50] The implicit structure is to be revealed in order to explore the hidden meaning of reality. Structuralist method emphasised that "*the true nature of social phenomena as relational*

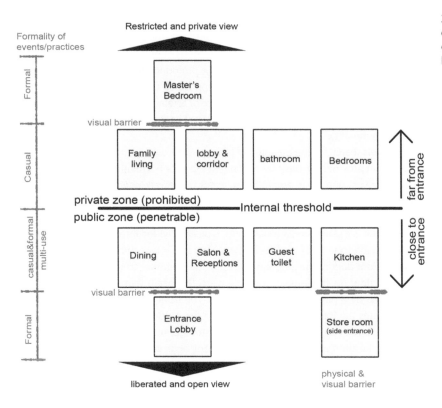

2.4 Spatial organisation of closed plan houses in Cairo

systems of meaning is to be sought in structure which lies somehow behind or beneath the phenomenal world of appearances".[51] Everyday practice emerged as a proper field for investigating this proposition: where meaning is embedded in everyday activities and rituals. Anthropologists such as Emile Durkheim[52] and Levi-Strauss developed this approach for cross-cultural and cross-disciplinary investigations ranging from religion, psychology, kinship to other cultural aspects of real life models.[53] Every action, interaction or practice is to be seen in terms of a model or structure that is either static or slowly transforming. Levi-Strauss, in particular, had configured a method to analyse real life practices as representations of underlying structures of social and cultural phenomena that are applicable to human beings, and which result from the negotiation of oppositional relationships of male–female, nature and society, inside–outside, dark–light. In this perspective, social practices need to be viewed through the lens of the cultural structures which constitute these comparative categories.

For anthropologists, the study of architecture is to be conducted as a search beyond the appearance of the object. In *Discipline and Punish*, Michel Foucault analysed the construction of the judicial techniques and apparatuses of the nineteenth century as a system exemplified in the typical array of the same cellular space.[54] For Foucault, objects can reveal the universal structure and its subsequent rules that are applicable to all situations. Pierre Bourdieu,[55] the *blissful structuralist* as he ironically called himself in *The Logic of Practice*, questioned the credibility of an objectivity that is totally dependent on rigid universal rules without the support of subjective assurance. In seeking evidence on how things get done and how a particular set of practices is experienced,

Bourdieu deviated from the structuralists' desire to impose system and coherence on the practices of men and women and their products. Instead, he looked at everyday practice: investigating spatial and temporal settings of homes and the way the home is structured using local cultural references. His profound analysis of the Kabyle house was, hence, constructed around conceptual opposition that was tied in to the system: the cooked and the raw. The tangible presence of artefacts, commodities and physical structures symbolises and refers to the abstract cultural order of values and the interpretive morality.[56]

During his search for the meaning of practice, Bourdieu introduced the *habitus*[57] as a notion that provides the basis for explaining the society's relationship to a structure[58] as *"an acquired system of generative schemes objectively adjusted to the particular conditions in which it is constituted"*.[59] In this context, the idea of home is represented in the dwelling that is *"a fragment of a society"* and *"the reference for every metaphor"*.[60] The Kabyle house is a simple rectangular form, typically divided into two unequal internal parts: the stable, one-third, and the human living space, two-thirds, with a high ceiling. The large front door in the long east wall provides entrance to both spaces, while the small back door for the females is to the west. The beam that supports both spaces is supported by a pillar at the dividing walls. The front door is the public transition that separates the indoor secured space from the outdoor domain. Passing through the door is a threshold journey between the stranger outside and familiar inside. The contrast in location and size between the front door and the back door represents the binary oppositions of front–back, large–small, male–female, and east–west; typical structuralist binaries. However, the house is divided into two spatially distinct parts, the lower and dark part where animals and the water vessels are kept, as opposed to the upper part, where the fire, grain mill and the woman's loom are located. The half wall to the left of the door is where vegetables are dried and where the man sleeps.

In this sense, the house is ordered to reflect the two parts of the family, the male and the female. The representations of the male and the female are highly emphasised; the main beam is the male, with the pillar as the female. Whilst the outside of the house is the male, the inside is female, and is itself structured into upper (male) and lower (female). Pierre Bourdieu's work developed evidence on the relationships between the practices and the impact these practices and meanings have on the organisation of the house. For example, privacy was a social practice that was represented in the thresholds. A practice needs space (spatial) and requires duration (temporal). To cross the boundary is a transitional practice that requires the transformation of status.

> *The transitional periods have all the properties of the threshold, a sort of sacred boundary between two spaces, where the antagonistic principles confront one another and the world is reversed.*[61]

Bourdieu, however, did not discuss the territory of courtyard nor identify to which world it belongs. It is, certainly, the central venue of transition; the manifestation of the outside male within the inside female. He did not tell us how the courtyard mapped the mutual interconnections between different practices, and how it contributed to the inclusive environment of the home. It is an overlapping venue where different arenas merge, and what makes the whole organisation work. Nevertheless, Bourdieu pioneered the investigation of dynamics of power relations in social life and how these relations translated into different forms of spatial relationships and conditions. In this sense, his work had tremendous impact on the understanding of the architecture of the home and his studies have become essential reading for architectural researchers of the late twentieth century.

Architects, on the other hand, have drawn much on the work of anthropologists to investigate the notion of home and local culture and the way they implicate the spatial condition of living. Amos Rapoport, through his investigations of the meaning of the built environment, uncovered relationships between the spatial form of a physical environment and its human understanding.[62] Leading projects that linked architecture to sociology and other social sciences such as anthropology, psychology, and ethnology, he set himself the task of studying buildings as symbols, focusing on *"the relation of signs to the behavioural responses of people, that is, their effects on those who interpret them as part of their total behaviour; this, then deals with the reference of the signs and the system to a reality external to the system – in a word, their meaning".*[63] The home, or probably the house, emerges in Rapoport's view as a combination of different elements that are either fixed elements (walls, fences, lawns) or semi-fixed elements (vegetation, furniture, …) that communicate symbolic messages and cultural information. It has four basic forms of organisation: space, time, communication and meaning. Human activities within the home represent an organisation of communication; hence, *"who communicates with whom, under what conditions, how, when, where, and in what context and situation is an important way in which communication and the built environment are related".*[64]

Rapoport's investigations, however, relied heavily on the findings of psychological and behavioural research conducted in different contexts and situations, with the aim to achieve universal conclusions. They were always missing that aspect of in depth understanding of spatial or materialistic components of each situation that needed individual scrutiny.[65] His analyses failed to investigate the holistic situation, in which all those elements work/interact together to articulate a particular social environment. In terms of the scope of this book, Rapoport's work, nevertheless, in tracing the social influence of the built form and the extension of social domains, had liberated the practice of home from the limitations of the physical boundaries of the house:

> Social boundaries, of course, are not necessarily spatial or physical but, once again, their perception, which must precede understanding and behaviour, is helped by clear and unambiguous markers – noticeable differences of all kinds. This is, of course, related to our earlier discussion about boundary markers as objects (fixed feature or semi-fixed feature), boundary-marking rituals (non-fixed feature in time or space), doorways and thresholds, and so on. All communicate meanings, the basic function of which is to reinforce basic cultural categories. Thus the whole notion of indicating boundaries by means of noticeable differences to delineate social groups, domains, and their spatial equivalents, and to define entry or exclusion, becomes very significant.[66]

Rapoport work draws, tentatively, an outline of a spectrum of social spheres in domestic environments. In defining boundaries that define territorial limits, he underpinned the diversity of social condition that represents private, semi-private, semi-public and public situations and venues.[67] It is essential that distinctiveness be communicated, and that involves the interaction of people in a particular social space, forming a complex organisation of social spheres rather than the traditional private/public segregation.[68]

Home as Social Practice

Home cultures and the way space is reproduced by the everyday came to the fore of architectural studies in the last quarter of the twentieth century. Grounded in the understanding of the phenomenon of home and associated culture through narratives of everyday living patterns is the anthropologist's mean of analysis that made its way into architecture in the work of some architectural theorists during the 1980s. Roderick Lawrence was one of the advocates of this approach as we can understand from his book, *Housing, Dwellings and Homes* (1987). Lawrence made extensive use of ethnographic research in investigating the order of daily activities of domestic space, indoor and outdoor, in both Australian and British homes.[69] He suggested the study of *"the reciprocal relations between architectural, social and cultural variables in specific societies"*, which he considered as a reorientation of housing research.[70] He has revealed a variety of interpretations of the socio-cultural parameters that are specific to the context and are continuously changeable over time and space. For him, spatial and social orders are agents of change:

> Buildings of all kinds are handed down as a legacy from one generation to the next; they are one vehicle for the embodiment of social ideas. Yet if there is a spatial and social order of residential environments then that order, however it interacts with personal attitudes and social irregularities, is an agent of both stability and change. In architecture the relationship between building form, its use, its meaning and time is a transactional process between physical and behavioural factors … The relationship between habitat and resident is dynamic and changeable, and it includes factors which may remain unresolved over a relatively long period of time.[71]

While moving from the broadness of the universal model of cross-cultural studies we saw in Rappaport's work to the specificity of national culture, Lawrence's main weakness is in the way he summed up the broad and diverse housing of Australia and Britain under two distinct models, undermining the diversity of living models that exist in each culture, never mind the diversity of the national culture itself, which is another problem. This historical comparison of fixed cultural models is highly problematic in cultural studies. Rom Harré's concept of the circulation of the 'practical' and "expressive" in the social life, for example, contests this analogy. Harré suggests that *"social forms and individual cognition of these forms are highly unstable and in rapid flux. This continuation of change is not only a fundamental property of social institutions, but of the built environment as well"*.[72] Eating habits, food preparation and dining rituals, as fundamental to home organisation, are agents for change that keep changing and informing in the idea and practice of homes, and as a consequence their architecture.

In the theoretical construction of this book, the cultural practice of everyday living, as informative about the construction and consumption of space at home, will form the prime methical approach. However, the specific contextual settings of Cairo and its old quarters represent peculiarities that do reflect on the way culture can vary even within the same context. In fact, it is the distance the researcher keeps from his subject that reveals the level of details and accuracy he would investigate. Also, to approach a context at a certain moment in time without investigating its past development would be a mistake. Practice and culture change with time, leaving traces that govern what comes next but infused by memories of what happened and was learned in the past.

Whilst actors and generations occupy the space of a home loaded with traditions, they superimpose their fingerprints on these traditions and transfer only their own version to the following generation, who will do the same. Through the path of history, community is defined and culture is practised. Hence, I would argue that only by a thorough investigation over a long period of time, of the meaning of home and practice of its spaces, can the culture of home and the meaningfulness of its architecture be captured.

This is an attempt to avoid the misleading assumption that one model of housing could represent one nation or culture (such as England or Australia), while removing the studied model from its immediate context. In a sophisticated socio-cultural political context such as the hawari of Cairo, it is difficult, or perhaps, inappropriate, to generalise the mode of housing in a particular period of time. Each harah has its own range of buildings and houses as well as social and cultural order. This difference grows from one zone to another, constituting different worlds, not only different situations and settings.[73] By following the development of homes in a particular harah through an extended period of time, insights about changes in the relationships between buildings, activities and rituals would enable us to understand the homes of Cairo. Food preparation and dining, socialising, meeting friends are considered as essential activities that impact on the spatial order and organisation of domestic spaces and environment.

Home as a Set of Rituals

Anthropologists seek to understand space organisation through the order which activities, practices and rituals impose on domestic domains. They look at certain kinds of repeated unfolding of events in an institutionalised context and in a regular sequence.[74] Victor Turner's *The Ritual Process* and Mary Douglas's *The Idea of a Home* remain two prominent texts that have focused on understanding the archaeology of social practice. They revealed the relation between the social reproduction of space and the inherent meaning of practice. Turner, for example, shed light on the *rites de passage* and the power the threshold poses in arranging the state of the individual between two different situations.[75] Entering the home is a journey through a threshold, which involves three stages; departure from public character (separation); ambiguous identity through the journey (margin); and joining a new state of settlement within the home territory (union).[76]

In this sense, the threshold is a *journey*, a passage though time and space between two different situations which is expressed through ceremonial events.[77] In the spatial organisation of home, the entrance zone (the threshold) and doors (the border) have all the qualities of these rituals, where, as Bourdieu mentioned, the world is reversed, new rules are implemented and a new order of superiority is at work. The covered inside becomes visible, and the open outside world becomes isolated behind the walls.[78] The publicly recognised social position is reconfigured through this threshold to the new privately recognised one.[79] A political leader is a father at home, and a male servant becomes the ultimate authority in his household. According to these qualities, the space is reproduced through every practice. In this way, a culture-specific relationship between a space and its users is revealed; i.e. the architecture and the habitual practices are brought together.[80]

Crossing the thresholds is a by-product, a transitional journey. Such liminality is expressed clearly in the Cairene architecture of both the house and the mosque.[81]

The entrance of both building types is configured carefully to represent this ambiguous situation of transition. First, the entrance is not a door, rather it is a series of spaces, that are relatively dark, enclosed, and provides indirect accessibility to the illuminated indoor space. The gradual fading of daylight at the entrance is a simulation of decay from the outside-in. Gradual progression through the spaces, however, allows the indoor light to emerge gradually in an expression of the emergence of a new reality. The entrance starts with a huge front door, a visual barrier to the entrance hall. Then, an inclined and indirect low-ceilinged corridor with no windows conveys the idea of loss in time, space and status. This corridor opens, finally, via a small hole in a corner onto the huge open space of the courtyard, which is usually wider than the external street. Examples can be seen in Sultan Hassan madrasa and mosque and Bayt al-Suhaimy in Old Cairo. New Cairene houses have simulated such practices in different spatial arrangements. The relatively horizontal journey of old houses has become vertical in the contemporary apartment buildings. The liminal zone of new homes starts at the building entrance, leading to a dark, tight stairwell that in most cases lacks windows and is open to the sky to allow light to emerge as you move upstairs.

Architecture, accordingly, and as per Susanne Langer, becomes a *virtual ethnic domain*,[82] which is a visual and spatial experience of structured domesticity that encompasses a set of analogous processes, like music (analogous with time), photography (analogous with two-dimensional space), and sculpture (analogous with three-dimensional space). While architecture is able to provide the multi-dimensional experiences of space and time, the home corresponds those experiences to the human actors in a structured and synchronised organisation. The idea of home lies in its organisation of space over time; in other words, in its capacity for memory or anticipation.[83] The home is an institution that is capable of anticipating future events through coordinating actions and planning consumption in an accurate and efficient response to memory and the regularity of experiences.[84] Rearranging the furniture, changing the usage of a room, is a process of anticipating future changes in the temporal order of activities.

This order produced what I called *"part-time space"* in small Cairene apartments: that is, a corner of the living room carefully decorated and formally furnished to represent the required image when others pay a visit to the family. In the absence of such visits, that space is an active family gathering space. This concept of part-time space highlights the need for public spaces which represent the public image of the household to strangers in the home, the supposedly private domain. Other indicators of anticipation of future activities appear at home, such as a cupboard full of products that have their annual use planned out (winter and summer clothes). Capacity of storage, storm windows and extra blankets are a response to the memory of severe winter, while shading devices and water tanks are a response to the summer droughts, and so forth.[85] Storage of these products is done according to the frequency of use. This quality of space-time arrangement is a cognitive plan for the common good of the home.

To read about daily activities of inhabitants is much like analysing actors' movement in a theatrical display, or what Turner called *"social drama."*[86] If the home is a kind of theatre and social drama is a kind of metatheatre, then rituals become the performance of a complex sequence of symbolic acts.[87] It is a *"transformative performance revealing major classifications, categories, and contradictions of cultural processes"*.[88] Hence, space organisation of the home is a theatre with frontstage, backstage and decoration, which could be arranged according to desired performance, the perfection of which is a result of the fine integration and movement of actors through space and time. This sense of scenery and drama is extended to the public space, which is considered as the stage upon which the drama of communal life unfolds. The streets, squares and parks are dynamic spaces that host such everyday social dramas.[89]

This approach was brought into architecture by Bernard Tschumi, looking at it as a spatial discourse associated with time, action, and movement. Architecture, for him, could be perceived as dynamic space: in flux rather than fixed and enduring.[90] *Space*, *movement*, and *action* (*event*) comprised the *semantic dimension* of what architecture really meant.[91] However, the implementation of this approach in architectural practice remained superficial and resembled a style of drawings and thematic concepts rather than semantic research through which the architect investigated the pattern of the rituals/practices taking place in the building, not to say the home.

The Home Reconstructed

While architecture has always been rooted in the domain of serving people's everyday life needs, this fact has always been taken for granted, without attempts to understand how such a mutual relationship between the space and everyday activities was constructed in reality. Sociological and anthropological investigation of the home had, by the middle of the twentieth century, given us insights into the organisation of home and inherent meaning embedded in its form and components. The anthropologist's approach, for example, investigating the home as ritual processes, detects the quality of the domestic environments and traces the semantic organisation of space and time. The threshold is an experience that freezes people's identity and recognised state in time and space during their *rites de passage*. In this sense, the threshold performs in a similar way to that of the hidjab. The journey through the liminal zone suggests a process of transformation that requires architectural treatment of spatio-temporal moments; in other words, designing the event rather than the space. Internal organisation, on the other hand, reflects the set of rituals that is expressed though the opposition of front–back, private–public, male–female and the barrier between them. The human movement, furniture arrangement over space and time relate to the occasion and its repetitive patterns. However, in the anthropologists' work we are left blind: as the accurate and supporting drawings that put the reader into the exact visual context are lacking. Drawings in this sense are tools to explore the nature of orders and hidden semantic systems.

The architecture of home, hence, was recognised as a reflection and expression of the organisational complexity of space and time, manifested in daily practices. However, there have been significant developments in architectural research, encompassed in two major stages. The first was the cultural analysis of environments introduced by Rapoport during the 1970s and 1980s: which fragmented the built environment, including the home, into groups of image forms that retain significant cultural signs and meanings. The second was the comprehensive analysis of the spatial organisation of the home based on the social practices, with the house as a set of quarters, as introduced by Roderick Lawrence in 1987. The construction of home as a set of rituals tends to investigate the home as a story, as a journey into a comprehensive and complex set of social relations/interactions between people and people, people and space and people and time.

NOTES

1   Alexander, *The Production of Houses*, p. 24.

2   Chandhoke, Transcending Categories: The Private, the Public and the Search for Home. In Mahajanand and Helmut (eds), *The Public and the Private*, p. 186.

3   S. Phillips, *The Christian Home*, p. 16. "By the physical idea of home, we mean, not only its outward, mechanical structure, made up of different parts and members, but that living whole or oneness in which these parts are bound up … By the moral idea of home, we mean the union of the moral life and interests of its members".

4   Alexander, ibid.

5   Bower, Territory in Urban Settings. In Altman, Rapoport and Wohlwill (eds), *Environment and Culture; Human Behaviour and Environment*, p. 179.

6   Birkeland, p. 139.

7   Douglas, The Idea of Home: A Kind of a Space. In *Social Research*, 58(1), pp. 287–308.

8   Nevett, *House and Society in the Ancient Greek World*, p. 4. It had operated in practice as a social institution in domestic domains that included the head of the oikos (usually the oldest male), his extended family (wife and children), and slaves living together in one domestic setting.

9   Gazda and Haeckl (eds), *Roman Art in the Private Sphere: New Perspectives on the Architecture and Decor of the Domus, Villa, and Insula*.

10   Pomeroy, *Women's History and Ancient History*.

11   *Oxford English Dictionary*. Online edition, as it appears on 25th June 2013.

12   On the other hand, dwellings are used to denote the primitive, non-professional and mainly vernacular production of housing, while residence, shelter, and family are associated with the functions which humans require within their living space.

13   *Oxford English Dictionary*, ibid.

14   Ibid.

15   Hakim, *Arabic-Islamic Cities: Building and Planning Principles*, p. 95.

16   Ibn Manẓur, Muḥammad ibn Mukarram, *Lisan al-Àrab*, vol. 2.

17   Ibid., vol. 5.

18   Hakim, ibid.

19   Ibn Manẓur, ibid., vol. 13, pp. 211–12.

20   Clapham, *The Meaning of Housing: A Pathways Approach*.

21   Kemeny, *Housing and Social Theory*.

22   Schneider and Till, *Flexible Housing*, pp. 41–3.

23   Kemeny, *The Myth of Home-ownership: Private versus Public Choices in Housing Tenure*.

24   Schneider and Till, ibid.

25   Harries, *The Ethical Function of Architecture*.

26   Kemeny, *Housing and Social Theory*.

27   Saunders and Williams, The Constitution of the Home: Towards a Research Agenda. In *Housing Studies*, 3(2), pp. 81–93.

28   Kemeny, ibid.

29   Douglas, ibid.

30   Harries, ibid.

31   Olwig, Cultural Sites: Sustaining a Home in a Deterritorialized World. In Olwig and Hastrup (eds), *Siting Culture: The Shifting Anthropological Object*.

32   Saunders and Williams, ibid.

33   Chandhoke, ibid.

34   Tietze and Musson, Recasting the Home-Work Relationship: A Case of Mutual Adjustment? In *Organization Studies*, 26(9), pp. 1331–52.

35   Nippert-Eng, *Home and Work : Negotiating Boundaries through Everyday Life*.

36   Cultural norms extend, in some cases, to their religious and in some cases mystical beliefs, especially in pre-industrial and relatively traditional environments.

37   Bahloul, *The Architecture of Memory: A Jewish-Muslim Household in Colonial Algeria, 1937–1962*, p. 28.

38   I used this term to reflect the intimate relationship and association between the inhabitant and his home. It is interpreted from Homer, and some dictionaries refer to the symbolically similar relationship of a pigeon that is trained to return home.

39   Lawrence, *Housing, Dwellings and Homes: Design Theory, Research and Practice*.

40   Hoodfar, Survival Strategies and the Political Economy of Low-income Households in Cairo. In H. Hoodfar and D. Singerman (eds), *Development, Change, and Gender in Cairo: A View from the Household*, pp. 12–13.

41   Lawrence, ibid., p. 114.

42   See Turner, *The Ritual Process: Structure and Anti-structure*; Bourdieu, *Outline of a Theory of Practice*; Douglas, ibid.

43   This situation appears in several countries. Middle-class families in either Egypt or the UK, for example, tend to have minimal kitchen spaces, as family members or the households are used to taking meals separately. This is a result of the long working hours, working shifts and the travel distance between home and work. In such cases, spaces for either entertaining or for children get bigger.

44   Anti-ritualist anthropologists who considered the ritual as a form of worship which is being imposed on secular cultural practices. See Douglas, *Natural Symbols*, pp. 2–3.

45   Douglas, *Natural Symbols: Exploration in Cosmology*.

46   Ibid., p. 162.

47   I did the field work over a three-year period (2007–11) in the hawari of Old Cairo. I had participated in some of those meals.

48   Interview with Dina Shehayyeb, a specialist in the socio-cultural context of the hawari of Old Cairo.

49   Jenkins, *Pierre Bourdieu*.

50   Saussure's work, adopting the epistemological tradition of realism, distinguished between the logical and grammatical structure of the language and the everyday practice, while giving the former the influential position of being the appropriate domain for the analysis of meaning. See De Saussure, ibid., p. 9.

51   Ibid.

52   De Saussure's perception of language is a "system of signs" that is at the same time a social institution. That system is a "science which studies the role of signs as part of social life". This is what was called semiology (signs), which was influential in architecture theory through most of the twentieth century and commonly adopted linguistic theories of signs and symbols. The latter was based chiefly on the assumption that there are innate conventions which enable buildings and their parts to convey meanings in a similar manner to language. See Lawrence's *Housing, Dwellings and Homes* (1987: 48).

53   Levi-Strauss, *The Elementary Structures of Kinship*.

54   De Certeau, *The Practice of Everyday Life*, p. 46.

55   He called himself a "blissful structuralist" in the Logic of Practice. Jenkins, ibid., p. 32.

56   Jenkins, ibid., pp. 32–3.

57   Latin word referring to a habitual or typical condition, state or appearance, particularly of the body. Jenkins, ibid., p. 74.

58   De Certeau, *The Practice of Everyday Life*, p. 58.

59   Bourdieu, *Outline of a Theory of Practice*, p. 95; Jenkins, ibid., p. 74.

60   De Certeau, ibid., p. 52.

61   Bourdieu, *Outline of a Theory of Practice*, p. 130.

62   Rapoport, *The Meaning of the Built Environment*, p. 178.

63   Rapoport, ibid., p. 38.

64   Rapoport, ibid., p. 181.

65   Most of the examples and the references he used utilised information and findings from such research. Several of those cases' findings are taken for granted, even though investigations took place in particular situations and contexts, and could not guarantee the same effect in combination with other elements.

66   Rapoport, ibid., p. 170.

67   Rapoport, ibid., pp. 181–2.

68   Rapoport, *Identity and Environment: A Cross-cultural Perspective*, p. 12.

69   Lawrence, ibid., p. 96.

70   Lawrence, ibid., p. 80.

71   Lawrence, ibid., p. 51.

72   Lawrence, ibid., p. 51.

73   Abaza, *Changing Consumer Cultures of Modern Egypt*, pp. 23–5.

74   Moore, *Law as Process*, ibid., p. 43.

75   Turner, *The Ritual Process: Structure and Anti-structure*.

76   Turner, ibid., pp. 94–5.

77   Turner, ibid.

78   Bourdieu, *Outline of a Theory of Practice*, pp. 130–31.

79   Refer to the Van Gennep third transitional phase in the liminal zone: aggregation; and Turner, ibid., p. 94.

80   Blundell Jones, *Social Construction of Space*, p. 62.

81   O'Kane, *Domestic and Religious Architecture in Cairo: Mutual Influences*, p. 149.

82   Langer's philosophy of architecture as *"virtual ethnic domain"* lies at the centre of her theory of signs and symbols, in which she distinguishes between the ingredients of a culture and its image. It is the task of the architect to supply the virtual image. *"The architect does not merely fill a given space with buildings. The given space is inevitably transformed into a new kind of dimension. The architect while manipulating actual space creates a place that is the image of a culture's world: a virtual ethnic domain"*. Donohue, *A Heideggerian Reading of the Philosophy of Art of Susanne K. Langer*, p. 31. For more details about Langer's philosophy of architecture see: Langer, *Feeling and Form*, p. 100.

83   Douglas, *The Idea of Home*, ibid., p. 294.

84   Douglas, ibid., p. 294.

85   Ibid., p. 294.

86   Turner, *Social Dramas and Stories about Them*.

87   Turner, *The Anthropology of Performance*.

88   Turner, ibid., p. 5.

89   Carr, *Public Space*, p. 3.

90   Bernard Tschumi in an interview by Khan and Hannah, *Performance/Architecture*, pp. 52–8.

91   Khan and Hannah, ibid., p. 53.

3

Architecting Homes: A Practice in Question

FROM SPATIAL DESIGN TO SOCIO-SPATIAL ARCHITECTURE

The home is a socio-spatial system. It is not reducible either to the social unit of the household or to the physical unit of the house, for it is the active and reproduced fusion of the two. The home is the most basic and simple of modern socio-spatial systems. Functionally (i.e. in terms of both its social and spatial organization) it is indivisible.[1]

In the introduction to his journal paper, *The Social Construction of Space*, Peter Blundell Jones accused strategies of architectural practice, during the twentieth century, of being inadequate for the prime mission of architecture as an everyday building activity. Those included functionalism, post-modernism, classicist and rationalist fundamentalism, enthusiasm for high-tech, and finally stylism and symbolism. He argued that their *"influence on works of high-architecture seems generally to be drifting ever further from everyday building as they become increasingly the vehicles for displays of individual virtuosity demanded by a market in images, and less and less concerned with habitation"*.[2] The centre of his argument is that even though buildings have always been artifice, they became, by the end of the century, more artificial and disconnected from their inhabitants and their social, political and economic processes.[3] Buildings lost their content and meaning and therefore were reduced to the capacity of mere objects. This realisation of the apparent misconception of the mission of the architect became popular at the turn of the twenty-first century and led to continuous calls for alternative practice. Termed by some as *alternative praxis*, *everyday architecture*, *spatial agency*, or *socio-spatial architecture*, all refer to the reunion of architecture with buildings' meaning and content.[4] They called upon the architect to put aside his/her personal ego of being the maker and join other and equally important players in making buildings meaningful and relevant.

In the previous chapter, I liberated homes from their physical boundaries and objective nature, searching for the implications of practice and rituals for spatial order. They were explored through the way people actually live and organise their lives at home, revealing the implicit meanings of such a complex spatial-temporal environment.

The question now is how can architects work beyond the physical determination of space to enhance the socio-cultural dimensions of buildings? How to shape and form social and spatial practices? This chapter, therefore, targets this particular issue, with the aim of elaborating on a tentative hypothesis that architects need to work (if the context is similar to the hawari of Cairo) with structured everyday homes, described by Steven Harris as *"irregular and open, inexact and conceptually fragmented".*[5] Adopting his words; there is no perfect order, no grand scheme.

Architecture as a practice tends to focus on the search for the optimum way to attend to people's needs (physical, psychological, spiritual, political, cultural and social). These needs, however, are perceived by architects differently according to time and context. In the medieval landscape of cities and towns small parts of the built environment have been affected by architecture as a discipline of professionals (architects).[6] Architectural production was, mainly, divided between the formal (by architects building palaces, basilicas, and churches/mosques) and the vernacular (by master builders building houses, shops, and markets). In the modern era, on the other hand, architects work largely as professionals, following an extensive raft of regulations that define people's needs according to political agendas and strategic planning.[7] The distinction between medieval and modern practice, the vernacular and the professional of any time, lies in the process of production and how inhabitants' needs are communicated between the architects/builders and users. Hence, architects are prone to perceive their mission to be elitist in nature, called by Spiro Kostof "city architects" and "special people", maintaining a wide distance from ordinary people's everyday lives and environment.[8] Such an elitist position is seen to compromise any revolutionary proposals for everyday architecture that might consider everyday activities and habits as essential for a sensibly built environment.[9]

During the post-war period, when the ordinary people's fundamental needs, such as housing and neighbourhoods, were set to drive the architects' agenda, the production of homes moved towards the mass-production of housing: with the prospective residents considered as subjects with no role in the decision making process. Although this approach may have been justified by the period and situation, it proved problematic in the long term. As a result, design informed by architect-user interaction and mutual negotiation was lacking in the majority of architectural productions during the twentieth century. In response to the dominant paradigm of commodification and consumption of houses, the twenty-first century has witnessed a renewed demand for more socially integrated architecture.[10] The teaming up of users with architects in Lucien Kroll and Peter Hübner's work represented a powerful approach to socially-based architecture that, with time, proved successful, as we have seen in Stuttgart student hostel.

Recent studies have considered the home's organisation, structure, and psychology, but have not considered the role of architects in its production. Sociologists such as Saunders and Williams invested much effort in calling for a new agenda for the constitution of home, in terms of the internal dynamics and space organisation, without considering the role of those who form its enclosures. Architectural consideration of that role, however, came from architectural theorists such as Amos Rapoport, Christopher Alexander, and Roderick Lawrence: who deliberated over various facets of the process but had a limited effect on actual practice. Moreover, the cognitive model of the notion of home as a complex socio-cultural environment implies the existence of a barrier between architects (as creators/elaborators of spatial settings) and the production of home as an environment driven by its occupants.

Here, I wish to revisit the role of the architect in making homes, raising two principal questions: Is the architect the creative designer of homes, or is she or he an architect who is aware of the distinctive socio-cultural context and its dynamics? In this book, I argue that architects need to understand the intrinsic mechanisms of social spheres and activities of homes if they are to be able to develop successful and integrative socio-spatial solutions that are dynamic enough to deal with temporal or social change, or even to develop new forms of living when needed. It is up to the architects to stay locked in specific typologies of housing, while societies and inhabitants keep developing new urbanity with forms of homes that are beyond the professional standards or requirements. The architect's duties, hence, move beyond the traditional mission of addressing the needs through the creation of space, to be more innovative, to be part of the living experience that is much more than the physical characteristics of spatial configurations. This chapter intends to offer a tentative argument about how architects deal with the sociocultural phenomenon of home.

The Act of the Architect

In today's world, architects are bound by two forms of institution; the formal organisation of the profession (regulations that control the practice in a systematic way, such as codes, standards and specifications) and the informal socio-cultural context within which the building will be produced (social analysis). The former seems dominant and expanding while removed from the context. Despite the repeated claims of architects regarding the suitability of their response to and inspiration from the context, their designs remain purely personal decisions or preferences of the architect him/herself, while real users, who test and judge that suitability, in reality, have absolutely no input. Even awards for buildings are given by experts and awarding agencies who are far removed from the project site, making few visits, largely talking to clients and in very rare cases may talk to users. But, under no circumstances were they part of the living experience of the building.

Architects struggle to situate themselves between two polar characters: architect as a *creator*, or architect as a *collaborator/facilitator of designs*. The former means the building reflects the architect's personal creativity and personality while the meaning belongs to his/her own concept and ideological thinking.[11] In this sense, the architect is like a surreal painter in being the only one who knows, for sure, the meaning behind his creation,[12] while others are left to make their own interpretation.[13] The latter, on the other hand, means the product is a group effort, a process, in which the architect is more like an elaborator. The building and its meaning, then, reflect the values and personality of the involved actors and the dominant culture of the community.

Outside this scale of creation-elaboration roles, other perspectives on the architect range from mere professional, designer, to theorist and reformer. Jeremy Till, for example, refers to architects as technocrats whose technical knowledge and expertise elevate them above mere building.[14] In this respect, the architect's knowledge, training and expertise provide him/her with essential tools to improve the built environment through construction and detailed creativity.[15] Others choose to design falsely and abandon the discipline and conventional techniques in favour of a purely sculptural intent, with abstract architecture that is detached from its place, history and any cultural roots of memories which it could contain.[16] The work of deconstruction architects such as Zaha Hadid and Frank Gehry is widely perceived as exemplars of such a Surrealist/ sculptural approach.[17] Le Corbusier, on the other hand, was a theorist who sought any available political support to get his plans implemented,[18] while the Bauhaus architects were real reformers of practice. As a creator, the architect maintains a good deal of control, responsibility and authorship which has been translated in recent years as copyright over designs.[19]

The term "architect" means, in many disciplines, the mastermind behind a particular process (political ideology, policies, new technology, or even a business deal). In that respect, the architect could be a politician, economist, activist, or even a systems designer.[20] However, the sole duty of the architect is recognised widely as that of addressing clients'/users' needs through the medium of the built environment while protecting the general public against the dangers of poor and insensitive buildings.[21] However, this duty and those needs require precise definition and clear objectives. Modern architects, according to their social agenda, believe that architecture is about forms of industrial technology which, in their view, bring about a revolution in the lives of ordinary people. Kenneth Frampton, a supporter of this view, sees standardised forms of industrial production replacing the customary practices of folk buildings.[22] Those new forms, according to Frampton, would benefit the working class people by facilitating patterns suited to modern culture and technology. In short, the architects' mission should be to create meaningful standardised forms. This view, however, reduces architecture to its physical characteristics and spatial rationality (raising the house above the home). In addition, it did not realise that those standard patterns meant superimposing [standard] lifestyles on people whose needs might be different. Systemising architecture led to systemising people's lifestyles and practices, which was later rejected and marked the failure of the modernist movement.[23]

In contrast to this view, there was a long rooted perspective, during much of the twentieth century, of the architect as a maestro whose main role is orchestrating the spatial design with the involvement of the users at an early stage.[24] Christopher Alexander, Peter Hübner, and Hasan Fathy are well known architects of this approach. They looked at producing homes as a social process, incorporating collaborative work in which they were coordinators and elaborators. While Fathy's work in rural Egypt emerged as early as the 1940s, maintaining the architect's mission to rationalise local and traditional forms,[25] Alexander's work in Mexico during the 1970s and 1980s helped the users to plan and build their homes at every stage.[26] Peter Hübner's work in Germany came, at the end of the twentieth century, to provide this approach with a more contemporary life and image, utilising community members in the development of design concept and in construction.

Homes, hence, require particular skills, training and awareness of the socio-cultural contexts, mechanisms of interaction and the way locals elaborate their own barriers and connections spatially and mentally. Such understanding requires social analysis of the users and the context in order to grasp social tensions, conflicts and real life situations. Dealing with everyday matters and ordinary people necessarily requires an architect to operate in a way that reflects awareness of the social context and impact of his/her work. In this instance, the size, scale and site are ideas only in the user's mind, not on the site plan. However, the architect does not have ultimate control over such extended environments; rather, he needs to cope with its parameters to elaborate the best environment from the user's perspective.

The context of the Old Cairo represents a particular example of this type of community, which requires special skills that elevate the architect from the designer of objects into a thoughtful thinker who combines talent with the basic social knowledge about what is going on. Adding a single building or a house in any harah in Old Cairo has enormous impact and could be more destructive to the existing order and culture than would be the case in a modern neighbourhood. For example, if the architect could not comprehend or be part of the social change taking place in that context, he or she could damage existing patterns of interaction and activities. Current failures include repetition of superficial imitation of classical-style buildings in search of a coherent image of the past, or the most absurd arrogance of installing an entirely modern glass façade on a building, which is considered alien, overexposure and in opposition to the local notion of privacy. We, hence, need to think more progressively about the broader role of the architect, not the architect who is *"appropriate for some building tasks but not for others"*. To quote Thomas Dutton's statement about what architects need:[27]

> *Our strength and skills (the architects) lie in our capacity to generate spatial form and architectural expression for institutions, peoples and groups, but we tend to be one-sided: we reproduce and manifest the practices and spatial requirements of the dominant culture. What we need to do more, is link with struggling cultures already present to help create spaces of cultural transformation: spaces of resistance that are linked to transformative cultural practices. It is at the intersection of space and culture that architects can make significant contributions to urban social change.*

But what requires investigation is whether such skills and awareness could fit in the conventional process of design, or, as I would suggest, an alternative process of *"architecting homes"* is required. In this respect, architecting homes is a complex process of collaboration, mutual interaction and production. This leads me to think about how the architecture of home (as socio-spatial practice) is being practised today, raising the question: does the architect *design* a home or does he *architect* it? Is architecture about designing or architecting and what is the appropriate action for mobilizing the architect's comprehensive processes to reach a rational and humanised solution for human needs?

The Dispute of Design: Architecting or Designing

Brian Lawson started his book *How Designers Think* by confronting the word design; considering it as *"a generic activity and yet there appears to be real difference between the end products created by designers in various domains"*.[28] Design is a term which has often been tied to the various applied arts and engineering fields and production *either as a* noun (product), a verb (act) or as an adjective (job description).[29] As a verb, *"to design"* refers to *"the process of originating and developing a plan for a product, structure, system, or component with intention"*.[30] As a noun, *"a design"* is used for either the final (solution) plan (e.g. proposal, drawing, model, description) or the result of implementing that plan in the form of the final product of a design process.[31] Being defined so broadly, there is no universal language or unifying institution for designers of all disciplines. No limitations exist and the final product can be anything ranging from socks, cars, fashion designs, to web designs and charts. Moreover, virtual concepts such as corporate identity and cultural traditions and rituals could be designed (such as a restaurant menu, uniform and dining style). On the other side of the image is *design* as an everyday activity being practised by everyone who arranges an office, room or decorates a home in a particular style.[32]

The process of design varies from the mathematical and systematic analysis of a structural engineer calculating loads on a beam, to the unpredictable and indeterminate process by which a fashion designer produces his collections. Within such a wide meaning and process of design, architects have to balance the two extremes (imagination and unpredictability on one side and technical knowledge and systematic application on another). Christopher Alexander considers the process of design as *"the process of inventing physical things which display new physical order, organization, form in response to function"*.[33] Any architectural production requires imagination and creativity along with technical knowledge and expertise. Then, to what extent, does design as a verb relate to the action of the architect in a distinctive context like the Cairene harah?

Architecture, in contrast to design, is a situated process that takes place within a specific site within a certain socio-cultural context that could never be isolated or limited in influence. According to Thomas Dutton, it *"is never capable of completely reproducing its own existence, for it is a primary medium for dominant institutions to manifest forms and images through which their power will be communicated and legitimated"*.[34] Architecture is not like social sciences which limit their scope of inquiry to constructing subjectivity; rather, the built environment is a form of inter-subjective relations that is being generated and entrenched.[35] To practise architecture is to elaborate the environment that governs such social interaction and communication. Linda Hutcheon argued that in architecture by *"its very nature as the shaper of public space, the act of designing and building is an unavoidable social act"*.[36] She, moreover, confirmed that architecture reinstates a dialogue with the social and ideological context in which it is produced and lived. Accordingly, successful architecture requires the understanding of the nature of such environments and considers the everyday lives and social norms of its occupants.

Such qualities of successful architecture seem absent in the broad process of design, which has to be specifically tailored to suit the complex process in which homes are produced in the hawari, encompassing the elements, listed by Lawson, of imagination and creativity or technical knowledge and expertise. Rather, it includes socio-cultural awareness and collaborative work. Hence, the mechanisms of socio-spatial architecture cannot be sufficiently considered as a design process. Rather, it is architecting, the prime act that is linguistically described as the action of the architect, and by definition recognises the complexity of architecture. The verb *"architect"* means *"to design [specifically] a building"*, hence, *"architected"* means *"designed by an architect"*.[37] The process is, then, *"architecting"*, which is significantly relevant to architecture and includes, in some cases, the limited scope of designing buildings as mere objects. What is surprising is why such a genuine and relevant verb has been absent in the discourse on architecture, in which writers never stop declaring the special nature of their creations.[38]

To justify the relevance of architecting to homes, I shall undertake a brief comparison between the two processes. *To design* is a determined and systematic process by which the designer has clear objectives and a clear and predictable product, with no clarity about its characteristics (the detailed features). It is a process led by a designer, individual or a group, to whom the rights of the authorship belong. Some designs are made without a need but are based on a personal initiative, with or without implications for others (artwork such as furniture, posters, and sculpture). The end product is primarily a physical object (a device, facility, software, image, artwork) that is being, basically, consumed by people in different ways. When a place is designed by a designer, it is lifeless, unreal and is so filled with the will of its maker that there is no room for its own nature.[39]

The value of the design lies in its innovation and it should be new and modern for its time. Humans (users/clients) are consumers who do not have real opinions in its production. As such, a designed object is short-lived and becomes outdated quickly by newer versions/products.[40] The action of the architect, hence, when simplified and stripped of its contextual complexity, turns out to be a design. The modern architect (termed by Alexander a designer), for example, relies more and more on his position as an artist and on personal idiom, and intuition to relieve some of the burden of decision making that is essential to coping with the complicated information (socio-cultural, economic, political, structural) he is supposed to organise and deal with.[41]

To architect a home (in Old Cairo, for example), on the contrary, is a process in which humans come as a priority, central characters around whom the building is constructed. Architecting (especially homes) is a generative and open-ended process, which could give results far different from the initial predictions. Architecting a space is always a need; no one architects a space out of his personal initiative; otherwise it is a negative addition (idle and destructive to the context). Therefore, its prime purpose is to satisfy particular social needs through spatial articulation. The product of architecting is an environment, a habitable space and place in time, whose value is more than its physical characteristics or spatial features. As a result, the product is an essential innovation (at least for its users); it could be a reproduction, development or, in some cases, refurbishment. Therefore, the product is one which is flexible and enduring and could be developed and reproduced in sustainable usage.

While the action of "designing" a building conceives of the designer as the principal creator, the act "to architect" means the architect is a composer of collaborative group work, including different players (the user, the community, builders, financer, institutions and others). An architect, for Schneider and Till, is an agent, and architecture is a spatial agency that teams up all agents (architects, users, builders, owners) involved in the production of a building to face up to their social responsibility for the future fate of the building in which their roles are influential.[42] Therefore, architecting homes requires the essential involvement of the user, who will help the architect to elaborate and enhance the suitable living and habitable environment. The architect, on the other hand, needs to get out of his office and spend time on site to be able to comprehend the local processes of mental construction of homes and their relative social boundaries.

The Rise of Socio-spatial Architecture

> ... [there] has been a call for socio-spatial practice as a form of resistance. It is a practice grounded in counter-cultural struggle as a means to establishing a counter-spatiality ... the resistance manifested in this socio-spatial practice use the only form possible, but it can link intellectuals and communities, culture and politics, architecture and change, theory and practice. It can also recover history and culture as the agency of human subjects, an achievement that is not possible through idealism, cultural detachment, or scenography ... Resistance as a socio-spatial practice is a way of describing, by our actions, a particular approach to the world and more specifically to culture, that allows us, as architects, to connect our skills and experiences of cultural groups to a larger critique of society.[43]

The call for re-establishing architecture as a field of socio-cultural investigation as well as practice has been extensive and prominent in recent years. Since modern architecture failed to impress the ordinary people with its economical production of houses as homes of the future and the lack of deep social understanding of the postmodern movement,[44] the production of homes (as opposed to housing) seems to have been a gap in the architectural production of the twentieth century. However, a few trials by individual architects have been devoted to such inquiry with the aim of re-discovering the contemporary home as a complex domain in terms of architectural reproduction.

At the centre of the pattern language theory of Christopher Alexander is the *"living architecture"*. Alexander believed in the association between events and spaces (similar to Bernard Schumi). Accordingly, spaces created by architects play a fundamental role in making sure that a basic pattern of events, which gives a certain building its character, keeps on repeating over and over again, throughout the space.[45] Accordingly, the architect, through this socio-spatial architecture, could contribute significantly toward a more alive urban environment. According to Henri Lefebvre, *"The transformation of our present society into a humanist one can take place as an urban revolution, that is, as a revolution of spatial design"*.[46]

This led to the emergence of a refined approach in which the input of the architect is paralleled by the input of the user in a process of mutual intervention and association. Architects need to spend more time in the immediate context of their production, trying to build socio-cultural understanding of people's lives and their everyday practices. To be trained for such missions, architects need to learn from past architectural experiences: in which architects used to work as active members of the production situation as well as being knowledgeable and influential. These lessons, therefore, should not be limited to our recent history, but should go as far back as the history of architecture itself, and focus especially on the production of homes in medieval cities (as we shall see in the Cairene harah in the following chapters). Such training and experiences, however, need to be situated in the modern life and to address the increasingly changing nature of everyday practices and associated technologies.

In this chapter, the intention was not to achieve what is called the *socio-spatial architecture*; rather it raises the issue, defines its structure, and makes different attempts to deal with it. So far, in this book, I have recognised the home, defined its parameters, and now I am exploring various attempts to architect it. It is the perspective of the author that there are no golden rules or guaranteed models of architecture. Architecture remains as a field of exploration and inquiry through individual experiences. Now, I shall refer to the work of three architects representing different social and cultural contexts in three chronologically ordered attempts at socio-spatial architecture in the twentieth century: Hassan Fathy (New Gourna project, Egypt, 1940s), Christopher Alexander (Mexico City, 1970s), and Peter Hübner (Germany, 1980s–90s).

Throughout their work, the culminating experiences and investigations of such socio-spatial practice were informed by the association between the architect and his subject users to produce a resident/human-generated architecture. However, I do not consider any of those cases a success or a failure. Rather, I shall work out a path of development of an enhanced experience of socio-spatial architecture. Such development and progression in making homes bring us evidence of how socio-spatial architecture does exist and can keep breaking new ground, demonstrating potential to become the driving force behind architecture of the home for the twenty-first century that provides a successful social and cultural fit.

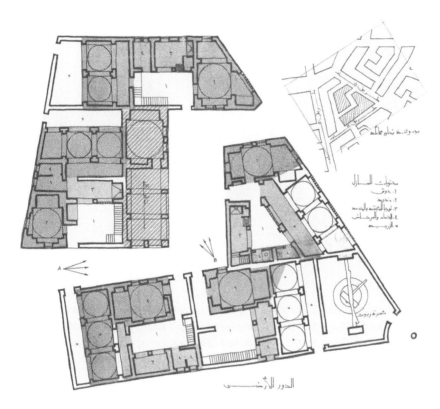

3.1   New Gourna
Village: plan
of *badana*

Courtesy of RBSCL of
the American University
in Cairo, Hassan Fathy
Collection.

ARCHITECTING PEOPLE'S LIVES

Hassan Fathy and Architecture for the Poor (1940s)

Hassan Fathy's central argument that architecture has been associated with people's culture and customs is apparent in his book, *Architecture for the Poor*. Demonstrating such association, he wrote: "*Every people that has produced architecture has evolved its own favourite forms, as peculiar to that people as its language, its dress, or its folklore*".[47] Architecture, for Fathy, is an exemplar for cultural production, in which producing a building is a process of integration between human intelligence and surrounding nature.[48] In this process, suitable spatial organisations and forms are invented to suit the environmental and ecological conditions as well as retaining social meanings. He believed that local materials and craftsmanship could provide the optimum and affordable architectural solutions while helping people to integrate with their homes and giving them the skills to adapt, extend or upgrade them. His approach to housing villagers, as it appeared in his 1945 New Gourna Project, has been hailed as a "Superb" example for indigenous vernacular architecture, and his book, "*Architecture for the Poor*", was considered among the best development handbooks of the twentieth century, while some considered his work as an exemplar for "*sustainable architecture*".[49]

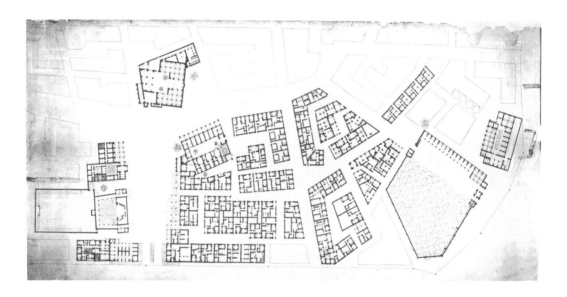

3.2 New Gourna
Village: master
plan of the village
Courtesy of RBSCL of
the American University
in Cairo, Hassan Fathy
Collection.

Fathy was a strong advocate of the community-based architecture and its social roots, while criticising the dominance of western architecture in Egypt. He asserted: "*It is not yet understood that real architecture cannot exist except in a living tradition, and that architectural tradition is all but dead in Egypt today*".[50] He tried, however, to elaborate on the traditional elements, such as the courtyard, the *Malqaf* (wind tower) and *mashrabiyya* (wooden lattice window) in modern forms.[51]

His approach to architecture was to learn from the past and produce for the future. Therefore, he could be considered the initiator of the resident-participation approach in architecture of the time. He got villagers to participate in designing, planning and building their homes while giving them the chance to express their needs and set their requirements.[52] Accordingly, every home should be different and the architect should deal with every case on its merit.[53] This was the central basic concept of the New Gourna village, which was built around the idea of "*building as a social process*". In contrast with the mass production of housing in Europe of that time, New Gourna's homes were built one by one, and every house was different in size, form and orientation.

Fathy understood the basic structure of the community: which was based on Badanas (clusters of small groups, micro society) that represented the basic family groups within the larger community structure that was composed of four tribes.[54] The *badana* was a shared outdoor space, around which the groups of 4–5 houses were arranged and attached. While recognising the working of the spatial arrangement as an actual social unit with its requirements of enclosure, security and intimacy, Fathy writes: "*streets giving access to the semi-private squares of the different badanas were made deliberately narrow – no more than six meters wide – to provide shade and a feeling of intimacy, and included many corners and bends, so as to discourage strangers from using them as thoroughfares*".[55] Accordingly, this semi-private sphere *par-excellence* included sitting areas, spaces for washing and cooking, so women could socialise while cooking and washing (a deeply rooted daily social practice). This sphere was defended through its size and control over non-axial access and was transferred into a male gathering venue in the evening when the villagers returned from work. Due to the nature of Nubian life and the use of immediate outdoor spaces as part of the house, the semi-public sphere inside was not an essential space.

New Gourna and other works of Hasan Fathy, on the other hand, have their own problems of limited application and the superimposition of particular forms, which turned out to have different meanings to the residents.[56] However, the philosophy of Fathy has been appreciated as a genuine approach that recognised the complexity and social meaning of architecture (especially that of the home) as early as the 1940s.

Christopher Alexander and the Production of Houses

While Fathy's approach and work were widely published and were subject to continuous debate during the late 1960s and early 1970s, Christopher Alexander was undertaking his experimentation in housing production in Mexicali, Mexico. He tried to put his thesis on pattern language to a practical test. He believed that the pattern language, as a set of entities of which any building is constructed, is a coherent and sensible mental language, which any individual (professional as well as non-professional) can use to build his own sensible and meaningful physical environment. For him, the pattern is the unit of the world and the truth about living.[57] Alexander was motivated by the idea that places are determined by events and activities taking place in them. Every meaningful space has "a quality without a name"; a quality that is circular (continuously repeated) and exists in our buildings only when we have it in our selves.[58] Accordingly, the person who lives in a building is the only one who can elaborate suitable language (from the suggested patterns) which makes him feel alive and human.[59]

The proposed language works as a set of a few spatial orders to be used for building homes. It sets out the possible alternatives, from Alexander's perspective, to order spaces, settings, boundaries and/or elements. As we notice in the patterns of Alexander, the outdoor space is part of the home environment. He constructed, rationally and logically, the essential link between the hierarchical levels of the built environment, connecting the home to the larger community and the city. His pattern language acknowledges that no pattern is an isolated entity, as no building exists in isolation from a context.[60]

In Alexander's theory, to feel at home you need to own it, so you can build and maintain it. Accordingly, the idea was to transfer the practice of architecting homes from the architect's hand to the non-architect residents, with the support of the architect-builder and funded by small government loans (for materials). Alexander suggested the "Grassroots Housing Process" in Mexicali as an alternative process to the conventional professionally led process of building housing and developing communities.[61] In Mexicali, the government asked Alexander to run a project for low-cost housing and every group of five families was given a collective plot of land in which a family could lay out their own house as they wished. Therefore, every house was different. Alexander and his team worked to support, but not to influence decisions or to promote forms or organisation.[62] The residents, depending on the pattern language, were to build their own home as they practised it every day.

The project started with a cluster of families (five families); and every family was provided with a trained builder for skilled jobs. Internal planning of the house followed what Alexander called "the gradient of intimacy", which means most public spheres are closer to the entrance and the more private are furthest from the entrance.[63]

The cluster plan and
open space (semi-private
sphere). (After Alexander,
1985).

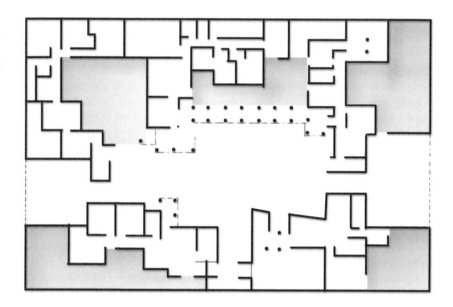

Shared outdoor space was part of the semi-private sphere and there was negotiation between families regarding associated activities and interaction in the interface spaces and shared areas.[64] The semi-private sphere was the dominant interaction social domain of such a small group, where the only actual borders between them were the ineffective plot edges (Figure 3.3). The internal private and ultimate private spheres were retained right at the rear of the enclosed structure; *"Within the private part of the house, we place a definite realm, a separate area where man and wife have their domain"*.[65] Less private were children's rooms, which were relatively connected to outdoor play areas. In the articulation of the transitional semi-private sphere, the five homes had what Alexander calls *"a beautiful family of entrances"* that opened onto the shared semi-private space. Porches were used to connect the indoors and the outdoors: which rendered a larger area liveable and made the outdoor area around the building habitable, and effectively increased the active living area of each house by 30 per cent.[66]

The resultant forms, however, were disappointing, to both Alexander and the developer, to the extent that the project was halted after one year, when the first cluster (instead of the intended 30 houses) was completed, and soon the site was deserted. Alexander expressed his disappointment, when reporting that *"the buildings were so traditional in appearance, and the naïve, childish, rudimentary outward character of houses disturbed them extremely"*.[67] While tagged by Alexander as a scientific experiment, the project was informative in terms of testing the intimacy between families and the homes they construct, the sense of community through collective production of houses.

Yet, there is apparently one problem which Alexander overlooked in his theory of patterns. People usually gain clear ideas about homes and social spheres by practising them, but, on the other hand, have very limited ability to transfer those ideas into a meaningful spatial organisation. The experiment, I would argue, confirmed the importance of the architect as a translator of everyday practice into suitable socio-spatial settings and environment. In order to get people to build their own homes, they need to learn how to architect their social spheres, which is not easy in what may be a once-in-a-lifetime job. The crucial thing is that the parts of a house also need to make a whole. Asking residents to simply locate their rooms in terms of a puzzle-like game of preferences does not help to provide a suitable and sufficient home. People need a tradition of building for themselves, not trying naïvely to invent a home overnight.

Peter Hübner and the Shoe-making

The ability to make one's own home is deeply rooted in people, almost genetically programmed. Building for oneself and with others is one of man's primal activities.[68]

Trained as an architect with an interest in the building process as an industry, Peter Hübner worked for 15 years on industrialising building processes by developing a new generation of prefabricated modular panels that suit different purposes. After spending much time on approaching building as an industry, he rediscovered the social facet of the building process during subsequent self-help/self-build projects that he initiated with Peter Sulzer. His approach to building avoided the imposition of the architect's signature: "*the dead-hand of uniformity*", and involved the users in the specification and the construction of a building. In Hübner's work, the participatory approach to building as a social process is essential to the success of any habitable building (similar to Lucien Kroll in Belgium,[69] Ralph Erskine and a number of architects working with the Segal method in the UK).[70] For such architects, the design process is "*a voyage of discovery whose end remained unpredictable, and it produced a building whose anarchic and anti-hierarchical image flashed across the world*".[71]

The majority of the buildings he commissioned were generated as an organism that would keep on growing indefinitely: "*The world has suffered enough from finished architecture*".[72] It is about creating community and therefore should involve its members: who are capable of making it a successful experience. Buildings are produced by an essentially social process that encompasses emotional ties between the occupant and the occupied which have existed for a long time; a normal process of human beings who are inclined to personalise the things around them.[73] The physical appearance and the uniformity of form are not as important as the experience and the power of the practice. As shanty towns grew unexpectedly and coincidently in response to the changing needs and development of the inhabitants, architecture, as well, should respond to such intrinsic dynamics of natural growth.

This approach was materialised through his renowned work at Stuttgart University's student village, "Bauhausle", where, in 1981, first-year architectural students, under the supervision of Hübner and Sulzer, planned, designed and constructed their hostel and rooms around a core building. The Segal structure was employed in that core building to accommodate the central toilets and kitchens as the shared public services and turned out to be the only conventional part of the whole site.[74] Each student sketched his/her room while consulting the professors and a hired master builder. Final designs, upon completion, showed a large diversity in the forms and interior arrangement adopted by each student. While the core building was the start and was distinctive in scale and character, the rest were of hybrid nature: from vaulted to pitched roofs, from single to double levels, from linear to semi-square arrangements.

While the building was not a single family home, it had attended to the collective needs of its shared tenancy as a large home, combining the notions of the house and the community within a building of limited size. This appears in the hierarchical arrangement of the social spheres. The core and central part resembles the semi-private sphere, in which all members meet and share everyday activities such as eating, laundry, cooking. It is equally accessible to everyone, while each room (the private sphere) requires permission to enter. However, every room has a small *table* and a two-seater couch, ready to receive a guest/neighbour (a potential semi-public sphere). Some rooms have two levels, of which the higher level is the bed space (the ultimate private). Semi-private spheres are well set off by the natural landscape, contained by open courtyard forms to the west and the east. This includes extended terraces and covered corridors. The full width and height glass panels ensure visual connectivity and control over the local public sphere from all sides.

The surprising uniqueness of the project is a product (unlike Fathy and Alexander's examples) of the successful and efficient organisation of the social spheres; a natural real architecture of the everyday. The important achievement of this experiment lay in the sense of the community it created and still sustains today.[75] In sum, Peter Hübner has attributed most of his work to his understanding of the social practices: which are sufficient to create a workable building and meet the basic need of its users. The understanding of this social process should be accompanied by a process of interaction during the design and the building, which will create an indeterminate stimulus that people need in their lives. Therefore, the architect, according to Hübner's perspective, should move away from design and more and more towards people.

ARCHITECTURE AS A SPATIAL AGENCY

Parallel to our three selected examples of twentieth-century architects, Tatjana Schneider and Jeremy Till looked at architects as agents of change in a kind of architectural practice they branded as *spatial agency* across the twentieth century. They displayed four examples of architectural practice from similar periods in Europe, which, according, to them, made "*a case for architecture as a socially and politically aware form of agency, situated firmly in the context of the world beyond, and critical of the social and economic formations of that context in order to engage better with them in a transformative and emancipatory manner*".[76] Those include MUF (established 1994), the London-based Feminist architects committed to public sector projects; the Moscow-based practice Obedinenie Sovremennykh Arkhitktorov (OSA; established 1925), who consider the architect to be an organiser (not maker) of buildings. Similarly, the Spanish architect Santiago Ciruged questioned the notion that the architect is the author and thereby the sole recognised designer. In his perspective, the architect acts as a catalyst of change for an unspecified period of time.[77] Finally, the New Architecture Movement, represented by publications such as SLATE, argued that architecture could not be separated from its political implications and social obligations.[78]

Architects in this spatial agency, as represented in these practices, deliver a different view of the architect's mission that is definitely not that of the knowledgeable expert. They all challenge the nature of the contemporary professional practice's reliance on the design and delivery of buildings which hold little relevance to the process of making home. Instead, architecture, or possibly the "spatial agency" for Schneider and Till, opens up to dynamic continuity and open-ended elaboration on inhabitants' lives, rituals and activities, even after the building has been taken over. This, in fact, lies at the centre of the process of making homes, as a process without end, but a continuous transformation and development. Architecture does not finish with the very moment of handing over a building to the users/clients. Rather, the transformation and development of its socio-spatial environments and domains starts to take effect.

Making Homes: From Experiments to Real-life Practice

When architecture considers the real life practices at the design stage, the immediate impact on design will appear in the indeterminate nature of the complete form. Users' actions and interaction with buildings are not predictable; however, the dynamic nature of human-building interaction should give space for alternative developments and improvements. Therefore, such dynamics should be reflected in the organisation of social spheres which appears in the spatial organisation. As a result, careful attention should be paid to the in-between spaces, and the overlapping nature between the public and private spheres. In this sense, irregular forms and non-repetitive order are by-products of the self-build experience, and the collective forms of different human experiences.

As we understand, socio-spatial architecture is a long and complex task, which appears to involve an extended and unpredictable process of interaction and association between the residents/occupiers and their desired homes. Getting families involved in the design process, even though proven essential, does not guarantee the habitability and suitability of the produced homes, both socially, and culturally. The examples of Hassan Fathy and Christopher Alexander showed us two unsuccessful cases. In the former case, families had their own mental and socio-cultural images of the home which contradicted those of the architect. The families were, moreover, opposed to the idea of moving to other sites which were not relevant to the economic base of their community (trading historical artefacts). In the latter case, despite initial success and emotional links with the new homes, involved families were left on a deserted site: which hindered any potential for sustainable life or community. The development of this approach could be seen as follows:

1. Architecting homes requires the involvement of families/residents during the design stage in considering the home as a mental idea and notion of life, before constructing it physically. Students in Hübner's case were trained to do this task, while in previous cases families were not suitably qualified or trained.
2. Community support and a sense of continuity and stability are important. That sense was strong and supported by students and enthusiastically by the university in Hübner's case. In Fathy's case, the families were reluctant to move, while in Alexander's case it was an artificial community that was easily fragmented.
3. While Hübner was a member of staff in Stuttgart (member of the community), Fathy and Alexander were invited as experts and were alien (celebrity) characters among the families and residents.

Indeed, these examples were remarkable in terms of architecting homes over the course of the twentieth century and have been explained extensively in different volumes. Our purpose here, however, is to recognise how this practice and the elaboration of social spheres developed during that period. In the light of our discussion, they appear to be experiments at an early stage in a complicated and lengthy process of understanding the architecture of home.

NOTES

1   Saunders and Williams, *The Constitution of the Home*, p. 83.

2   Blundell Jones, *The Social Construction of Space*, p. 62.

3   Ibid.

4   See for example: Schneider and Till, *Beyond Discourse* (spatial agency); Harris, *Everyday Architecture; Dutton, Cities, Cultures and Resistance* (socio-spatial practice).

5   Harris, *Everyday Architecture*, p. 4.

6   Kostof, *The Architect*, p. 3.

7   Similar to modern capitalist societies and strategies of privatising the built environment. For more on this issue see: Schneider and Till, *Beyond Discourse: Notes on Spatial Agency*, p. 100.

8   Ibid.

9   Haig Beck, the editor of Architectural Design (1976) as cited in Berke and Harris (1996), *Architecture of the Everyday*, p. 85.

10   Berke and Harris, ibid., p. 3.

11   Examples of this type could be seen in Charles Jencks, *The Iconic Building*, where he works out an analysis of the growing phenomenon of building as an *"amazing piece of surreal sculpture and an understated insertion into the urban fabric"*. Surprisingly, Jencks claims that those iconic buildings are driven by social forces, a reproduction of the modernist claim that standardisation is socially based and reinforced. It has become apparent that to justify any particular architectural agenda, no matter how distant from the general public, architects and critics link it to the social needs. This claim leaves us wondering, what are those social needs that they are talking about?

12   Lefebvre, *The Production of Space*.

13   See Charles Jencks' exploration of different interpretations through critiques of Frank Gehry's Guggenheim Museum in Bilbao (1993–97), in which the building was seen by some as fish, artichoke, mermaid. Everyone tells his own story of the building. See Jencks, *The Iconic Building*, p. 9.

14   Jeremy Till refers to architects' abilities in detailing, which helps to refine complex conjunctions through the application of technical and aesthetic judgement. In this perspective, architecture is a technocracy. See Till and Wigglesworth, *The Future is Hairy*. In line with this is Christopher Alexander's principle of "The architect-builder". See Alexander, *The Production of Houses*, p. 63.

15   Jean Prouve's ways of folding metals and creating machines for living in is one of the clearest examples of this view. See Kroll, *The Complexity of Architecture*, p. 21.

16   Kroll, *The Complexity of Architecture*, p. 21. Zaha Hadid's work is usually a good example.

17   Jencks, *The Iconic Building*, p. 7.

18   This was apparent in his *La Ville Radieuse*, which marks Le Corbusier's increasing dissatisfaction with capitalism and his turning to the right-wing Vichy regime, when, on receiving a position on a planning committee, he designed plans for Algiers and other cities.

19   Christine Murray discussed the copyright of architects' designs which, according to her, leaves architects without legal rights to protect their creativity. See Murray, *The Architects' Journal*.

20   For example: Moore and Slater, *The Architect*. About one of the masterminds (Karl Dove) in the George W. Bush administration; the job title "systems architect" is a familiar reflection of this theory.

21   Spector, *The Ethical Architect*, p. 5.

22   Berke and Harris, *Architecture of the Everyday*, p. 89.

23   This substantiated the emergence of post-modern architecture as a revolt against modernist ideologies. David Harvey, in *The Condition of Post-modernity*, referred to Charles Jenks' consideration of the Pruit-Igoe housing demolition in 15 July 1972 as the symbolic end of modernism. See Harvey (1989), *The Condition of Post-modernity*, p. 39.

24   Hertzberger, *Space and the Architect*, p. 9.

25   Fathy, *Architecture for the Poor: Experiment in Rural Egypt*.

26   Alexander, *The Production of Houses*.

27   Dutton, ibid., p. 3.

28   Lawson, ibid., p. 4.

29   Lawson, *How Designers Think*, p. 3.

30   *Cambridge Dictionary of American English*.

31   Ibid.

32   Lawson, ibid., p. 3.

33   Alexander, *Notes on the Synthesis of Form*, p. 1; Alexander could be named as the father of pattern language, which has been influential in the development of software engineering. Therefore, he could be responsible for introducing architecture to such a technical field, which resulted in software architecture and the growing demand for software architects.

34   Dutton, ibid.

35   Bickford, *Constructing Inequality*, p. 356.

36   Hutcheon, *The Politics of Postmodernism*, pp. 180–81.

37   *Oxford English Dictionary*.

38   I presume that it was part of elevating architecture from its socially-grounded practice into the higher-profile cultural and elitist domains of applied arts during the nineteenth century. This claim, however, requires further investigation and analysis that is not in this book scope.

39   Alexander, *The Timeless Way of Building*, p. 36.

40   This by no means denies any value of artworks or lasting designs, generally. However, the most appreciated and valued designs are, after a short time, being considered classic work.

41   Alexander, ibid., p. 11.

42   Schneider and Till, *Beyond Discourse: Notes on Spatial Agency*, p. 99.

43   Dutton, *Cities Cultures*, ibid., pp. 3, 8.

44   Or its fake unity with culture. Linda Hutcheon referred to postmodernism's failure to offer genuine historicity, and considers social, historical and existential reality and discursive reality when it is used as the referent of art, including architecture. See. Hutcheon, ibid., p. 183.

45   Alexander, *The Timeless Way of Building*, p. 92. As an example of his living patterns, Alexander gives a detailed description of a young boy living within a family in a traditional community: where homes and workplaces are combined in the same area, and explains how such association makes it alive in contrast with the dead communities which separated workplaces from homes. See p. 108.

46   Lefebvre, *The Production of Space* (extracts).

47   Fathy, *Architecture for the Poor*, p. 19.

48   Fathy, *Urban Arab Architecture in the Middle East*, pp. 42–3.

49   Panayiota, *Hassan Fathy Revisited*, p. 29.

50   Fathy, *Architecture for the Poor*, p. 19.

51   An example of this work is his residential compound in Baghdad, in which he designed linear blocks of apartment buildings arranged around closed (but large) courtyards.

52   Kholosy, *Hassan Fathi*, p. 85.

53   Ibid.

54   Fathy, ibid., p. 70.

55   Ibid., p. 71.

56   Panayiota, *Hassan Fathy Revisited*; Taragan, *Architecture in Fact and Fiction*.

57   Alexander, *The Timeless Way of Building*, p. 91.

58   Ibid., p. 62.

59   Alexander, *A Pattern Language*, p. xvii.

60   Ibid., p. xiii.

61   Gabriel, *The Failure of Pattern Languages*, p. 337.

62   Alexander, *The Production of Houses*.

63   Alexander, ibid., p. 186.

64   Alexander, ibid.

65   Alexander, ibid., p. 190.

66   Ibid., p. 184. However, the families could not see the point of such semi-shared space, as it did not add effective closed space to the house. However, Alexander imposed this element on each house while hiding its cost as overheads (p. 186). This highlighted the problem of relying on inhabitants to take serious decisions.

67   Ibid., pp. 355–6.

68   Hübner, *Shoes, Not Shoe Boxes*, p. 337.

69   Kroll, *The Architecture of Complexity*.

70   Guy and Farmer, *Reinterpreting Sustainable Architecture*, p. 146.

71   Blundell Jones, *Sixty-eight and After*.

72   Blundell Jones, *Peter Hübner*, p. 45.

73   Hübner, *Design Schools and Powerhouses.*

74   Blundell Jones, *Peter Hübner,* p. 14.

75   The residents of the 30 rooms formed a committee who used to interview all new applicants to make sure they were able to cope with the obligation and responsibility of being a member. The waiting list has always been long, and the building is always well maintained by its residents, without outsiders' interference. The experiment proved successful with time, and has attracted students from different fields of study over the past 20 years.

76   Schneider and Till, *Beyond Discourse*, p. 98.

77   Ibid., p. 106.

78   Ibid.

4

Making Homes in Cairo:
Constructing the Socio-spatial Architecture of Home

THE SOCIAL CONSTRUCTION OF DOMESTIC SPACE

Old Cairo is a unique context that represents a special structure of the home, both in its public and private spaces. It disputes our understanding of the home as an exclusive private sphere situated within the boundaries of the house. In the harah, domestic activities are not bound by physical boundaries of each house. Rather, domestic rituals can take place outdoors and indoors, in collective settings as well as in individual arrangements. As such, Roderick Lawrence's division of houses into four distinct zones[1] (front stage public, back stage private, zone of clean activities, and zone of dirty activities), holds little significance in explaining the phenomenon of such local communities and their socio-spatial patterns. While Lawrence's classification clarified the complexity of domestic activities in individual units, it was a rigid spatially-based typology that fell short of denoting the elasticity of domestic social practice that keeps changing its sites and venues at frequent temporal intervals. At certain and occasional times, social practices flip the home around, turning public into private and vice versa (examples include a home wedding ceremony, partying, family gatherings and national or seasonal festivities).

The hierarchy of spaces actually impacts on the rituals taking place in them. This hierarchy allows for activities to develop through a gradual transition from private to public domains as implied by building arrangement. Starting in the bedrooms, moving to home's shared living spaces, front of the house, local alleyway, and ending by the street and local thoroughfare, private activities develop in nature, defence, exclusiveness, and security. While everyday rituals such as eating, gathering, and even sleeping affirm this hierarchy, there are times when occasional rituals tend to alter this hierarchy in favour of attending to certain temporal needs. For example, the wedding rituals transform the exclusive private spaces, such as a bedroom and a house living space, into part of the public venue where female guests reside and fill them with the joy of exclusively female dancing and singing.

On the other hand, immediate outdoor space in front of houses or shops is recognised as an inseparable part of the property. Inviting neighbours and passers-by to drink or smoke in front of a house resembles a practice of rightful ownership and indicates control over private area within the public sphere. Furthermore, extending the effective protection domain for local residents beyond its gates/entrances becomes a manifestation of power and defence in the immediate proximity of the harah. Domains, hence, naturally extend beyond physical boundaries of spaces and buildings. Examination of practices, in fact, reveals a hierarchical organisation of privacy that is not manifested physically, but interpreted by local residents into fields of power, control and defence. In this sense, Lawrence's classification was isolated from external and surrounding forces that influence the domestic environment: either to expand outwards (if the outdoors is familiar homogenous and intimate), or to shrink indoors (if it is violent, strange and dangerous). This elasticity requires more inclusive categories that are not spatially bound by particular space. Rather, it requires a notion that envisages space as part of a larger social domain that is flexible to expand, shrink, or disappear according to temporal needs. If we are to understand how Cairo works as collective homes, domestic domains need to be re-structured into more comprehensible socially-based domains. I argue that the social sphere is an appropriate notion to represent such complexity both inside and outside the home. The complexity of home needs to be deconstructed into a set of socio-spatial domains that will help architects to understand and elaborate upon its everyday habitable life. The principal argument, here, is structured around the notion of social sphere as a comprehensible domain (both socially and spatially) of homes.

THE NOTION OF THE SOCIAL SPHERE

Social sphere is a notion that has been extensively, but implicitly, used to describe human activities, habitual practices and rituals, especially those taking place in groups. Humans create their own range of social spheres which control their behaviour and the way they act. The terms public sphere and private sphere are frequently used to distinguish the activities and social interactions within the house from those of the public spaces and modern media of communications. Public sphere, for example, has been recognised by contemporary theorists as a non-state arena of communicative interaction, a central space of opinion formation among the mass public.[2] Private sphere, on the contrary, connotes domestic environments as a private domain that includes different social situations whose limitations and nature change continuously. However, the distinction between public and private and their meanings change with time. What constituted a private sphere in the Roman era included slaves, teachers and servants who either disappeared (slaves) or we regard them today as independent workers and equal members of the public domain (teachers and servants).[3]

Jurgen Habermas found that the division of the social spheres is due to the intrinsic nature of both domains. He states: "*We call events and occasions 'public' when they are open to all, in contrast to closed or exclusive affairs – as when we speak of public places or public houses … . The private sphere evolved into a sphere of private autonomy to the degree to which it became emancipated from mercantilist regulation*".[4] He sees the two-fold social sphere of "public and private" as resulting from the separation between the state and the society.[5] In this context, social sphere is reduced to its political meaning, which defines two institutions of human life as subject to control, the public (outside homes), where state and media are in charge, and the private (inside homes), where the family is in charge.[6]

In everyday life, neither term is bound by the boundaries of homes. They are used in extended contexts to describe different types of human activities, everyday situations: the open and shared against the exclusive and intimate. Richard Sennett differentiated in *The Fall of the Public Man* between private and public lives. In public life, people act collectively without being the same: while they tend to carve a sense of familiarity in their private life.[7] As a result of the dominance of a predominantly public pattern in the modern world, individuals tend to act out of psychology of privacy. Such emergence of private within the political and public life affects both perception and the organisation of the social space in general.[8]

Both terms are adjectives of the social situation. Every action and activity of a human being is a social situation, whether individual, in private, collective or in public. The confusion emerges at the level at which the interactive situation could be described as private or public. Despite the extensive use of both terms, private and public, and their representation of social interaction, we rarely hear the term "social sphere": the basic structure upon which the distinction between both is constructed. Some scholars claim that "public" and "private" are useless terms or even dangerous categories that should be avoided, combined or replaced.[9]

Architects, on their side, often take the two-fold private/public for granted, without comprehensive reflection on daily activities and their association/dissociation with spaces. The public is outdoor, government managed spaces or buildings, while private refers to an enclosed, privately owned property (house, buildings, garden). Furthermore, they consider social space as a shared public space (even within private property), denying the private situation of being social.[10] They are used to considering the social space as a space of gatherings (public): which implies that the non-public/non-social is the domestic realm of the home.[11]

In contrast with this view, Anthony Giddens's notion of *"locale"* links the interaction to the setting in all situations. He sees social interaction as constituted by its spatial setting; location is part of the explanation of why and how things happen in the way they do.[12] Gidden's notion of "locale" works in line with Henri Lefebvre's understanding of the social space, which covers all types of spaces from a single room to a marketplace:

> *Everyone knows what is meant when we speak of a 'room' in an apartment, the 'corner of the street, a marketplace, a shopping or cultural centre, a public space, and so on. These terms of everyday discourse serve to distinguish, but not to isolate particular spaces and in general describe a social space. They correspond to a specific use of that space, and hence to a spatial practice that they express and constitute.*[13]

Moreover, Henri Lefebvre argued that the space could be read through three different approaches: mentally (psychological), physically (natural and visual) and socially (interaction). These three facets are interconnected dimensions of what he called the *"social space"* and which he sees as a social product, which *"incorporates social actions, the actions of subjects both individual and collective who are born and who die, who suffer and who act".*[14] Therefore, the social, as used in this book, is a term that covers the different situations of communal and non-communal activities. To depict the complexity of the social situation within a particular social and spatial context, the term *"social sphere"* refers to a relational domain that relies on the social interactions within particular spatial settings during particular moments in time (Figure 4.1).

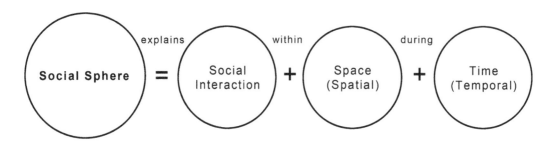

4.1   The notion
of social sphere

Social sphere in this discourse does not contradict the radical idea of space; rather, it links the real-life essence of activities to its spatial organisation. While some recognise home as a site for personal and familial production, others see it as a physical location with the structure and routines of social reproduction (eating, sleeping, sex, cleansing and child-rearing). The discrepancy appears in every attempt to limit the home physically. For example, while Neil Smith states that *"the borders of the home may be sharply defined, as in the walls of a structure of the markers of private property that include other private space such as garden or courtyard"*, he returns to qualify this statement by saying *"or they may be more fluidly defined as the space of the home fades in community space"*.[15] He realised that the home is a contested zone (internally and externally) based on struggles of authority (gender relations: men/women) and dominance (power relations: culture/ethnic).[16]

SOCIAL SPHERE OR SOCIAL SPACE

Henri Lefevbre, in his seminal book, *The Production of Space*, asks: *"Is space a social relationship? Certainly, – but one which is inherent to property relationships* [land ownership] *and also closely bound up with the forces of production* [which impose a form on that land]*"*.[17] Social space, in this perspective, contains a diversity of objects, both natural and social, including the networks and pathways which facilitate the exchange of materials, things and information. Furthermore, there is no single space to be defined in isolation from its context. Local appears in the public and the worldwide does not abolish the local. *"The intertwinement of social spaces is a law"*: which is devoid of any definition of a particular space in isolation from others. Considered or defined in isolation, such spaces are mere abstraction, not real phenomena.[18] Hence, space as an individual object becomes problematic and insufficient to describe the socially rich environment of homes. Social sphere, on the contrary, appears to be more inclusive and fluid. It represents the social situations as well as gender authority and power relations within its everyday practices and interactions.

Social sphere, therefore, could resemble the architectural manifestation of Anthony Giddens notion of *"Locale"*, which is basically *"simultaneously and indivisibly a spatial and a social unit of interaction"*,[19] in developing understanding of how architects can build homes in the hawari of Old Cairo. It has become evident that housing design failed to respond to extended socio-cultural dynamics of homes during the past century, moving towards the consumption of space in lieu of developing it as an interactive social domain.[20] To identify the difference between the SPHERE and the SPACE,

I shall undertake a comparison between home (sphere) and house (space) in what I call a socio-spatial equation:

Home becomes larger than house in old towns' alleyways and their relatively safe and controlled environment that allows private activities to extend outwards. People in those areas make less distinction between indoor and outdoor within the territory of homes. Here, the hawari of Old Cairo emerge as significant examples of this inclusive home, where private activities are welcomed and shared by neighbours within controlled entrances. Similarly, new gated communities of suburban compounds demonstrate their exclusive environment through barricaded gates and high fences. Philip Johnson's Glass House gave us another, but modern, example by utilising the surrounding natural environment as part of his glass house. While the physical space and motion of people stops at the glass wall, the home as an environment and social sphere extends outwards, protected by natural boundaries.

Home becomes smaller than house for individuals who tend to spend intimate time on personal interests such as writing and painting. Their home is the room in which they feel at home. It is the protected and spatially organised territory from which others are essentially excluded.[21] In Japan, entrance lobbies where shoes are kept tend to be lower than the rest of the house and these indoor spaces are excluded from the domain of the home. This space is part of the house, but not home, the highly decorated domain. Recently built care homes and shared tenancy properties are good examples of this type, where the resident feels at home when he feels in control of his/her environment. Outside this realm, every action and behaviour is subject to negotiation with strangers/others.

As a result of this comparison, one can depict the difference between the conventional notion of physical space and the proposed social sphere. While the space is a leading design unit for a house, sphere is, I argue, the appropriate tool in the domestic social environment, for the architect, when s/he is architecting a home. Spaces are becoming limited in their scope as primary components of a home. They are rigid, artistic, artificial creations and engineering productions of the architect rather than real-life experiences. Accordingly, the social sphere is introduced as an alternative setting; its inclusive nature produces real-life homes. In that sense, the space is part of the sphere, which includes types of activities, social settings and temporal flexibility that tolerate change over short as well as long periods of time. Social sphere could extend over several spaces, indoor or outdoor, depending on the context, rationale and the needs of its residents.

PUBLIC AND PRIVATE SPHERES: THE MYTH OF SEPARATION

> We exist in two realms, the private on the one hand, and the public on the other. The realms are distinguishable not only theoretically but because different behavioural codes and communications cues obtain in the two. One does not require the same characteristics, for example, from the person one marries as from the person one supports for public office.[22]
>
> Binary categorization of urban space as public or private ignores transitional spaces which may be influential in the formation of patterns of sociation.[23]

The statements above from Harvey Cox and Malcolm Miles respectively complete each other to provide a uniform understanding of how our domains of activities work in everyday life. While Cox recognises human and social needs in two distinct situations and the way relevant behavioural codes are at work, Miles acknowledges that two such domains don't work in isolation from each other. Rather, he sees each social domain, private or public, as interlocking with the other in what he calls *transitional spaces*. In other words, both realms are defined by their existence side by side in one inclusive pattern of social world, in which both overlap within certain zones, providing a space for transitional journeys.

Comparatively, Islamic medieval cities and their hawari have long been analysed, on the basis of archaeological information, as two isolated worlds: the indoor private sphere of women versus the supreme public sphere of men. The latest studies, however, have proved that this perception was not accurate. Social investigation of the hawari and domestic environments provides evidences that private and public spheres have always involved interconnected and overlapping relationships. The harem was a site of joy and social networking of women across families, while *al-mandharah* (male reception) was a busy gathering space for families.[24] Both are parts of medieval houses.

Our lives take place in, as Harvey Cox mentioned above, two realms, the intimate and closed, and the open and shared. They complete each other in every single action and the meaning of each is realised in relation to the other. However, those two realms are not rigid; they are flexible, interconnected and dynamic. Some of the intimate relations of ancient cultures have become public today and what is private in the evening could be public in the morning.[25]

Richard Sennett undertook a historical analysis of the private and public domains, concluding that they have developed along with the development of their social context and western cultural ideals. Public, in contrast with the family's private domain, *"came to mean a life passed outside the life of family and close friends; in the public region diverse, complex social groups were to be brought into ineluctable contact"*.[26] The linguistic meaning of each term is defined by its opposition to the other. The *Oxford Dictionary* defines private as *"Restricted to one person or a few persons as opposed to the wider community; largely in opposition to public"*.[27] It extends the meaning to the modern context when it describes personal ownership, paid for services as opposed to public/ state services. Public, on the other hand, is defined as *"the opposite of private; open to general observation, view, or knowledge; existing, performed, or carried out without concealment, so that all may see or hear. Of a person: that acts or performs in public"*.[28] Both terms are correlated with the conditions of behaviour and terms of belief, which change from one historical era to another. While Public emerged as identification of the common good in society during the 1470s, it developed quickly to mean manifest and open to general observation. Private in the same context referred to the privileged and high calibre governmental level.[29]

This is similar to its meaning in medieval Arabic, in which both terms, public and private (*Aa'mma* and *Khassa*), had principally social meaning, and represented the class hierarchy and stratification. *Aa'mma* (Public: literally means general) referred to the native people (mainly middle class), and *Khassa* (Private: literally means special) were usually the ruling class.[30] With the development of society, the social distinctions of these terms disappeared in favour of modern concepts of ownership and production.

The contemporary meaning (in Arabic and English alike) developed during the nineteenth century, upon the growth of cities and the development of networks of sociability.[31] Private ownership became the meaning of al-Khassa, while government owned/managed spaces and services became al-Aa'mma. During the twentieth century, and the expansion of formal political institutions to dominate the city's outdoor spaces and services, homes emerged as the protective shelter of the private lives of human beings.

People are more sociable the more they have some tangible barriers between them; and in the same way they need open public places whose sole purpose is to bring them together.[32] Such barriers help them to construct subdivisions of their habitual environment, which, consequently, facilitate the mental process of organising a hierarchy of social contexts. This hierarchy does not mean a discrete separation between the subdivisions (e.g. private, public, semi-private … etc.); rather, it helps the individual to situate his/her reactions and feelings into an intelligible order. This is not constrained by the radical private (indoor)/public (outdoor) divisions; rather they appear at every spatial and social level from the family to the society.[33] People create their own private domain within public spaces (personal e-mails, chatting online in the office, or a homeless man sitting on a corner of a pedestrian walkway/park), and can expose themselves publicly at home (a woman posing on her home's terrace).

When we consider homes at mealtimes, they are rearranged to suit eating rituals and situations. During breakfast, the family is mobilised in the kitchen and other rooms are abandoned. The breakfast table is arranged, and in most cases, exposed to direct daylight through an adjacent glazed window. There is a need to link breakfast to its temporal as well as psychological situation: a new beginning, a new day. If no guests are expected, it takes place in an informal setting in the private part of the home, the kitchen, the backyard garden, or a small dining table.[34] Formal meals, on the other hand, are associated with a front part of the home, presentable to guests. They are associated with reception, living, and dining, the front door spaces.[35] Formal dining at home depicts much of the restaurant meal formality. It is formal, presentable, honoured; and the contextual environment, including furniture, lighting and decoration, is enhanced in an attempt to craft a public image of the household. This was the situation among the hawari's high profile residents.

In contrast, in a tight working class home, where the cooking capacity was minimal, the kitchen was reduced to a small counter, and the breakfast table disappeared. Living space was merged with dining and sometimes kitchen spaces. The setting was made entirely different by merging different practices within smaller and multifunctional spaces for social and financial reasons. In such a context, the home became an active functional space with no formal dining or formal invitations/visits. These two models display different social contexts; the spatial arrangement of each is justified by and explains certain social practices (private, public or public-private). Each model represents either the inclusive organisation (public and private spheres) as in the former case or the exclusive (private only) organisation as in the latter case.

To provide distinct boundaries between the nature of indoor and outdoor activities, the private and the public, social scientists and consequently architects have applied a holistic approach of studying the phenomenon of home as a private sphere versus the public sphere of the street and the city at large.[36] Yet, this notion is not convincing when it is applied to different spatial contexts either in location (culture) or in time (historical).[37] The house as a single private sphere does not explain the inclusive typology of homes which includes spaces with similar characteristics to the public domain (formal salon, antique dining set), nor does it explain the private activities taking place outdoors (as in the hawari of Cairo). This is basically due to the two intrinsic labels, the private (indoor) and public (outdoor), which do not stand in an everyday context (parties, events, invitations …). Saunders and Williams argued that any home can have both public and private spheres: *"the specific use of rooms, allowing dwelling occupants to be together and separate, also enhances the regionalization of the dwelling in terms of relatively public and private spheres – the front and back regions. In that regard, the home, like the nation state itself, may act as an essential constitutive and reproductive element in conceptions of the public and private"*.[38] When the public sphere (activities and rituals) extends to the home's representational spaces, other thresholds and boundaries will be required inside the home (between the private and the public). These constructed thresholds could be physical or psychological, according to the social and personal settings organised by the residents.[39]

ZONES OF EXTENSION

Domestic spaces are dynamic in nature and keep developing and changing according to the need and situation. The significance of spaces' physical characteristics is limited when compared with active practices that take place in them, and which become routine or occasional rituals. While public spaces are tagged as outdoor, public practices and rituals extend indoors, replicating similar behaviour and features inside the private domain. Similarly, private activities and practices expand outdoors in the immediate context of the private, creating what are called semi-private spaces/ spheres. The nature of such extension/expansion is to be defined within its zones of influence (where private and public spheres are not bound by their physical/spatial settings), which prove more flexible and elastic than spaces. Therefore, the notion of the social sphere could be more inclusive and useful than social space, as it explains the complex nature of overlapping private and public activities.

Zones of extension are seen as intermediate domains that may accommodate activities totally at odds with their spatial designation. For example, the front lawn of a house is outdoor space with private activities and characteristics. The alleyways of old towns are extensions of the private spaces of the house.[40] Similarly, the balcony, the porch and the courtyard of Mediterranean and Middle Eastern houses in cities such as Naples, Barcelona and Cairo are hard to categorise.[41] The balconies are private spaces with private activities (put out plants, store fruit, hang up washing and exchange conversation with other balcony users) that are part of the public domain (residents' dress needs adjustment before commuting to it), while the courtyard is a public space implanted in the private domain. On the other hand, while reception and dining rooms are indoor spaces, they are arranged to reflect the abilities, prosperity and image of the household to the public rather than to its residents. We cannot categorise the front lawn as a public space; rather, it is a *semi-private sphere*, as it accommodates private practices and rituals in a public area. On the same principle, we cannot consider the reception area as private space; *semi-public sphere* seems more appropriate to describe public events and rituals taking place in it.

Amos Rapoport has considered semi-private space as it appears in the cul-de-sac alleyways: where all homes have visual access to it. It is relatively secure and this safe environment encourages collective activities.[42] However, he considered the space and its spatial characteristics to have psychological significance. This created an artificial division of one combined whole, which was later reduced it to its physical characteristics: making it easier for the modern movement and modern architects to deal with in their illusion of understanding and working with the home. Complexity is hated and simplicity is celebrated, even if it is artificial.

Contemporary architects of the twenty-first century, on the other hand, have started to appreciate the value of homes (quality of life, environment and context) and their relevance to almost all architectural production. They have realised the positive influence a good organisation of social spheres and the sense of home have on the quality of life within and around buildings and on the appreciation of the users. They tend to capitalise on combining the private and the public into one inclusive socio-spatial system with a planned hierarchy of spaces and activities. The architects Rogers Stirk Harbour planned both private and the public spheres of their Maggie's Centre in London, a care centre for cancer patients, to complement the inclusive organisation of social spheres within an intimate and comfortable environment. They did not attempt to define internal boundaries of privacy as they did with the harsh outside urban context. The architects transformed this public service into a private environment of a home-like lifestyle.[43] To do so, they softened public spaces with intimate pictures and images, such as a house-like dining arrangement, extending the internal environment to outdoor spaces, and considering the terrace as an essential part of the private domain. This Stirling Prize winning building for 2009 was protected from the harsh surrounding urban chaos of the frantic Hammersmith thoroughfare by a thick orange masonry wall, carefully planted gardens and groves of trees.

It is important to emphasise that the organisation of social spheres represents a comprehensive and inclusive domain within which games of privacy and public interaction are in continuous and dynamic interplay of dominance. As such, this text perceives space (buildings and urban; social or intimate) as the spatial arrangements to accommodate such social dynamics, temporal adjustment and continuous change in an inclusive elastic unity. The social sphere, in this sense, is a similar construct to the cell, which is a coherent unit of life with continuous action and flexible movement, producing and consuming energy. If movement and action stop, no energy is produced/consumed and the cell dies. Similarly, if games of interplay cease between private and public, the social sphere is frozen, no longer active, and the home is reduced to the object form of the house, a dead physical space.

Therefore, making a home requires an enhanced organisation of spatial form and relationships that consider layered social activities and dynamic interaction. Space, hence, needs to have the capacity to perform according to the changing order of activities and rituals, on a daily, weekly or annual basis. By integrating private and public domains into the flexible and inclusive notion of social sphere, architects have the flexibility to plan their interconnected spaces and hierarchical order of interaction in much extended settings. This involves inclusion of public outdoor spaces as part of their building environments and indoor private spaces as part of the public world. Conventional perspectives of social interaction consider only two basic domains, the private and public. However, social spheres, in reality, are comprehensive and overlapped domains that include zones of extension in which semi-private and semi-public zones accommodate transitional journeys between the intimate and the shared (Figure 4.2).

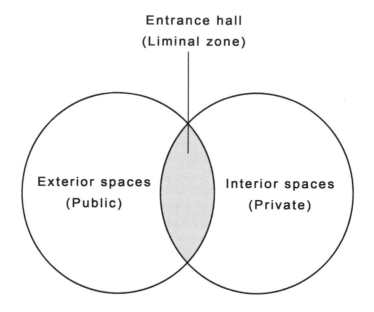

Entrance hall
(Liminal zone)

Exterior spaces
(Public)

Interior spaces
(Private)

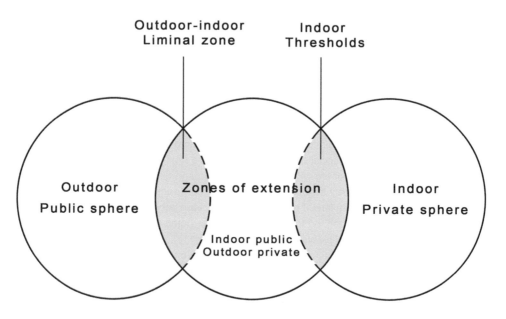

Outdoor-indoor
Liminal zone

Indoor
Thresholds

Outdoor
Public sphere

Zones of extension

Indoor
Private sphere

Indoor public
Outdoor private

4.2   Private/Public boundaries:
from liminal zones to the complex nature of overlapping social spheres
After Sibley and Lowe (1992) and Abdelmonem (2011).

BUILDING THE SPECTRUM OF EVERYDAY SOCIAL SPHERES

Instead of using the binary categorisation of the private and public spheres, homes need to be understood through a refined spectrum of social spheres that governs their socio-spatial order and organisation. This spectrum includes the basic activities and their spatial order, indoors or outdoors, which are extremely relevant to the hawari of Old Cairo. The proposed spectrum emerged as scale of privacy with open (ultimate) public and ultimate private spheres at its extremes.

A. Public Spheres

Open-public sphere is the urban site that brings together the various social structures within society. This sphere stitches different urban components and units (communities, neighbourhoods, districts) together at central state-managed venues, spaces, institutions, and services, which do not belong to a particular locality. It is a place that is, in Sennett's terms, "*a human settlement in which strangers are likely to meet*".[44] It offers respite from constraints of the locality. It is where everyone is a stranger, without social responsibility or roles. This sphere contributes to the shared public image and character of society that shape people's feelings and perspectives.[45] It is far from natural; rather, it is cultural and ideological.[46] This category matches what the majority of social scientists and theorists, such as Jurgen Habermas, Manuel Castells and Richard Sennett, consider as the public sphere.[47] Architecture of this sphere takes shape over a long period of time and develops out of dominant cultural and architectural perspectives. Al-Muizz Street, the central thoroughfare of Old Cairo, for example, represent a chronological development of Islamic Cairo, and its religious and cultural significance is entirely different from Cairo's downtown, with its predominantly classic European-style buildings of the late nineteenth century (Figures 4.3 and 4.4).

Local-public sphere is defined within the perimeters of a single urban unit (community, neighbourhood etc.). It is a domain for social and political gatherings which emerges through shared socio-cultural, spatial contexts, and in some cases, interests.[48] This sphere is situated within a locality and represents its social organisation and cultural norms. It constitutes defensible space within which it is possible to perpetuate cultural practices which are relatively immune from penetration, dilution or destruction from the outside.[49]

B. Zones of Extension

The semi-private sphere extends outside the traditional private sphere, into the public space. It represents the power, ownership rights and influence of the private owner in his/her immediate outdoor space. It works as outdoor transitional zone between the traditional private and traditional public. It replicates the rules and social systems of the family in the local public space. Doorsteps of all workshops in Old Cairo are reorganised to host a collective meal followed by communal tea-drinking at breakfast and lunch times.

The semi-public sphere is essentially part of the private sphere that is exposed and uncovered to the public (guests, visitors). It works as a transitional venue between the private and the public. It shares several activities and features of the external public sphere; however, it is controlled by the household. This sphere includes front of the house activities, such as those taking place in reception space, living room, dining room, front door gardens, and front lawn.[50]

C. The Exclusive Private Sphere

Common-private spheres include family living, internal circulation, kitchen and services spaces and associated activities. This sphere lacks a defined physical form as it extends between the semi-public sphere and the ultimate private. It works as a transitional zone between the semi-public and the ultimate-private. It is not accessible to strangers, but accessible for all members of the household at any time.

Ultimate-private spheres are, culturally and socially, recognised as sacred and defined by activities of sleeping and intimate intercourse. It is at the end of the zones, paths, and the final destination of movement. It is marked by minimum interpersonal but intimate communication and interaction, such as sexual intercourse between partners, breast feeding of babies, and discussion of private issues. In sum, social functions are marked by high intimacy and enclosure.

This classification, however, is meant to be provisional and explanatory of the hierarchical arrangements of social spheres of the hawari of Old Cairo. It is not a rigid model that is applicable to every case. The spectrum of social spheres of Old Cairo's hawari, for example, is entirely different from that of London or Manchester's apartment blocks. The hierarchical arrangement of spheres differs from one city to another (based on the dominant ideology, urban planning, and architectural practices). However, any socio-spatial organisation of a home does include this spectrum, either fully or partially. Some spheres shrink and others expand according to the context: which produces diverse models of socio-spatial organisation of homes. City centre apartment towers, for example, experience neither the local public sphere nor the semi-private one. Once people leave their apartment door, they have neither private zones nor local public sphere. They move from the exclusive private to open public sphere, where they are not distinguishable. In most cases, these apartments are tight enough, (for economic reasons) to afford a semi-public sphere inside the home. As a result, this model of experiences could represent the most reduced hierarchy of social spheres.

In contrast, the hawari enjoy an extended organisation. Houses normally have semi-private spaces, and in some cases private access. The local public sphere is powerful and local members are familiar with each other and used to meeting on a daily basis (in the mosque, in the alley, in the market, at the food shop, retail store). The organisation of old-quarter communities, such as Old Cairo, provides a unique experience of overlapping of social spheres within very tight spatial arrangements.

THE BOUNDARY, THE BORDER AND THE THRESHOLD

At the core of the proposed notion of the spectrum of social spheres is how people enter/exit from one social sphere to another. By what process does such socio-spatial transition take place? This leads us to discuss the way in which the social sphere is mentally constructed and spatially defined and to suggest three types of crossing point: the boundary, the border and the threshold. Each of these forms has its own meaning and structure based on the social norms and practice in its sphere. But how is the proximity and distance between spheres established and ordered according to these crossing points? How can we define the end of one sphere and the beginning of another and the consequences that crossing over have on individuals' self-adjustment?

4.3   Open public sphere of Old Cairo (Share'i al-Muizz, North end)

Note: Character predominated by mosques, minarets and artefact workshops.

4.4   Open public sphere of downtown Cairo (Ahmed Orabi square)

Note: Character predominated by massive and classic European-style public buildings and radial street networks.

While the boundary resembles the hardest edge, which is constructed for defensive and protection purposes (not essentially physical: every man has his own boundary), the border provides a permeable barrier, which allows overlapping and self-adjustment, psychologically and socially, for the coming change of status. While the boundary is sharp and immediate, the border is smooth and transient. The threshold, on the other hand, represents the journey, process of admission/exit, the rite-de-passage of crossing spheres; it might not involve physical movement or transition: rather, it incorporates temporal events, social practices, such as parties, festivals, birth and death ceremonies, which change the nature of the social sphere within the same physical characteristics. It is constructed through a set of rituals as detailed earlier in the work of Pierre Bourdieu, Mary Douglas or Victor Turner.

Boundaries at home are established at every point that constitutes a significant change in the degree of privacy (from the ultimate-public sphere to the local-public, or from the semi-public to the ultimate private spheres). This is evident at old communities' narrow entrances (usually monitored by local residents) as well as modern gated communities, where boundaries are manifested physically through high fences and guarded entrances.[51] It is constructed between the inner (pure) self and the outer (defiled) self. To this extent, gated communities are a spatial manifestation of the need to separate the self from the polluted and unidentified city (the ultimate public sphere).[52] As such, gates are boundaries that construct and manifest social relations: and, in this case, mark the segregation between the outside and inside. Gates make boundaries more visible and psychologically salient.[53] This physical manifestation of separation is used to make a [artificial] place-identity. This identity is created by investing architectural qualities of style and form to reflect the way the inside contrasts with the outside: a strategy in which gates are significant spatial and social elements.[54]

Boundaries are seen as a fortification for defence and protection against intrusion. A boundary is essentially a guarded territory, whether natural, like those guarded by prides of lions, or physical, such as the military camp.[55] The boundary represents in the ecological structure, according to Sennett, the cell wall, while the border is the cell membrane. The boundary does not allow penetration of any kind and should be visible, physical, made of porous materials, and demonstrate strength.

On the other hand, the border allows interaction, absorption and transition, and should be permeable and transparent. The border could be a floor step, pattern, or decorative frame that forms the edge without fencing it. It is a site of exchange. Sennett referred to the ecological border as the site where organisms become interactive, such as the shore-line of the lake, where the land and water overlap, with no defined edge or limit. Borders could be transparent, non-physical, psychological, or social, and facilitate fine tuning of a human being's behaviour when moving between two zones/spheres. Another stimulating and interesting theory about the border comes from the French psychoanalyst Didier Anzieu, who sees the human skin as the border between the familiar and the foreign.[56]

The border appears within a larger homogeneous context, but represents a transitional situation, where an individual's status changes, such as moving between the semi(s) spheres: semi-public to semi-private, from semi-private to internal private.

Accordingly, border resembles the socio-cultural code of practice that is neither physical nor visible. However, Modern architects of the twentieth century played the game of transferring the boundaries of the pre-modern buildings into borders. They were keen to ease the fortification of the house boundaries into transparent but impermeable borders.[57] In that sense the solid walls of traditional houses were turned into full width glazed doors/screens.[58]

DEFINING SOCIAL SPHERES IN CAIRO

One of the serious problems that face architects in Cairo lies in the lack of interest and knowledge of how people use their spaces, indoor and outdoor, private and public. The uninterrupted flow of activities, informality and intimacy of social interaction, tightness of indoor spaces, and the reliance on outdoor spaces in everyday interaction are among the many factors that make the architect's mission to accommodate the user's needs within a closed and limited building almost impossible. This stems from architects' inclination to conform to conventional design processes and adopt a specific programme within the fixed boundaries of the site. Hence, it is crucial to understand the way the spectrum of social spheres in Cairo is structured and at work, if we are to understand the nature of people's interaction and everyday activities in this part of the city.

Old Cairo, in particular, represents an interesting case in terms of the spectrum of social spheres. Well-defined by its old walls and limited in size, Old Cairo is spatially protected from the disorganised urban chaos (fabric and traffic) outside the walls, consolidating a distinct and homogeneous urban character. The walls defining the boundaries of the hawari's open-public sphere are characterised by al-Muizz thoroughfares with the bustle of human traffic, busy stores, and rising minarets. The shared public image is predominantly medieval Islamic, with popular indigenous culture and traditional food, drinks and coffeehouses. Thus, strangers and tourists, heavily present in this area, are easily recognised and monitored.

Such activities, open interaction and image of the open-public sphere are broken at the hawari's entrances, gates or junctions. At the usually tight entrances of the hawari, busy stores and well-attended coffeehouses disappear and small trade and industrial workshops emerge, with tighter security maintained by shop workers. The harah's entrance is the only threshold within a solid, thick, and impenetrable mass of back-to-back buildings and houses. Unlike the old walls, the hawari's in-between shared boundaries are neither spatially nor visually defined. The thresholds of the old city and the hawari, however, are carefully spatialised and designed to drive the visitor through the *rite-de-passage*, in which he/she experiences a decline in their status and power as the daylight decays in the shadows of the city/harah's narrow and dark gates/entrances. Such experience prepares the visitor to submit to the power and control of local culture at its sphere of dominance (within the city or within the harah) (Figure 4.5).

Within the harah, the central alleyway joins ground level spaces of different houses to provide zones of extensions that overlap the private with the public social spheres.

4.5   Boundaries of public spheres in Old Cairo

Emphasis of control through scale and darkness in the entrance of Sikket Berjwan.

Private domestic activities may take place in the harah if space in the house is too limited to host them. On the other hand, some public activities (such as gatherings, celebrating an occasion, and socialising with neighbours) take place in houses when space permits. While boundaries of the city and between the hawari pose strength and power, internal boundaries between different houses and the central alleyway or between semi-public and semi-private spheres are blurred and. largely imperceptible The *semi* zones act as transitional zones, with no fixed venues for either private or public territories.

4.6   The interplay of privacy in the central alleyway of haret al-Darb al-Asfar

Inside homes, private spheres are dynamic and flexible, intimate spaces like bedrooms are open to the rest of the family during the day. They could be used for studying, to host friends and family members at times. They could become public, when relatives and friends stay for the night, or at the time of weddings and parties in which all spaces are open to public activities.

The boundary between the hawari and old city public sphere or between the old city and the outside world defines the hierarchical order of the social spheres, while this order is intentionally blurred when it comes to internal divisions and activity within a single harah. This could be interpreted as supporting the notion of the harah (as perceived by its residents) as a secure and solid community and a home with a spatial capacity to adjust itself to changing needs in the short and long term. Good protection from outside threats and intrusion, internal integration and homogeneity could be seen as the principal assets and compulsory requirements of the Cairene harah as a home with an inclusive and dynamic order of social spheres.

NOTES

1   Lawrence, *Housing, Dwellings and Homes*, pp. 106–7.

2   Bickford, *Constructing Inequality*, p. 356.

3   Gazda, *Roman Art in the Private Sphere*, p. 4.

4   Habermas, *The Structural Transformation of the Public Sphere*, pp. 141–2.

5   Perri 6 et al. in *The Future of Privacy* adopted a similar view when they recognised that the social domains were in conflict in the late twentieth century due to the continuous intrusion of the private sphere by the spread of surveillance systems. See Perri 6 et al., *Future of Privacy*, pp. 21–2.

6   Bickford, ibid., p. 356.

7   Sennett, *The Fall of the Public Man*.

8   Tonkiss, *Space, the City and Social Theory*, pp. 25–6.

9   Kilian, *Public and Private, Power and Space*, p. 115.

10   In Terence Riley's book *The Un-private House*, any house that is not a closed and isolated domain is described as non-private. Private emerged, therefore, as a natural characteristic of the home which could be ignored by the architect. However, the book does not tell us whether the un-private house is habitable or not.

11   Miles, *After the Public Realm*, p. 256. A large number of architectural books refer to community buildings and public services as social spaces and social buildings. See for example: Blundell Jones, *Peter Hübner*.

12   Saunders and Willams, *The Constitution of the Home*, p. 81.

13   Lefevbre, *The Production of Space*, p. 16.

14   Ibid., pp. 26, 33.

15   Smith, *Homeless/Global*, p. 104.

16   Ibid.

17   Lefevbre, *The Production of Space*, p. 77.

18   Lefevbre, ibid., p. 86.

19   Saunders and Willams, ibid., p. 82.

20   Schneider and Till, *Flexible Housing*, pp. 36–8.

21   Storey, *Territory*, p. 170.

22   Cox, *Sociology and the Meaning of History*, p. xvi.

23   Miles, *After the Public Realm*, p. 255.

24   Keddy (harem activities).

25   Gazda, *Roman Art in the Private Sphere*, pp. 4–6.

26   Sennett, ibid., p. 17.

27   *Oxford English Dictionary*.

28   *Oxford English Dictionary*.

29   Sennett, ibid., p. 16.

30   Refer to al-Jabarti's chronicles for frequent use of these terms to signify the difference of social positions.

31   Sennett, ibid.

32   Sennett, ibid., p. 15.

33   Such hierarchy appears within the bedroom, the most private space, where there is a sofa/chair where sons /daughters are allowed to sit (internal public), whereas the bed is restricted to the married couple. Similarly, it appears in the street, where a grocer retains the adjacent area to his door for his personal activities. For him it is private frontier. Similar situations could be seen in the parks (lovers' corner, immigrants' corners … and so on).

34   Roderick Lawrence found that dining in the kitchen is an essential activity in England and Australia, while respondents rejected the idea of inviting friends to eat there: which means eating in the kitchen is a private activity associated with family meals. See Lawrence, *Housing, Dwellings and Homes*, p. 104.

35   Ibid., pp. 105–6.

36   Lawrence, R., *Housing, Dwellings and Homes*; Sibley and Lowe, *Domestic Space*.

37   Rapoport provided us with different categories of spaces in homes, such as semi-private and semi-public, in the cross-cultural analysis of his thesis on "non-verbal communication" and its implications for Architecture. See, Rapoport, *The Meaning of the Built Environment*.

38   Saunders and Williams, ibid., p. 82.

39   It could be a closed door, a movable curtain, difference in level which reflects the transition between two distinct worlds.

40   See Baker, *Rebuilding the House of Israel*, pp. 113–14.

41   Miles, *The Public Realm*, p. 256.

42   Rapoport, ibid.

43   Project information and details are displayed in the RIBA website: http://www.architecture.com/Awards/RIBAStirlingPrize/RIBAStirlingPrize2009/MaggiesCentre/MaggiesCentre.aspx (last accessed on 24th May 2010).

44   Sennett, ibid., p. 39.

45   Bickford, ibid., p. 356.

46   Miles, *After the Public Realm*, p. 255.

47   Habermas, *The Structural Transformation of the Public Sphere*; Castell, *The Power of Identity*, and *The Urban Question*; Sennett, *The Fall of Public Man*.

48   While this sphere has received less attention in the western world, it retains its power in the organisation of developing societies, to the extent, in some cases, of locality becoming a parallel state with its own institutions.

49   In this respect, Brixton in London, Harlem in New York City, are examples as relevant as a harah of Old Cairo. See Saunders and Williams, ibid., p. 86.

50   Refer to the Saunders and Williams, ibid. and Roderick Lawrence, ibid., pp. 162–70 classifications of home zones. The former called them public spheres of homes, while the latter called them front spaces and in some cases transitional zones.

51   Bickford, ibid., p. 361.

52   Ibid., p. 365.

53   Flint, *Behind the Gate*, p. 139.

54   Ibid., p. 169.

55   Sennett, *Sites of Resistance*, p. 227.

56   Skin in his perspective is a biological border that is a network of various sensory organs that register touch, pain, and sensation (warm/cold, wet/dry, painful/refreshing). It is an organ of perception, where mental life meets with its biological and social reality. While it is a protective cover for the human bodies, protecting our inner world from outside problems, which appear on its surface, colour and scars, the results of external influences, it is also a place of penetration, exchanges with others (absorption/evaporation of water) and transition (reducing, increasing inner temperature). See Lethen (1996), *Between the Barrier and the Sieve*, p. 301.

57   Ibid., p. 302.

58   This was described as a tendency among the dominant socio-political conditions of immigration, colonisation and emerging ethnically mixed societies. Traditional borders were violated at every level, which created a need for more transparent, permeable borders that allowed penetration. In this context, the argument that the international style had been developed to blur and confuse the sense of local identity seems convincing.

PART II
Homes of Old Cairo between
Two Centuries

5

Cairo and the Cairene Harah:
Structural and Contextual Consideration

A HOME IN THE CITY

The architectural history of Egypt is capable of being understood only when we take into account the successive waves of influence that flowed in the capital immediately following each new change of dynasty. Perhaps no country in the world since medieval times can show such an unbroken series of rulers of foreign extraction or such a spineless lack of popular resistance to them.[1]

Investigating the architecture of home in a Cairene context would essentially require an understanding of the structure of Old Cairo, its architectural and urban development. It would also require introduction of the urban unit of this area: the harah and how it manifests the idea of home to its occupants, socially and spatially to the reader. Cairo, similar to others, is a city whose form and urban image is constructed through a comprehensive integration of people's culture, traditions and activities within a distinct spatial order and architectural character. However, Old Cairo's structure of the hawari is far from being typical or repetitive in terms of patterns, order, social structure or even visual perceptions. Every harah in Old Cairo retains its distinctive characteristics and identity that are different from others. However, strangers and visitors, impressed by the homogeneity of the spatial structure and pattern, would not be able to see differences between one harah and another. The architecture of homes in the old city cannot be understood without underlining the character they picked up through these "*successive waves of influence*". This character develops with time and relates architectural and spatial forms to their particular socio-cultural and temporal situation. In this chapter, we shall come to know the way the congested urban patterns of Old Cairo were formed as a consequence of socio-economic and political situations. Immigration, wars, colonisation, disasters, and economic and political dependence are found to have exerted significant impact on Old Cairo as we see it today.

I do not attempt here to generate historical narratives of Cairo, which has been covered by extensive historical accounts and titles. Rather, the urban history and development of Cairo will be interrogated in relation to the idea and practice of home. I try to unfold the urban development of the city and the emergence of the harah as its central urban unit. This supports the understanding of the narratives and nature of the hawari as enduring homes that have their own characteristics, mechanisms and features. It, moreover, answers the question of why the harah is still strongly defended territory by its residents, despite the current deterioration of its physical and social structure.

ON CAIRO: AN OVERVIEW ON THE CITY'S URBAN DEVELOPMENT

The city emerged as an outline of formal organisation and structural pattern which required groups of people to add their input to fill in that pattern. Some cities, however, develop in a natural form of growth around nodes of settlements, with their character and spatial characteristics influenced and shaped by those who live in them. While the former form represents the conventional top-down city planning approach, the latter represents primitive, unplanned and informally evolved cities. A comprehensive study of Cairo's history and development demonstrates that this ancient royal city, with strict formal patterns, soon took a diversion to develop over-informal processes of infill addition, growth and congested urban fabric. Such a pattern of undetermined piece-by-piece addition was later followed by other layers of formal planning practices in the nineteenth and twentieth centuries to give the city its distinct form, character and multilayered structure of different lifestyles.

Cairo[2] or al-Qahira[3] (total area 453 km[2]) has retained its own distinct character as the principal and continuous capital of Egypt for the past 14 centuries. It has been the uninterrupted centre of power in Egypt since 641.[4] With its unique history, the current site includes four previous capitals which preceded Cairo: Memphis (Pharonic); Al-Fustat; Al-Qata'i and Al-Askar (Islamic). Cairo is, perhaps, one of the few cities that combines the treasures of Islamic architecture of the middle ages side-by-side with classic styles of nineteenth-century European architecture in addition to the international style of the 1950s–70s. That is why scholars tend to divide it spatially into three cities: Medieval-Islamic Cairo, European Cairo, and Contemporary Cairo.[5] Their Medieval-Islamic city developed during the period (969–1863) and closely approximates the boundaries of historical Cairo built during the Fatimid, Ayyubid, Mamluk and Ottoman periods. The medieval Islamic quarter represents the very tight and dense urban form, with active social interaction, activities and support. The European-style downtown area, in contrast, provides a true image and life style of the classic revival of Belle Époque, with a radial street pattern, paved wide roads, circular squares, plazas, boulevards and classic façades.[6] It represents a period that witnessed a domination by European architects, the rise of western capitalism in Egypt and the stark segregation in social class and life style between rich westerners and the natives, which Janet Abu-Loghud termed *the dual city*.[7] While it was the district of the bourgeois class before the 1952 revolution, it currently reflects the business and economic power of companies, banks, finance and even political establishments.

The site of Cairo has been significant throughout Egypt's long history. It was described by Fourier as the city where wealth, trade and science were conjoined with power, religious leadership and irresistible strategic location.[8] Representing the wealth and power of the Arabic and Islamic civilisation, Cairo was one of the world's significant meeting as well as conflict points. In justification of the French occupation of Egypt (1798–1801), Jean-Baptiste-Joseph Fourier said in his *Préface Historique, vol. I of Le Description de L'Egypte (1809–1828):*

> Placed between Africa and Asia, and communicating easily with Europe, Egypt
> occupies the centre of the ancient continent. This country presents only great
> memories; it is the homeland of the arts and conserves innumerable monuments;
> its principal temples and the palaces inhabited by its kings still exist, even though its
> least ancient edifices had already been built by the time of the Trojan War. Homer,
> Lycurgus, Solon, Pythagoras, and Plato all went to Egypt to study sciences, religion,
> and the law. Alexander founded an opulent city there, which for long time enjoyed
> commercial supremacy and which witnessed Pompey, Caesar, Mark Antony, and
> Augustus deciding between them the fate of Rome and that of the entire world. It is
> therefore proper for this country to attract the attention of illustrious princes who rule
> the destiny of nations". And he continues, "No considerable power was ever amassed
> by any nation, whether in the west or in Asia, that did not also turn that nation toward
> Egypt, which was regarded in some measure as its natural lot.[9]

Cairo is located at the point on the River Nile where it splits into two branches forming the river's delta to the north, and this allowed control of the northern part of the country. The site is on a plain that is framed by two heights, the Muqattam spur to the east and *Al-Ahram* (the Pyramids) hill to the west, which worked as natural boundaries of the site. The site was used by many dynasties that ruled Egypt before the Muslim conquest of Egypt in the seventh century.[10] Al-Maqrizi told us that *Salahuddin*[11] had to dismantle hundreds of small pyramids to build his own castle on the same site (south-east of Cairo).[12] The contemporary site of Cairo has been considered the regional centre of power (cultural and religious for both Arab and Muslim worlds) since 641AD: and this distinguishes it from other Middle Eastern cities, where different cities vied for supremacy in different epochs.[13]

Three Arab-Islamic capitals of Egypt have been located within the confines of contemporary Cairo: Al-Fustat (641–750), Al-Askar (750–868) and Al-Qata'i (868–969). Following the Muslims' conquest of Egypt in 641, Al-Fustat was founded as a military camp city to the north of Coptic Cairo and Babylon and subsequently became a regional centre of Islam and the home to the first mosque in Egypt: Amr Ibn Al-'As mosque.[14] When the Abbasid caliphate took over from the Ummayyads in 750, they moved their capital to the north to their military camp city, Al-Askar. In 868 and under the Tulunids, Egypt's capital was moved further north to another settlement, Al-Qatta'i.[15] However, neither Al-Askar nor Al-Qatta'i achieved the prominence of Al-Fustat, the then long established, popular and economic capital of the province (641–1168). By the end of the ninth century, Al-Askar had been merged with Al-Fustat following the growth of the latter, whilst Al-Qatta'i was destroyed by the Abbasids when they recaptured Egypt in 905. With the Abbasids' second conquest, Al-Fustat once again became the capital of Egypt.

Previous cities, along with Cairo, owed their existence to being home-like settlements for the triumphant army, located far enough from the existing capital city dominated by the natives.[16] In other words, all were *home away from home*. Albeit different in planning principles and forms,[17] they were divided into small quarters, each assigned to an ethnic, tribal or religious group of warriors along with considerable freedom to form their own home. Among the three preceding capitals of Islamic Egypt, Al-Fustat[18] sustained the greatest influence on the urban development of Medieval Cairo. The informality and tightness of the city alleyways, prevalent in the excavations of Aly Bahgat and Gabriel Baer,[19] were to evolve in Cairo following the relocation of the population of Al-Fustat to Cairo in 1168AD

The first informal development of the unplanned settlement of Al-Fustat emerged when the powerful tribes, who occupied the central quarters of the city and maintained strong influence, had developed and expanded their allotted plots to occupy the surrounding territories and fill open spaces. It took Al-Fustat five centuries to move from this socio-spatial tribal structure towards a uniform urban settlement with adequate distribution of services.[20] Central quarters developed into rich and active districts while those on the outskirts remained marginalised and were left to people of the lower orders. By the end of its lifetime, house planning and construction were developed to include three-storey houses, with some houses extending to six or seven storeys by 985.[21] Moreover, according to *Naser-I Khusrau,* there were buildings of 14 storeys in Al-Fustat by the eleventh century.[22] However, the excavations of Bahgat and Baer revealed to a large extent how the organisation of street networks and the arrangements of residential clusters were extremely complex.

Al-Qahira was no different from its predecessors in being a royal, or more accurately, a Forbidden City that hosted the royal Fatimid family, their ruling class and triumphant army. The new city, at the time of foundation (696AD), was called *al-Mansuriya*: which meant the victorious, and was built to be a royal home of Fatimid's Caliph *al-Mu'iz Ledeen Ellah*, whose palace occupied the central location.[23] Being a royal city, it recalls, as per Creswell,[24] similar arrangements in Beijing, and the Forbidden City which was founded three centuries later. The city was enclosed within high walls for protection and defensive purposes. No one was allowed to enter the city except the garrisons, high officials and princes. Therefore, *Al-Fustat* kept its importance as an active metropolitan centre for the state, while the new city was exclusively for ruling purposes.

The Fatimids placed their new capital and fortress to the north of all previous cities to control trade routes and defend their state from the Abbasid Caliphs in the north-east (Figure 5.1). The first construction was a palace and fortified walls of area about 1,100m x 1,150m.[25] City walls were of rectangular shape, with the four sides facing the four cardinals and eight gates, two on each side. The new city was based on very few planning principles, of which the dominant elements were the Caliph's two-part palace, built at the centre of the city, and the central north–south thoroughfare dividing the city into two halves. Twenty residential quarters called *Khitat* (small vacant quarters/land plots, singular – khitta) were distributed to different army groups on either side of the thoroughfare to the south of the city.[26] Those included Berbers, Jews, Greeks, Africans, Sudanese, etc. As well as the khitat, there were broad squares, gardens, and later, in 970AD, Al-Azhar mosque was constructed to lead the Shiites' religious education. At its foundation, Ali Mubarak estimated that 20 per cent of the city was made up of palaces, 10 per cent was garden, 10 per cent squares, 60 per cent khitta and harahs (local alleyways).[27]

The central thoroughfare was planned as an extension of the old trade route to Al-Fustat within the new city and was used as celebration venue at the time of the army's return from victory. The two palaces dominated the skyline in both scale and richness and, for the first time, the mosque became a secondary (not the central) element of the Islamic capital of Egypt. No buildings survived from that early plan except a few mosques: Al-Azhar (972), Al-Hakim (992), and Al-Aqmar (1125).[28] In contrast to the strict order of the central quarters, the khitat were relatively free of restrictions, giving their occupants more freedom to manage their settlement and distribution of houses, while some of them were fenced and gated for protection.[29]

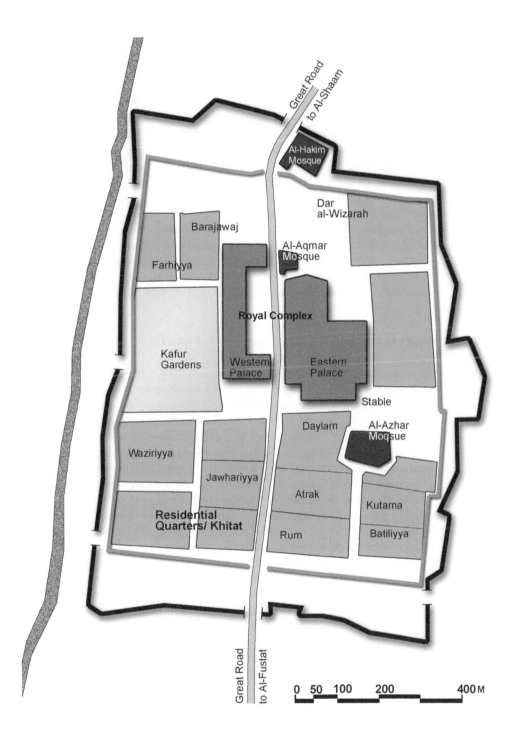

Great Road to Al-Shaam

Al-Hakim Mosque

Dar al-Wizarah

Barajawaj

Al-Aqmar Mosque

Farhiyya

Royal Complex

Kafur Gardens

Western Palace

Eastern Palace

Stable

Daylam

Al-Azhar Moqsue

Waziriyya

Jawhariyya

Atrak

Kutama

Residential Quarters/ Khitat

Rum

Batiliyya

Great Road to Al-Fustat

0   50 100   200   400 M

5.1   Early master plan of Fatimid's Al-Qahira:
original city walls (grey) of Jawhar (969AD), appears within Badr Al-Gammali walls (Black, 1078AD)
Developed after Sanders (1994); Al-Sayyad (1999).

They kept to the traditions and social structure and hierarchy of their particular group. While the tribal structure had dominated the distribution of houses in early times,[30] as time passed, each khitta began to be classified more by occupation. Evolving from something resembling the practice of home of close relatives to an economically based unit, each khitta moved from a tribal structure to a mixed and multi-occupancy societal unit. In the course of the twelfth–fourteenth centuries, they started to host workers, migrants and people from other groups to satisfy their local needs, services and economy. The emerging hawari were soon classified according to the dominant activity. For example, haret Al-Sukkariyyah, beside Bab Zuwaily to the south of the city and in the proximity of Al-Azhar Mosque, was the site where the sugar (*Sukkar*) trade was the dominant activity. Similarly, Al-Nahaseen was the location for coppersmiths. However, while *Haret Al-Yahud* (Jews) was so named because of the overwhelming majority of Jewish residents (97 per cent), it was known as the location for moneychangers (*saraf*) and the financial transactions in which they specialised.[31] Similarly, but less exclusively, *Haret Al-Nasara* (Christians) was known for its professional homogeneity. It was chiefly the home of clerks and public writers, with four out of 10 Copts registered as being specialised in this profession (according to the 1848 consensus).[32]

MEDIEVAL CITY AND THE FLOURISHING OF THE HAWARI

The medieval history of the city can be traced from the twelfth century, when the city was flooded with 200,000 migrants after the burning down of the already densely populated city of Al-Fustat during crusade wars. Following the dispersal of this massive wave of people to Cairo, the royal city began to experience the overpopulation that continues to this day.[33] The urgent need for shelters to house the newcomers along with shortage of available spaces and limited resources[34] (during the Crusade Wars) led to the phenomenon of construction as a response to need. This phenomenon occurred at two levels: building houses outside the fortification walls on the north–south spine of travel, and filling empty plots and spaces in inhabited areas. Cairo rapidly became a densely populated city, with a dramatic change in its character and urban pattern. The construction of houses outside the walls created peripheral informal zones with no planning order and depending on attachment of one house to another. Soon, they had founded their own popular markets and redistributed commercial activities within new quarters. That expansion was mainly longitudinal towards the south and the city walls had gradually lost their importance and were decaying structurally.[35] This expansion led *Salahuddin* to build another set of fortified walls to enclose the growing urban expansion.

The construction inside the city walls, on the other hand, introduced a pattern of infill and replacement that had changed the spatial structure of Cairo, introducing the dense and congested urban communities of the hawari. They were formed when several houses' backyards, gardens and open spaces were used for newly-built houses, leaving tight and non-straight alleyways to be the only urban voids. In the long term, small houses started to replace the large ones, through streets became dead-end alleys and linear streets received many additions on both sides, becoming a zigzag interior as described by Jomard in 1798.[36]

Despite such congestion, medieval Cairo represented the richest period of Islamic architecture in the east, supported by the enormous funds of the ruling elites, their families and merchants. The local skilled builders, funded by large investments from the elites, made the evolution of unique Cairene architecture possible.[37] This development is exemplified in the emergence of new building types and complexes such as *Khanqah*, *Wekalat*, educational and medical buildings (such as *Madrasah* [school] of Sultan Hassan and *Bemaristan* [hospital] Qaytbai). Magnificent houses such as Bayt Zaynab Khatoun, Bayt Al-Razzaz are well documented structures of the medieval period. In addition, new decorative structural elements, such as *Muqarnas* and *mashrabiyya*, appeared.

EUROPEAN CAIRO AND THE DUAL CITY OF THE NINETEENTH CENTURY

While Cairo retained its medieval character until the late eighteenth century, its new character as a modern city emerged with the coming of Muhammad Ali Pasha (1805–49) and the departure of the French (1798–1801).[38] The native and indigenous hawari were to be challenged by the new forms and organisation of the modern western culture and lifestyle. The already congested old city could not be developed or changed to suit the ambitious plans of the new ruler. Hence, modern Cairo, as he saw it, required addition and expansion but of a new type and with a different image. Muhammad Ali initiated new expansion towards the north of the city to Shubra, paving wide roads with European style palaces. Later, with his grandson, Khedive Ismail (1863–79), in power, Muhammad Ali's dream of European and modern Cairo was brought to reality. However, to achieve more rapid development of a modern city, modern quarters were planned and constructed to the west of the old city, causing a stark spatial and socio-cultural division between two distinct images, the old and the new. Each represented the character and personality of a distinct architecture, cultural context, and lifestyle:[39] the native Egyptian quarters and the Western European boulevards.

By the turn of the nineteenth century, the city was struggling to define itself between those two extremes. In this cultural and social confusion lie the routes of Cairo as a *Tale of Two Cities*.[40] Cairo's urban development since the mid-nineteenth century could be seen as a pattern of *replacement and change* to be achieved by replacing the traditional character with an imported modern one. At the centre of Ismail's dream, was a modern and attractive Parisian-Cairo, or what was called by Cynthia Myntti *Paris along the Nile*.[41] Ismail, amazed by his French experience (education, culture and lifestyle), founded the first Ministry for Public Works[42] with a mission to build modern Cairo, inspired by the 1867 *Exposition Universelle* of Paris.[43]

Nineteenth-century Cairo's urbanism was best emphasised by Janet Abu-Loghud's[44] description of 1897 Cairo as a dual city, where two distinct types of homes were introduced: the homes of the natives in historical Cairo and homes of the classy western strangers in the Isamilia quarter (now the downtown area). The former was a historically developed labyrinth with tight unpaved alleyways, while the latter was a newly-developed European quarter with radial wide paved boulevards and classic style palaces and apartment buildings (Figure 5.2). The classy quarter enjoyed western hotels, coffee shops, Greek restaurants, Chinese terraces, with municipal services such as sewer systems and street gas lights along with a European opera house (a copy of Milan's opera house). These facilities were made exclusively for Europeans and did not appeal to Egyptians, even those who could afford them.[45]

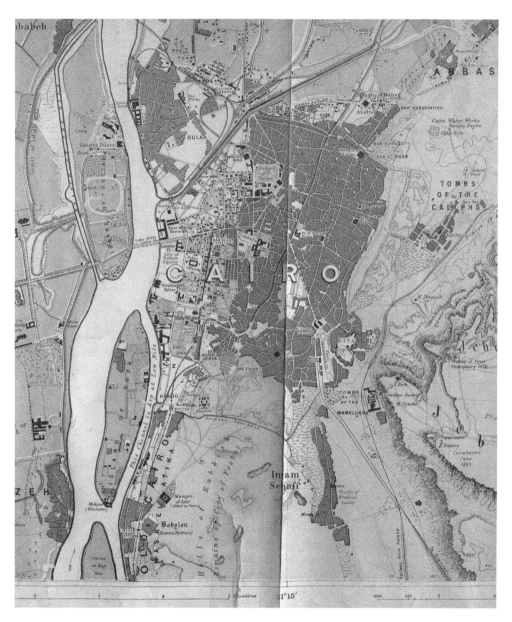

5.2   Cairo in 1882:
the old city is shaded while the emerging modern city appears to the West
British War Office (1882); Courtesy of Egyptian Geographic Society.

Nineteenth-century Cairo, hence, tends to be seen as a set of contradictions, derived from the notion of dualism: old/new; elitist/ordinary; traditional/modern; Islamic/ European; western/native. However, these contradictions later proved to be integrative in the historical continuity of the city's character and personality. In a similar way to ancient Cairo (initially formal but becoming labyrinthine), European Cairo had become, by the end of the twentieth century, congested and over-occupied by native Cairenes, with similar practices of informality, that turned the European dream somehow into a native environment.[46] In Cairo, it had become apparent that the processes of change and development were sufficiently powerful and overwhelming to overcome the initial intentions of its rulers and planners to have a formal and Western-styled city. Throughout this book, my interest is always to look at the traces of continuity and development, inclusive of basic stability and provisions of change. What began to emerge as an exclusively European-style city was later dominated by native Cairenes and was integrated into the multi-layered urban fabric of twentieth-century Cairo.

METROPOLITAN CAIRO OF THE TWENTIETH CENTURY

The various notions of the dual city lasted throughout the British occupation of Egypt (1882–1952), with the social division between the ordinary natives on one side, and foreigners and western-educated elites on the other, being spatially manifested between the hawari-communities (the baladi) and Afrangi[47] quarters. In contrast, the post-liberation layer of the city emerged with a clear agenda of equality and mission to showcase a flourishing future of the independent nation. Following the 1952 revolution and the formal recognition of Egypt as a republic in 1954, the new leadership adopted a locally-driven agenda with emphasis on equality, citizens' rights to free education and decent housing, as well as turning Egypt into an industrially based economy.[48] However, during the 1960s and 1970s, the city skyline received several iconic structures on the Nile frontage such as the 1960s public National TV building and the Tahrir administrative complex and 1979's Cairo Plaza Towers in Bulaq. In addition, similar trends were dominant in the representation of modernity during the second half of the century through individual additions of western-inspired architecture.

Cairo Urbanism of the post liberation period was marked by the ideology of Arab Socialism as manifested in a state-run programme for housing projects in 1958. This programme added modular urban zones to the city both to the north and to the west, such as al-Muhandseen, to the north of the western bank of the Nile and Madinat Nasr to the north-east of Cairo. The government invested 5–6 million Egyptian pounds in providing 10,000 units per year,[49] a number that was not enough to fulfil the needs of a growing working class population. One of the famous projects was the social housing of Masakin Zeinhom on the periphery of the old quarter, which was reduced in a few decades to a state of ruin. Therefore, the return to seeking private investment for this sector during the 1970s was considered as a sign of the failure of the implemented programme.[50] The poor and lower class groups, non-governmental employees were left with no option but to build their homes informally, which resulted in overcrowding in both the old city and informal communities on the peripheries.[51] As a proposed solution to the problem of overcrowding, following the 1973 war, Sadat's government (1974–81) started the development of the already planned new desert cities in an ambition to spread the population of Cairo[52] over larger plots of land in order to "*reconstruct the demographic map of Egypt*".[53] The cities of the Sixth of October, Al-Obour, Al-Sadat, and the Tenth of Ramadan[54] were all located on the outskirts of Cairo. However, these cities remained vacant until the end of the millennium when the private sector emerged to take the lead in injecting investment, developing industry, and attracting people to settle in high-class exclusive compounds.

The urban form of contemporary Cairo can only be explained as an evolving cultural urbanism that is fundamentally social in pattern and cultural in behaviour and requirements. This is a result of the increasing polarisation among social groups, difference in affordability, and the notion of exclusive communities. Such spatial division of the social ladder reflects starkly on the composition of the city as a group of nodes/pockets. Each node has a high-class or bourgeoisie community at the centre, surrounded by middle-class districts with lower but decent living conditions and the last circle is a group of informal communities at the periphery of this sector. Examples of those urban nodes could be seen in *Misr Al-Gadida* (north-east), *Maddi* (South), *Garden City* (Central), *Dokki* (West), and *Muhandseen* (north-west). The node, typically, includes the shopping malls, classy restaurants and places of recreation, such as hotels and sports clubs; the middle class areas include fast food restaurants, supermarkets and moderate-level shops; while the informal communities on the edge include popular markets, workshops and services and provide jobs and workers at the same time. Hence, we can appreciate that the contemporary urban form of Cairo requires a spatial attachment between zones of all social classes in order to survive.

OLD CAIRO: RATIONALE OF TIME, PLACE AND SPACE

The French occupation in Egypt (1798–1801) has always been marked as an influential transitional period in Cairo's modern history:[55] the climax of medieval urbanism and congested hawari as well as the end of isolation from the European modern urbanism which was soon to emerge in Cairo. In 1800, the indigenous revolutionary movement was at its peak. The revolt was a response to the massive power used by the French army to control the local population. While the whole period of the French occupation is widely accepted as the entrance to modernity, it was around 1800AD when radical change occurred and the modern character of the Cairene society began to be formed. During the conflict, the French attacked the hawari of Cairo, and demolished their gates: an action that was considered by the Cairenes, as *Al-Jabarti* described, as a violation of the local population's privacy and security that ultimately was to change their lives.[56] Their sacred space, the harah and home, was frequently invaded and their boundaries were breached. In response, local communities had to defend their hawari, leading to rebellion on a huge scale.[57]

The year 1800AD was selected as the point of departure for my enquiry on Cairo's architecture of home. First, it represents a historical defining moment in Cairo's urban history: when the harah's position as a principal urban unit would have been challenged by borrowed classical European forms which started to appear during Muhammad Ali's dynasty (1805–48AD). Second, by the time of the departure of the French in 1801, the old, traditional organisation of the city's urban settings had reached its climax and its hawari had become dense and saturated. The following decades witnessed the imposition of the then modern state's regulation of housing designs and features: which included building heights, colours, window types, and represented the early emergence of a new system of housing production.[58] Thanks to Edward William Lane and the Chronicles of Al-Jabarti, we have detailed accounts of social life in the traditional quarters during that period. Along with the survey and analysis of a few surviving houses of that period, this part starts with shedding light on the spatial praxis and social performance of space during the introduction to modernity in Cairo.

The hawari of Cairo as an administrative unit could be traced from 1800 through two principal courts. The first was *Al-Salihyah Al-Najmiyyah*, which was concerned with everyday legal activities such as buying, selling and divorce cases. Its records give us an idea of the names and affiliations of people involved as well as a description of the subjects' houses. The second was *Al-Bab Al-A'ali* court, which was concerned with the general trade and commercial issues of the whole city.[59] The documents recorded in the first court gave us information about house trading within the harah, while the second helped us to see how conflicts, especially concerning building construction, were resolved and how building conventions were developed. We could trace the roles that both individuals and social institutions played in the local system of house planning, design and construction from these documents.[60]

The historical accounts of nineteenth-century historians such as Al-Jabarti tended to focus on the ruling class and their counterpart merchants. The narratives of the ordinary Cairenes were almost missing. However, these narratives and detailed accounts of Cairenes of different social stratification appeared in Edward Lane's accounts, which have become the prime source of such information. However, Lane's accounts are criticised by Juan Campo as being the reflections of an *aristocratic orientalist* whose English lifestyle influenced his writing. While we are not able to verify his images of Cairenes' habits of that time, we can use his descriptions of events, images or activities in combination with archaeological evidence, archival records, and trace those habits through later accounts of Cairenes' lives. This approach enables us to reach a reasonable understanding of the social structure and behaviour in the Cairene harah of 1800.

Archival materials have been the major resources for studying this particular period of Cairo's history (reference can be made to Andre Raymond, *Cairo*; Nelly Hanna, *Habiter au Caire* and *Construction Work in Ottoman Cairo*) This contemporary account, on the other hand, has relied heavily on field work, which was undertaken in the harah during the period May 2007–September 2012. This period witnessed spatial changes in the form of demolition of a several residences and the construction of many others. Field work depended on both interviews and observation as effective tools of engagement with social activities of residents.[61] Both approaches traced issues of modernity in the structural transformation of Cairene society in general, and the local traditional families, called *Baladi*, in particular. My aim is to understand social actions and needs of people that determined architectural development and the modelling of home as a space. Therefore, while sociological and anthropological approaches are involved, the desired outcomes are terms comprehensible to wider audience, and particularly architects.

In her PhD about the social relationships in the hawari of Old Cairo, Nawal Nadim followed the relationship between the sexes in *Haret Al-Sukkariyyah*, one of the poorest hawari of Cairo. While she looked as the structure of the community and changes happening over time, she had a preference for a holistic view of anthropology.[62] However, such detailed investigation of certain families' structure and social relations cannot be considered as a reliable source when there is a need to reflect this information on the hawari communities as a whole. The relatively small size and complex social organisation of the harah cannot be subdivided into different elements to be analysed individually. Rather, the social structure and interaction of its units: the families, is the fundamental force that influences the social behaviour in this context.[63]

It is the arrival of newcomers: mainly rural migrants who flooded into the hawari of Old Cairo at the turn of the twentieth century, which has changed its social structure. Today, Nadim states *"the integration and sovereignty of the harah have nearly vanished. The term no longer denotes a quarter, but refers to the narrow alleys which are found in abundance mainly in the Old Cairene quarters".*[64] Acknowledging the changes in the harah's form, Nadim, nevertheless, ignored the fact that integration and sovereignty had remained visible characteristics in the form of the narrow alleys and their branches. They would hold the residents together on issues of cultural values, social norms, security and shared interests.

Particular social issues of the Cairenes, such as motherhood and gender, have been studied, but such studies have tended to cover the perspectives of the Egyptian society as one group, which is, to a large extent, misleading. The influence of spatial settings and particular contradictions among subcultures, even within the same society, is ignored in such studies. For example, Mervat Hatem has made generalisations on mothering culture in Egyptian families and the domestic role of mothers, giving much weight to their essential duty in raising children as a general notion of Egyptian society.[65] Generalisation of social practices risks missing the differences and sometimes the contradictions among subdivisions of the society. When we view modern districts, we find that most of the wives are working, and some of them have babysitters who perform, partially at least, a mothering role.

Other literature on social settings of the harah has tended to avoid analysis of the family and its internal relations and behaviour and stayed within the boundaries of numerical surveys, which fail to reflect the real image of the harah as a social realm. Mohamed El-Sioufi's *Fatimid Harah* and Yasser Elsheshtawy's *Urban Transformation: Social Control at Al-Rifa'l Mosque and Sultan Hassan Square* are examples of this direction. This gap implies the difficulty in studying and analysing domestic life that is very *private* and *sacred* and, therefore, overly mysterious. This meaning of the Cairene home as sacred space and the significant religious meanings in domestic space are explored by Juan Campo in his book *The Other Sides of Paradise*, which traces religious symbolism in Cairene homes. It is not acceptable for *baladi* families to discuss their daily domestic practices such as cooking, sleeping, fights with strangers. To tackle this problem, I developed different approaches that range from unstructured discussions and interviews, to full involvement in these families' daily lives. Farha Ghannam, in a similar fashion used strategies like living with families; joining in celebrations, parties, daily cooking and shopping. I developed similar venues of communications and participation in daily, seasonal or occasional events.

THE CONTEXTUAL SETTING OF CAIRO OVER TWO CENTURIES

Investigating daily life and spatial practice two centuries back, essentially, requires a review of the time configuration and setting of that time, otherwise some of the normal everyday practice would be confusing and unintelligible to us today. For example, practices in the harah in 1800 entailed certain activities taking place within a defined period of time, which changed over the day or over the week. During prayer times, public spaces became quiet and almost deserted because people were attending the mosque.[66] Time configuration, in 1800, was organised by the precisely timed ritual performance of prayer. Prayer times were periods of silence, freezing of activity, especially within the public spheres.[67] Following the sequence of five prayers per day, the resulting four periods were clearly defined through a space-time pattern of activities.[68]

Here I would refer to Pierre Bourdieu's notion of transitional periods that were influential as thresholds, *"a sort of sacred boundary between two spaces"*,[69] in the Algerian Kabyle communities of the 1960s. Moving between transitional periods was like moving between two spaces; rituals and activities change. Bourdieu stresses the subjectivity of time, which influences domestic environment, which in turn, is opposite to the objectivity introduced by Isaac Newton as *"absolute, true and mathematical with its own nature"*.[70] Subjectivity of time is inherited in the Cairene *hara's* culture, with private and public times configured to daily practice and associated rituals rather than the temporal precision of modern times.[71] In the Cairo of 1800, the night preceded day, as the eve of a new calendar day started following sunset after the last prayer (*Al-Maghreb*),[72] when people headed home.[73] Two main parts could be easily defined; the time for public gatherings, meeting and business, which was between Al-Fajr and Al-Maghreb prayers; and private time which was defined by night time, from Al-Maghreb until Al-Fajr. We can follow the sequence of daily practice based on Edward Lane's following descriptions[74] (Figure 5.3):

- Following Al-Maghreb prayer at the mosque, men headed home, spending night time in their harem qa'a with their children and wives. A light meal was introduced and the family would eat together.
- Sleeping time started after *Al-Ishaa'* prayer and ended with *Al-Fagr* (Dawn) prayer, after which men went home to take their breakfast with their families before heading to work.
- By mid-day, the third prayer, *Al-Dhuhr*, would come around and thereafter men went home for lunch and in some cases to take their nap.
- The last part of the day started after this rest, when the Cairene men went to their afternoon work and spent the rest of the day with their friends and fellows at the coffee houses, shops or markets until sunset and the start of Al-Maghreb prayer.

While such a programme of activities was masculine in nature, the routines of females were consistent with domestic productive activities at home.[75]

Even though time configuration did not differ on Fridays, social practices and space activities did change. *Al-Juma'a*[76] prayer on Fridays had to be attended at the mosque, following which men used to go to market and families sometimes to open parks.

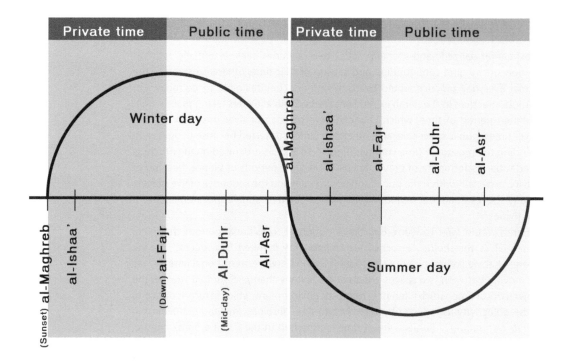

5.3 Configuration of time in the Cairene harah in 1800AD

Note: Width of private and public columns shows the dynamic nature of time from winter to summer.

Moreover, Friday was the day for celebration in both religious and mundane terms, especially weddings and engagements. The bride used to make her first arrival at her groom's house on a Friday eve.[77] Both Lane and Campo have considered the preferred time for a wedding as the eve of Friday (*Laylit ig-gum'a*), Thursday night. We thus understand that weddings took place at that time to add sacred meaning to occasional rituals, combining a religious assembly (Friday prayers) with a significant mundane and social gathering: the wedding.

When we move to a longer time scale, that of the year, then two Feasts (*Eid al-Fitr* and *Eid al-Adhha*)[78] and several *Mawlids* (religious figures' birthday celebrations[79]) would be considered. The two feasts: Eid al-Fitr and Eid al-Adhha, come at a particular Hijri calendar timing (end of Ramadan and 10th of *Zul-Qe'da* respectively) and were the most respected and celebrated events during that period. Daily routine was, then, subject to a significant change of timings and places. Feast days showed more communal and social practices which involved collective gatherings in public spaces. During Ramadhan, practice did change as all adult Muslims fast from dawn to sunset without eating or drinking at all. Fasting does not restrict eating and drinking only, but also sexual intercourse between couples, which meant that eating arrangements of both changed accordingly.

Religious rituals were also substantial to understanding the way the home was designed, organised and used. Conceptually, homes within their physical parameters (houses/dwellings) are sacred places in Islam. They have, like God's house (*al-Kaa'ba* in Mecca), been created by God, "*holy creations*".[80] Houses retain religious significance that is traced in Quranic verses[81] by Campo as: "*failing to acknowledge God's grace in such (house) mundane matters places people in the dangerous condition of denial (kufr)*". Hence, prayers could be performed at home as well as the mosque. However, there is no particular place at home where prayers should be performed other than facing towards Mecca and away from toilets (the house of evil in every home). Prayers were usually undertaken in living spaces within both public and private spaces: *al-mandharah* (the former), or the *harem* (the latter).

Domestic prayers were performed by Cairenes on a floor mat or carpet (according to financial affordability) which was laid on a stepped or protected floor plate. Prayers were performed on any surface and material (fabric) that was *Taher* (clean, purified, not to be touched by shoes), and away from dirty places (such as bathrooms). Both males and females performed the same rituals, but in segregated arrangement and females had to cover their whole body and their hair. Rituals performed at home were not limited to praying; in addition, reading Quran, inviting shaykhs for Quran recitals were common practices in the hara's homes, especially on Fridays or at occasions such as funerals,[82] or after sad occasions, when Cairenes sought to purify their houses from the devil. Quranic scripts spread over the walls in halls in several homes of 1800 were attempts by households to conduct themselves as sincere to their faith within their community.[83] They were features of halls in Bayt al-Suhaimy for example. Smaller houses of the lower middle class lacked these wall scripts, which, in my view, was due to the lack of interest in presenting themselves as of higher mundane or religious status. If we attempt to analyse this sequence of ritual and time configuration in domestic spaces in depth, we find that daily activities demonstrated vibrant movements among places of different social nature and setting. None of the practice can be defined in terms of time without variables of space and activity (such as gender division at prayer times), social status (between house quarters), and interpersonal modes of communication.

On the scale of community, some conventions were important to organising mutual interactions, in a way similar to accepted/informal legislation. These included building conventions, Islamic Shari'ah and social contracts. The social structure of the harah as an essential part of its formation had to be translated into a group of building regulations and conventional guidelines. They were basically "conventions" that were not written or formally regulated. Hence, they were informal standards of practice but used as a reference for resolving disputes. The power of those conventions was better seen in the rulings of the state's courts in Nelly Hanna's study of construction work and institutions of Ottoman Cairo, in which she declared that "*Among the most important functions of the courts with regard to construction work was that it saw to it that certain building and urban regulations were followed by people who were constructing, restoring or enlarging buildings*".[84]

Legal courts in their decisions did not refer to particular regulations; rather, judgements on protection of individual privacy and public good were based on Islamic law (the *Shari'ah*) in conjunction with already established conventions. According to Nelly Hanna, several cases had inspection teams, usually formed of professionals (builders/architects), advising the court on disputes and allegations that privacy of one man's house was violated by the windows of newly constructed buildings. One such dispute was a case where the local population had accused a newcomer of building a new house with a "projection" overlooking a closed street, which to them, constituted an intrusion of their privacy. Following the visit of the inspection team, the claim was accepted and the owner was ordered to remove any projection over that particular street.[85]

Some of the court's rulings set the basis for conventions. For example, an inspection committee recognised that window projection across more than one-third of a street's width was not acceptable, while others recognised that maximum allowed height should not allow the new build to overlook its neighbours' courtyards or roofs. In effect, this last rule had applied restraint on building heights for the whole city. While these restraints were applied to regulate the building processes on the basis of Islamic rules, building conventions retained a considerable flexibility in internal arrangement of spaces and ownership rights. People were allowed to change the use of their buildings as long as it did not cause harm to their neighbours. Moreover, it was common practice for people to rent the roof of their neighbours' houses to extend their own houses.

Building conventions of 1800 retained a dual strategy which combined restraints and flexibility. Both were associated with the social contract and Islamic rule of *no harm*. While restrictions were applied to building heights, projection, and harmful land use, especially within residential areas, flexibility was allowed in cooperative ownership of a building (by levels), its internal arrangements and the variety of land uses within a single hara. Building conventions in the hara of 1800, as such, had already been established for a long time through a case-by-case approach, to maintain national coherence and a safe environment and to find new solutions to emerging problems: "*The building regulations in practice in Ottoman Cairo were based on religious law, specifically on the percept that one was not permitted to harm one's neighbours by building on a public thoroughfare and obstructing passage, by intruding upon the neighbours' privacy, by depriving them of light and air, by disturbing them with noise or smelly activities*".[86]

NOTES

1   Jairazbhoy, *An Outline of Islamic Architecture*, p. 143.

2   *Cairo* was first recognisd by the European researchers through the French scholars who joined Napoleon's expedition (1798–1801), and called it Le Kaire, translated later into English to became Cairo (Behrens-Abouseif 1989). The original Arabic name of the city was and still is *al-Qahira*. However, this name was not commonly used by the public as they knew their capital for centuries as Misr al-Mahrusa which means "Egypt – the Protected". It was the common name at the arrival of the French at the turn of the century. It continued to be used until the twentieth century. Until today it is used with the exclusion of the term "the Protected".

3   *Al-Qahira* (Cairo) is an adjective of the Arabic verb (قهر) (*Qahara*), which means to defeat someone or something valiantly with ultimate power; "the Victorious". However, this word is, in Islam as in Arabic language, recognised as a holy adjective of God and not to be used for human beings (Ibn Manzur 1232–1311, 1956) (the Holy Quran). The first name given by the founders was "al-Mansuriyah" (the victorious), which has a similar meaning but resembles the victory of human beings over each other. The Caliph al-Muiiz Ledin Ellah changed the city name from "al-Mansuriyah" to "al-Qahirah" on his arrival in the city four years later. He, most probably, wanted to give a glorious name to his new capital, especially when we know that their previous capital, of the Fatimid dynasty before Egypt, was named "al-Mansuriyah" as well (Creswell 1952).

    Currently Cairo means: al-Qahira al-Kubra (greater Cairo). It has three sub-municipal levels, al-Qahira governorate (east side of the Nile); al-Giza governorate (west side of the Nile); and al-Qalubiyyah (the north of al-Qahira governorate).

4   Behrens-Abouseif, *Islamic Architecture in Cairo*, p. 3.

5   Fahmi, *Global Tourism and the Urban Poor's Right to the City*, p. 159.

6   Myntti, *Paris along the Nile: Architecture in Cairo from the Belle Époque.*

7   Abu-Loghud, *Cairo: 1001 Years of City Victorious.*

8   Edward Said, quoting Jean-Baptiste-Joseph Fourier's 1828 introduction of *Le Description de l'Egypte*. See Said, *Culture and Imperialism*, p. 33.

9   Jean-Baptiste-Joseph Fourier, *Préface Historique*, pp. 37–8.

10   Such as Memphis, the capital of the old Kingdom of Egypt until 2200BC.

11   Salah Uddin Bin Ayoub, the ruler of Egypt (1174–93) and the founder of Cairo Castle.

12   Al-Maqrizi, 1959, "al-Khitat al-Maqriziyah" (al-Mawaiz wa al-itibar bi zikr al-Khitat wa al-Athar), Beirut, Part two.

13   Behrens-Abouseif, *Islamic Architecture in Cairo*, ibid. Cairo and Egypt were under control by foreign rulers until the 1952 revolution and subsequent independence in 1954. These rulers included the Arabs, the Fatimid (North African), the Ayyubid (Kurds), Mamluks (Asians), Ottomans (Turks), Muhammad Ali's family (Albanians), and the French and British occupations. Every dynasty and ruling elite have left their fingerprints on the urban features and character of Cairo. However, during all these dynasties Cairo was the centre of regional control and power.

14   Butler, *The Arab Conquest of Egypt*, p. 341.

15   Ibid., p. 342.

16   At the foundation of Al-Fustat, the structure of the city was mainly dominated by houses that occupied all the quarters except the central one. These houses were mainly tents, huts and primitive mud structures used for residential purposes. Even the central mosque was constructed using palm trunks as columns, with half trunks and reeds for ceilings. For more details on Al-Fustat's structure see Kubiak, Al-Fustat.

17   From informal branch-like alleyways of Al-Fustat, to the perfect rectangular form of al-Qata'i.

18   Al-Fustat (the capital of Egypt 641–969AD, and destroyed in 1168) reached its peak in the twelfth century, with a population of approximately 200,000. It was the centre of administrative and population power in Egypt, parallel to the formal rulers' capitals, until it was burned down in 1168, following the claims that it could be defended against Crusade attacks.

19   Kubiak, ibid.

20   As reported in Al-Maqrizi, Al-Khitat; and K.A.C. Creswell, *A Bibliography of The Muslim Architecture of Egypt.*

21   As reported by al-Muqaddasi during his visit to Al-Fustat. See Al-Muqaddasi, Ahsan al-Taqasim, pp. 316, 329.

22   Abu-Lughod, *Cairo*, ibid., p. 19.

23   Creswell, *The Muslim Architecture of Egypt*, p. 22.

24   Creswell, ibid., p. 23.

25   Ibid.

26   Raymond, *Cairo*, pp. 39–40.

27   Ali Mubarak, Al-Khitat al-Tawfiqiyah, p. 206.

28   Al-Akmar, which lies in al-Darb al-Asfar at the heart of the city, was the first stone building to be built in Cairo, demonstrating an ecclesiastical appearance with interior Persian arcades and decorative Kufic texts. The other two were rebuilt several times at later dates and have become iconic structures of historical Cairo today. See Briggs, *The Fatimid Architecture of Cairo*, p. 190.

29   Raymond, ibid., p. 40.

30   Where the leader was at the deepest and most protected point of the group, followed by the other most senior members and so on.

31   Fargues, *Family and Household in Mid-Nineteenth-Century Egypt*, p. 30.

32   Ibid.

33   Abu-Loghud, *Cairo*, ibid.

34   Egypt was struck by famine for a few years prior to the burning down of Al-Fustat, which, with the crusade attacks and the required military preparations, used up the wealth of the nation and left the rulers without the budget for any construction activities. However, the Crusade never reached al-Qahira or Cairo.

35   Al-maqrizi, Khitat mentioned that the Persian traveller Nasir-i Khusrau stated that there was no evidence of the existence of city walls at the time of his visit to al-Qahira during the eleventh century. From Nasir-I Khusrau, Schefer's ed., pp. 38–57; transl., pp. 145–59.

36   Jomard, *Description*.

37   Examples include Sultan Hassan mosque and Madrasha, Sultan al-Ghuri khanqah and Wekalat, as well as various maristans and hundreds of houses and mosques.

38   Said, *Orientalism*. Edward Said considered the French Occupation of Egypt the point at which imperialism was to be experienced and the other, the Orient, was to be explored. The researchers of eighteenth-century Cairo, on the other hand, stressed that the modernisation project started five decades earlier during the Mamluk rule of Ali Bey and Muhammed Bey Abu al-Dhahab (1760–75). See for example: Crecelius, D. (1981), *The Roots of Modern Egypt: A Study of the Regimes of Ali Bey al-Kabr and Muhammad Bey abut al-Dhahab, 1760–1775*, p. 169.

39   Stanley Lane-Pool simply pictured this notion of the two worlds while describing the sounds of natives' night celebrations which reached his room in the western hotel in European Cairo from the heart of the old quarter, and which gave him the sense of the One Thousand Nights stories behind the walls of the old medieval city. See: Lane-Pool, *The Story of Cairo*, p. xx.

40   The title of one of the most important studies of Janet Abu-Lughod, later developed into the most exhaustive study on Cairo by the same Author, *Cairo: 1001 Years of the City Victorious*.

41   Myntti, *Paris along the Nile: Architecture in Cairo from the Belle Epoque*.

42   Raymond, *Cairo*, p. 270.

43   To guarantee the outcome, Ismail, had employed the landscape architect of the Exposition, Barillet-Deschamps, to draw the plan of European Cairo, which remained dominant in the city's downtown.

44   Ibid., p. 98. This model of dual cities was adopted by later scholars who analysed the urban setting of Cairo during the nineteenth century.

45   Raymond, ibid., p. 274.

46   The tone of regret of the vanishing of this European dream was loud in the Egyptian bestselling novel *The Aycoubian Building* by Alaa Al-Aswany, in which the author praised the European history classy life style of European Cairo and its antique buildings, while regretting their deterioration after native Cairenes (mainly army officers) took over buildings and apartments abandoned by their European occupiers following the 1952 Revolution. See: Al-Aswany, *The Yacoubian Building*.

47   Baladi, means local country people, or people of the lower order. It has been used by higher-class western-educated elites to describe the native majority, who were not educated and did not have a sense of modern lifestyle. Afrangi, on the other hand, meant foreign, and was used by the natives to describe foreigners and western-educated elites who adopted western morals and lifestyle. For more detailed definitions, see: Abaza, *Changing Consumer Cultures of Modern Egypt*, p. 12; Early, *Baladi Women of Cairo*, pp. 53–4.

48   The new regime managed a comprehensive state-led social housing programme on the periphery of the then crowded city (which reached 2,320,000 people in 1947 and reached 6,400,000 people in 1975) (UN, *World Urbanization Prospects*, 2007). In addition, it aimed to overcome the increasing problems of unemployment and poor living conditions in the growing squatter communities over the edges of the old quarters. Time proved that those plans were too ambitious to implement.

49   Feiler, *Housing Policy in Egypt*, pp. 295–312.

50   Ibid., p. 298.

51   Shubra district, to the north of Bulaq's industrial area, expanded towards the north with an industrial pool of many factories in Shubra-Al-Khaima to satisfy the state's industrial ambitions, and later required additional housing for migrant workers who sought low-cost housing in the neighbouring old district of Bulaq.

52   It is worth mentioning that Cairo's population by the year 1976 had reached 6,690,000 people. This figure was discussed by the media as a huge challenge to the administration of the city. The introduction of those cities was celebrated by the media as a sign of a new future for the nation.

53   Feiler, ibid., p. 299.

54   All those names made reference to the victory of the Egyptian Army in the 1973 war. It could be claimed that the regime wanted to emphasise Sadat's central role in development as well as the victory in the war, with the Egyptian media of the time calling him "*the hero of war and peace*".

55   Crecelius, *The Roots of Modern Egypt: A Study of the Regimes of Ali Bey al-Kabr and Muhammad Bey abut al-Dhahab, 1760–1775*. Minneapolis, MN, USA: Bibliotheca Islamica, p. 169. See also, Said, *Orientalism*.

56   Al-Jabarti, A'ajeb al-Athar fi Al-Tarajim we al-Akhbar.

57   This revolt pushed the French army to lay siege to Cairo, culminating in an assault on the city on 21 April 1800. Later, in June 1800, General Kleber, the French commander in Egypt, was stabbed to death in his palace by a Cairene student from al-Azhar University. The French had to leave the city in 1801 after a controversial period of conflict and insults that challenged both sides: the Cairenes and the French Empire.

58   Abu-Loghud, *Cairo: 1001 Years of City Victorious*, p. 65.

59   Hanna, *Habitier au Cairo*, pp. 31–3.

60   Lawrence, *Dwellings*, p. 66.

61   In the period from May 2007 to April 2008, 23 interviews and around 20 video and tape recordings were undertaken by the research team.

62   Nadim, *The Relationship between the Sexes in a Harah of Cairo*; Al-Messiri-Nadim, *The Concept of the Hara.*

63   Sibley and Lowe, *Domestic Control, Modes of Control, and Problem Behaviour.*

64   Al-Messiri-Nadim, *The Concept of the Hara*, p. 319.

65   Hatem, *Toward the Study of the Psychodynamics of Mothering and Gender in Egyptian Families*, p. 288.

66   Muslims have five prayer times a day when they should attend the mosque. In the practice of 1800, a man would be criticised for doing business during prayer times. Those not attending the mosque would tend to disappear to their homes or other places. See Edward Lane, ibid., "Chapter Three: Religion and Law", pp. 65–110.

67   For description of everyday life of the Cairene in 1800, we have limited resources, of which, Edward Lane was found significant in terms of clarity and detail. We relied on his description of timely activities, which we assume would not normally be subject to misinterpretation by him as an Orientalist.

68   John Urry (1986) has emphasised the spatio-temporal nature of social relations, associating social practice with its context of space and time.

69   Bourdieu, *The Theory of the Practice*, p. 130.

70   Stephen Kern, *The Culture of Time and Space*, p. 11.

71   Ibid., p. 15.

72   Daily Prayers: *Al-Maghreb* (complete sunset); *Al-Isha'a* (Complete dark; almost 1–1:30 hours after Maghreb); *Al-Fajr* (Dawn); *Al-Dhuhr* (Noon); *Al-Asr* (mid-day; mid-afternoon time, 3–4 pm).

73   Lane, ibid.

74   Lane, ibid., p. 141.

75   Nikki Keddie, *Women in the Middle East: Past and Present*, p. 52.

76   Al-Juma'a prayer is the same as al-Dhuhr (noon) but, has to be made at mosque in a collective setting and include a speech by a religious scholar. It is like the Sunday worship in Christianity.

77   Campo, *The Other Sides of Paradise*, p. 111; Lane, ibid., p. 161.

78   Eid al-Fitr is an annual celebration which marks the end of the fasting of the holy month of Ramadan, while Eid al-Adhha is the celebration of Pilgrimage day in Mecca, when Adult Muslims have to slaughter lambs, just as Ibrahim (the prophet) did instead of slaughtering his son Ismail at the place of the Pilgrimage.

79   These events are not recognised by most orthodox rites of Islam; however, sufis used to organise massive celebrations, along with associated recreational activities, which most of the uneducated rural population fled to Old Cairo to attend.

80   Campo, ibid., p. 14.

81   Campo's explanation of Quranic verse (16 Nahl 80–83), p. 15.

82   Lane, ibid.

83   Personal conduct appropriate to a sincere believer is the crucial requirement for attachment of religious significance to such status (Campo 1991: 15). Rich households tried to manifest such sincerity of faith in the Quranic scripts on the walls of their halls, mainly in al-mandharah, where they received their guests.

84   Nelly Hanna, *Construction Work in Ottoman Cairo*, p. 10.

85   Al-Bab al-'Ali 248, 47, 1167/1753, pp. 28–9; Hanna, ibid., pp. 13–14.

86   Hanna, ibid., p. 10.

6

The Harah and the Embodiment of Home

THE HARAH: A HISTORY OF HOME

As social practices configured effective zones of action, in terms of private/public (as we saw in working-class British houses), inside/outside (liminal zone of Turner), male/female (Kabyle house), these zones were determined by particular physical and spatial characteristics, reflecting their social transition from one domain to another. The extent of a home is, hence, defined by the boundaries residents construct around what they believe to be their unchallenged home. These boundaries could be physically constructed by ownership rights in the case of fences, lawns, backyard gardens or walls,[1] or be mentally and cognitively organised to extend beyond the ownership lines, for example, to the local street, the neighbourhood and community.

Homes, as such, are extended outwards or inwards, following patterns of activities and socio-cultural intercourse within the immediate context. Amos Rapoport emphasised such influence: "*because the living pattern always extends beyond the house to some degree, the form of the house is affected by the extent to which one lives in it and the range of activities that take place in it*".[2] Shrinking inwards reduces the effective area for the private practices to particular part(s) of the house, where the individual feels at home. Outward extensions are usually associated with more secure and active outdoor social environments (traditional Cairene communities), while inward shrinkage implies division of the house into more than one home (home workers, writers … etc.). As outward expansion represents a variety of models, here, we tackle the two extremes. The residents of a home in a Cairene hara easily define the boundaries of their home according to the local socio-cultural code and custom; the local alley is essentially a part of their home, as the front lawn of a British detached house is for its household. However, it is not an exclusively their home; rather it is the bigger home which includes other families. The harah's public space, hence, attends to the requirements of a home: inviting guests, taking meals, partying, socialising with the family and neighbours.[3] It is an open space that is an addition to the closed domestic space. Home extends beyond the house and its boundaries appear when the local codes cease to work, e.g. at the junction with larger streets.

Modern gated communities of high-profile social classes represent the modern form of extended homes. They spread out around Cairo either in the new satellite cities or at the periphery of the old urban zones. The new compounds have, in contrast with the hawari, a typical house model with a systematic arrangement that reflects the wealth of their elite residents. Their particular cultural and social codes, albeit entirely different from those of the *hara*, end at the high fences and guarded gates which form the major threshold of elite distinction.[4] In this example, the rituals and socio-cultural codes of practice differ dramatically on the two sides of the fence/gate. For example, informal football games are a popular street activity outside the gates, while inside it is considered an offence and prohibited.[5]

Islam and Islamic ideology dictate everyday life and ideology of Muslim Cairenes, what makes the home of a Muslim a "moral centre of life", within which daily praxis of the good Muslim should follow particular rituals. Stepping into home with the right foot, eating while sitting and with the right hand are examples of rituals that mark the practising Muslim and give the domestic space a ritual purity which defines the home as a sacred place, a domestic paradise, or part of God's house on earth.[6] However, the lack of these rituals turns the home into a house of repellent evil.[7] In that sense, home is the ultimate place of rituals. While the Ka'ba in Mecca is God's home, the mosque is the home of Islam. The home is conceived ritually on a wide-ranging scale. So, while the individual's home could be a sacred space (if fulfilling the requirements of purity, cleaning and associated practices), an Islamic territory where Islam is the primary religion is called Dar al-Islam (the house of Submission). It is where Muslim believers are protected.

On such diverse scales, the idea of home has been translated spatially by virtue of religious rituals, morals, security and protection. Spatial organisation reflects such influences and inherent ideology through location of particular spaces at certain places/distances within the house. The toilet is a dirty place which is considered evil (it should not be in the path of prayers directed towards Mecca). The floor of the living space should be covered with mats for prayers and shoes should be taken off outside the space as at the mosque. If there is not enough transitional/lobby space within the house, shoes should be taken off at the entrance door, to maintain the purity of the house, so it is suitable for prayers at any time. In this ideological context, the harah emerges as an intermediate practice of home, where security and protection are guaranteed and privacy is combined with shared activities and interests of the local community.

While the home as an idea and a practice is my concern in this book, it is essential to recognise the narratives and development of the *hawari* (the local homes) in Cairo since their emergence in the city's urban forms. In Cairo, with its extended history, nothing ever starts from scratch. Former capitals and human settlements have imposed themselves on Cairo since its foundation. Despite the foundation of Cairo as an isolated walled royal city with distinct form, it is hard to believe that it was not inspired by the then busy and popular metropolitan centre of al-Fustat. At least this appears to be the case in the initial division of the residential quarters into sets of khitta similar to those of Al-Fustat. Therefore, here, I follow the development of the idea of home as residential environment and urban form in Cairo.

The earliest Arabic term for a residential settlement in Egypt is *Khitta*. Khitta literally means a plan,[8] and refers to a district or a quarter that was assigned to a certain group of people.[9] It emerged as early as the foundation of Al-Fustat, when ethnic and tribal groups were assigned plots to settle down and form a permanent home for their tribe/group.

The khitta was usually divided into the smaller *mehallah* (place), which later would become the harah.[10] Later, each khitta was developed and formed inner services and local markets. While the distribution of *khitat* (plural of *khitta*) among different tribes as zones was traced by Kubiak in his book, "*Al-Fustat*", he was unable to define their particular spatial forms or sizes. The khitat formed a reasonably regular road network along the south-north *al-Muizz* thoroughfare. During the following two centuries of the Fatimid dynasty, the khitat, which were relatively empty by 973 as mentioned in the travel accounts of Ibn Hawqal,[11] were becoming filled with structures and houses, but did not compromise the formality of the city plan or challenge the pattern of street networks.

The situation was entirely different by the thirteenth century. The formally marked and organised khitat were transformed into crowded and dense hawari. Nizar AlSayyad has indirectly attributed the transformation of the Fatimid khitta to the medieval harah, which remains as a dominant urban unit today.[12] AlSayyad followed the change of the city's urban form in the accounts of two famous travellers: Abd Elatif Al-Baghdadi (from Iraq), who visited Cairo in 1193, and Mohammad Ibn Sa'id (from Morocco), who visited it in 1243.[13] In the 50-year period between the two accounts, the Ayubbid took control of the city with the goal of changing Shiite Cairo into a Sunni state. Al-Baghdadi noted in 1193 that the city was dominated and characterised by palaces occupied by notables, detached structures and wide streets, while in 1243 Ibn Sa'id described it as crowded and claustrophobic, filled with twisting, irregular streets and tall buildings.[14] The houses, according to Ibn Sa'id, were mostly attached and traffic jams and bottlenecks were experienced along the alleys, which were irregular and hosted mixed activities.

This change was owing to the basic political process; the Sunni Ayyubids (the dynasty which followed the Fatimids) were determined to erase the remnants of their predecessors (physically, socially and culturally).[15] The city was filled with unexpected numbers of people within a very limited time. To accommodate the growth and newcomers, gradually, private activities, businesses and new homes were built in public spaces and streets. The organisation of regular quarters then changed under the mounting pressure to narrow alleyways with dead-ends. The maze-like closed lanes dominated the city from the medieval age until the arrival of the French in 1798AD.[16] Unfortunately, there is no detailed record of such change in the urban settings. However, the process of change appears to have occurred gradually as we understood from the archival and historical evidence.[17]

THE HARAH: IN LANGUAGE, MEANING AND URBAN HIERARCHY

The idea of the harah was developed from the earlier khitat whose domestic functions were extended to include the social and economic activities in the public sphere of the city in general.[18] Despite the fact that initial subsections of the city were of equal size, the hawari's flexible nature as a community allowed it to expand or shrink in size and population. As an urban unit, the Cairene harah had a complex nature whereby interrelated spatial and social structures complemented each other and integrated in harmony in the absence of formal organisation or authoritarian control, which created an old mystery about how it came to take on such elastic interrelated forms, and how and when it moved away from the regular and formal plan into such maze-like dead ends.[19] The harah is defined according to three aspects: socially as a group of people unified by ethnic or occupation characteristics and segregated from other groups in the city; politically, it is a unit of administration and control; while physically, it is a subsection of the city, with limited access, controlled gates and walls, which were barricaded during confrontations and could be closed at night.[20] It is not a rigid setting; rather, it is a dynamic and interdependent entity. It has changed some of its elements, modified activities, functions and structures to adapt to changing demands and eras. Even though it has been described as a self-contained unit, which reflects the geographic as well as the social climate of medieval Cairo,[21] its hidden dynamics retained the cultural values and social phenomena of family-based Islamic morals and traditions.

The urban hierarchy of medieval Cairo came to recognise the harah in terms of a solid mass of buildings surrounding usually dead-end lanes, that are branches off the main thoroughfare called *Qasabat al-Muizz Ledin Ellah*.[22] The street network included, in addition to the harah, secondary streets, perpendicular to the thoroughfare, *Durub* (local streets sing. *darb*), and *Utuuf* (small dead-end turns, sing. Atfa). In Arabic terms there are five levels of road hierarchy, four of which are applicable inside the city (Share'i, Harah, Darb, and Zuqaq / Sikkah/ Ataffat), and one is for transportation between cities (*Tareeq*).

The *harah*, as a term, referred originally to local streets branching off from Share'i (street with open access from both sides).[23] Linguistically, it is derived from the word (حيرة - حير) (*hayyara-heirah*), which means confusion caused by a lot of people and houses in tight spaces. Therefore, Ibn Manzur states that every group of houses adjacent to each other may compose a harah.[24] *Harat* (multiple of harah) was the term later developed to refer to quarters of particular group/community characteristics or identity (handicrafts, trade, religion, or ethnic origin), such as *Haret al-Nahhaseen* (copper workers), *al-Yahoud* (Jews), *al-Rum* (Greeks), with the harah's smaller branch of lanes being called *darb*.[25]

Share'i (Main Street), is considered in the original Arabic to be the great street, with the public moving through and buildings with doors opening directly on it. It could be understood in terms of the main route for heavy traffic through the city, which means the basic layer of town planning. As it is exposed to high levels of public movement, it should include mosques. Therefore, we can understand that main axes and thoroughfares were called Share'i in Cairo, like *Share'i al-Gammaliyah, Share'i al-Muiiz*. This name is still used today; however, it includes different levels of the city's fabric.[26]

Darb (smaller harah, Plural: *Durub*) (by-street, lane) (Lane 1956), and mentioned in traditional language as local roads with gates on both sides, and sometimes coming to a dead end. In many areas, the darb is a central spine to the harah, such as the main darb in haret al-Darb al-Asfar. Basically, it is a private level of the alleyways with limited external access. In some cases it is purely residential and in other cases it includes workshops, *wekalats/Khans* (sing. *Wekalah*, closed market complex). From its meaning and practices it resembled a self-contained domestic unit with night security.

Zuqaq/Atffat/Sikkah, are the smallest streets and consist of dead-end branches with no direct access to outside the harah or Share'i. They are subdivisions of the Darb.

The harah, as defined above, as a local street with a surrounding mass of buildings holds comprehensive social and cultural meanings. Socially and culturally it is recognised as a group of people who practise certain values and social contracts specific to them. As such, the harah represents a shared and distinct identity-model and, sometimes, habits or character relevant to the place. In that sense, a residential harah is entirely different from commercial or industrially-based ones in terms of habits and traditions. That is why most of the hawari of Old Cairo are named after the profession, ethnic background or even religion which is considered significant to the local occupants and represents them and their norms collectively.

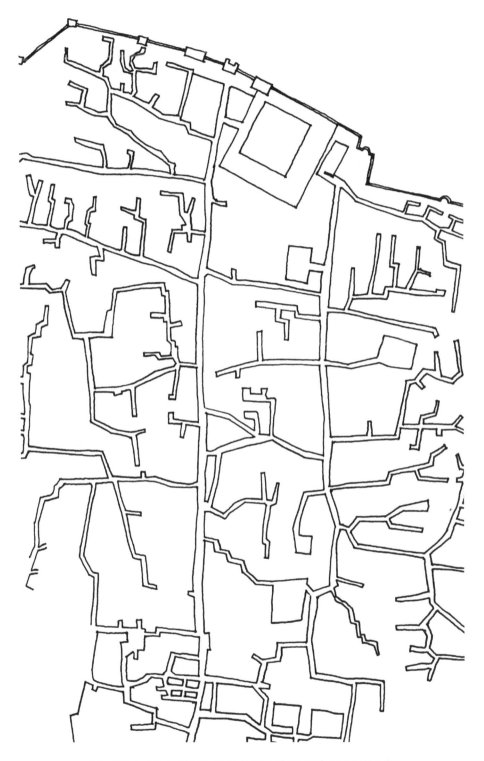

6.1   Street pattern and hierarchy in the old city as they appear today

Spatially, the harah has been defined differently in respect of the formal administrative purposes and informal daily practice of the locality. Administratively, Old Cairo was divided into quarters, and each quarter was divided into smaller units. Most of the contemporary hawari are part of a unit (subsection of a quarter). Each has boundaries that define a regular form, aligned with surrounding streets/alleyways, such as the formal boundaries of haret al-Darb al-Asfar. However, in practice, the harah's boundaries are defined in terms of its spatial organisation of shared public sphere (alleyway) with defined entrances/gates, surrounding residential buildings, social structure, distinct cultural identity and local security. People recognise their harah neighbours according to whether or not the harah alleyway is their access to their houses.

Nasser Rabat emphasised the importance of the harah's inherent meanings as a cultural and social unit, describing it as *"the city's best unit to reflect the Islamic Shari'a's (law) principles of privacy, sight barrier and gender segregation"*.[27] He explained how the harah reflects a place to which the occupants belong and it is the place and space they protect and defend. Harah was a local defensible sphere to its local population, including lifestyle, inner spaces, houses and shared spaces as their private property. In addition, the harah becomes a reference for identity. Most of the Cairo historians (e.g. Al-Maqrizi 1250; Ali Mubarak 1883) defined the character of Cairo as a set of hawari (harats), listing them in terms of their recognised identity, based on occupation, religious or even vicinal ties.

The harah, in essence, represents the kind of residential quarter which, according to Aldo Rossi,[28] is *"a group of elements that connect and integrate and develop within the city instead of being imposed on it by external forces, whatever administrative of political, in order to form a distinguished unit physically and socially"*.[29] It is in that sense a *community*, which, according to Richard Jenkins, is *"a powerful everyday notion in terms of which people organise their lives and understand the places and settlement in which they live and the quality of their relationship"*.[30] It is about responding to the humans' fundamental needs in everyday reality. It is a *"collectivity"* that is more than *"the sum of its individuals"*.[31]

People in the hawari have mutually influenced each other in terms of defining their identity. While the residents are locally identified with their harah/darb name, some harahs are identified in reference to particular distinguished residents (e.g. *Haret beit al-Qadi*, the harah of a judge's house). Al-Maqrizi (1250) and Ali Mubarak (1883) used to relate the name of the harah to a single person's name (a distinguished figure). On the other side, individuals are recognised by the characteristics of their harah, such as morals, traditions, and sometimes, toughness, strictness, or even generosity and hospitality. Interviewed residents of one harah "al-Darb al-Asfar", similarly, considered themselves as one family, which is indispensable for everyone. Daily activities such as eating breakfast, taking tea and coffee, and sometimes lunch, which involve individuals from different families/households, are frequently taking place in collective settings in the harah's public space.

The connection between the social and spatial characteristics of the Cairene harah was best described by Stephen Kern as a *"path"* that is *"closed by masonry"*. This perception negates the form of modern neighbourhood by turning the outdoor open path into a closed interior space.

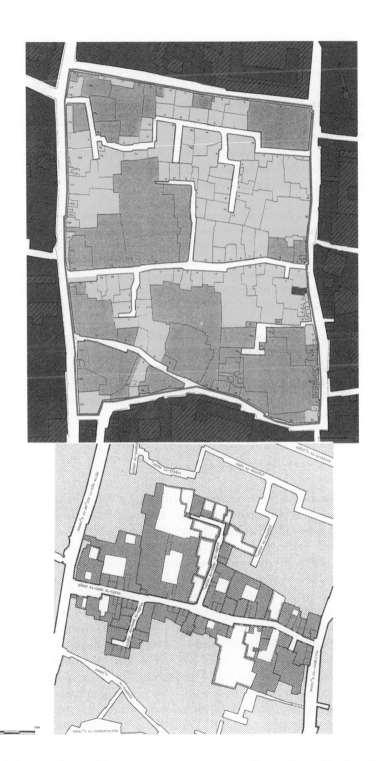

N

0   10   50m
1:1000

6.2   Haret al-Darb al-Asfar: between formal administration and the reality of practice

Notes: a) Formal boundaries of haret al-Darb al-Asfar.
b) Locally recognised boundaries of al-Darb al-Asfar. The dark-filled areas are the houses which are considered part of the harah's community in reality.

> *The Egyptians conceived of space as a narrow path down which the individual soul*
> *moves to arrive at the end before ancestral judges. Their most distinctive constructions*
> *are not buildings but paths enclosed by masonry.*[32]

He shows the path to be the closed protected space similar to homes. It is the actual home which represents the harah's social phenomena physically and spatially. Jamil Akbar, in contrast, questioned the chronicled order of the formation of the harah: "*whether the linear composition of the city has influenced and composed the social organization? Or what happened was the opposite?*" In his view, social units are the principles on which the city developed:

> *first cities were primarily composed of a tribe or ethnic groups units, which means*
> *social units in principle, however, this linear structure changed with the growth of*
> *population; vacant spaces were filled, and structures were attached to each other in*
> *smaller units … . Some houses expanded onto parts of their neighbours plots through*
> *the buying/selling process. Therefore, some houses conjoined two smaller plots,*
> *each of which belonged to two different groups. Accordingly social organization has*
> *changed as a result of physical change of the built environment.*[33]

The complexity of the urban fabric in the old city is a testimony on how spaces, social patterns and time were equally responsible for the interweaving nature of the settings in the old city. Any attempt to isolate any factor in an attempt to understand and analyse the others proved meangingless and a misjudgement. With all the difficulty involved, the synchronised time-space-social is the only unit of analysis that could justify and explain that setting.

THE ARCHITECTURE OF THE HARAH

Throughout its history, the architecture of the harah is marked by its flexibility in adapting to changing needs and conditions. Early harahs were dominated by simple and regular building forms and houses that were set freely in open space with wide roads and straight pathways.[34] Such regularity has changed over time to irregular forms, caused by the process of filling in vacant and left-over land. Connection to the harah's central and active spine and pathway was essential and crucial; some of the backyard plots acquired for building purposes had lengthy, indirect, tight entrance corridors to the harah. Some were single houses while others included wekalat (such as wekalet al-Tuffah; apple wekalah connected to Shar'ei al-Nahhaseen), mainly in commercial hawari. Such strong desire to connect to the public space emphasises the centrality of this venue to every resident.

Building arrangement around the central spine and openings overlooking it are of no less importance than a single house's space organisation. Rather, house construction in the harah, traditionally, used to pay much attention to neighbouring or facing buildings to avoid the harsh penalties imposed by Islamic law (Sharee'ah), which in some cases led to the demolition of the whole structure.[35] The master builder-architect had to design the house both internally (i.e. around a courtyard) and externally (to avoid social or privacy conflicts with neighbours). Such arrangement of houses gives an unmistakable sense of privacy and inclusion. The houses are attached to each other, providing a solid wall with door openings and high level windows.

6.3   Architectural form of the Cairene harah

Such attachment and enclosures stress a strong sense of a solid-void contrasting relationship and the borderline between private and shared activities. Gaps are only found at the atfa, or Zuqaqs junctions. Such solid form and continuity form, in Kevin Lynch's terms, distinct visual perceptions.[36]

Even though the hawari houses (mainly large ones) were orientated towards the inner saha (courtyard), there were major openings and decorated facades which were much opened and connected by the shared pathway. Cairene houses along the harahs were and still are full of external large openings; however, these are always raised to the highest possible level, described by Edward Lane in the nineteenth century to be higher than a man riding a horse.[37] This connectivity required extraordinary attention to exterior windows: which were made of perforated and highly-decorated wood screens called *Mashrabiyya*, allowing internal privacy with views to the outside and good ventilation during summer.[38]

Services and a number of small retail stores are usually found in a comparatively wide space called *Ouqda*.[39] Such spaces can be either internally or externally placed. The internal ones usually exist in an intermediate portion of the harah's main access and at the junctions between different paths as in Zuqaq and atfa. Such spaces are called *Saha* (plaza or open space) despite their limited size and scale and are considered the centre for social activities and the gathering point for public events. The external Ouqdas are placed mainly at the harah's entrance, with more services relevant to the public passers-by, such as food shops, grocers, retail stores, workshops, and in some cases cooked food shops. Most of the harahs in the medieval Cairo were gated with narrow arched gates, which used to be closed at night for protection against strangers, robbers, and foreigners. However, by 1798, most of these gates had been destroyed by the French army: or were destroyed soon after their departure. After several cycles of demolition and rebuilding during Muhammad Ali's era, Cairenes eventually ceased to rebuild the gates and adopted other strategies to maintain security, which will be discussed later.

Architecture of the medieval harah is characterised by continuous solid stone walls with high level windows at ground level and higher level projected floors with approximately 1.00-metre exposure into the harah's void. High-level decorated Mashraibyyat and wooden *malqafs* (wind towers) dominated the scene. The contemporary harah's ground level is, however, dominated by shops/stores and workers, while top level projections have become balconies and the mashrabiyya have been replaced by the *sheish* (louvered wooden windows). While medieval houses were organised around central courtyards, contemporary houses are externally-orientated single family apartments in multi-storey buildings.

In some hawari, mosques and wekalat are significant buildings due to their relatively massive scale, style and modular façade design and decoration, as seen at share'i al-Nahaseen (previously mentioned). These buildings, with their wide decorated entrances and heavy and rich ornaments, give interesting views among the typically long and less decorated houses' facades. Those buildings are still emulated today by modern architects aiming to restore such significant elements of the built environment in Old Cairo (an example is the Feda complex in share'i al-Nahaseen) (Figure 6.4).

Similar to the continuity of the hawari patterns in Cairo from previous capitals in Egypt, medieval houses of Cairo were also inspired by their predecessors, mainly in Al-Fustat. They were described by the Persian traveller Nasir-i-Khasrau, in the tenth century to have houses of four, seven, eight and even eleven and fourteen storeys in residential complexes hosting several families and arranged beside each other with large surrounding gardens. He had described the city domestic environment in utopian terms stating that no cities in the world had such glory.[40] Excavations of the early twentieth century revealed some of their organisation and typologies as courtyard houses with internally-oriented rooms and no external windows.[41]

6.4   Feda complex in share'i al-Nahaseen

Fearing fire or demolition similar to that which occurred in Al-Fustat, Cairo houses, buildings and complexes were kept to a maximum height as low as three or four levels. Cairo inherited courtyard houses for people of the higher order while people of lower orders used to reside in small apartment buildings, or collectively in residential complexes such as Rab' and Wekalat Qunsuh al-Ghuri, in Khan al-khalili and wekalat bazara'a in al-Gammaliyyah.

This divides residential buildings of medieval to pre-industrial Cairo into three types, which appeared to dominate the urban scene until the end of the eighteenth century, exemplified as: al-bayt, al-Rab' and Wekala, and the Palace. The scale of each of these residential types of structure was in accordance with their location and their occupants. Al-Rab' was a residential complex composed of multiple rented apartments. Each of these was composed of one main space with a small room for services, all arranged beside each other around an internal courtyard in a similar manner to contemporary hotels. Occupants were workers, artisans or foreign trade merchants who stayed in Cairo for business and to sell their products. The *wekalat*, on the other hand, included open markets on ground level, while upper levels were residential wings, such as those in *Wekalat* and *Rab' Qunsuh al-Ghuri* (1504–05AD) which remained functioning until the twentieth century.

Al-bayt and the Palace could be categorised together as both are different types of private houses. However, there were obvious differences between the *Palace* (the house of rich merchants, Ulama and ruling elites) and *al-Bayt* (the house of ordinary middle class people or maybe merchants). The difference can be seen mainly in the size, capacity of rooms and decoration. While Bayt Al-Razzaz[42] and Zaynab Khatoun houses were large, highly decorated, palace-like houses, Mustafa Gaa'far's house was of the average bayt type as we shall see in details in later chapters

The contemporary harah has added two new typologies: the multi-storey apartment building and the multi-occupancy bachelor accommodation (where young professionals and workers rent a room or a bed in a shared room). The latter, in particular, was not allowed during the medieval and pre-industrial harah, and emerged only early in the twentieth century. However, neither have distinct characteristics and the bachelor accommodation type usually consists of large apartments that have been subdivided into shared units/rooms of multi-occupancy (as many as nine per room in some cases) for better value and profit.

Between the city and the house, the harah resembles the intermediate organisation of home that works effectively to link a group of people to their place and location with a strong sense of belonging and privacy. It, however, does not emphasise or promote internal separation and division. The harah's organisation is understood as a reflection of the internal organisation of the house that accommodates more than one family. While the harah maintains a strong boundary with the outside world, it eases internal boundaries between its internal components: houses. The harah's boundaries with the outside world are composed of thick masses of back-to-back buildings and tight and controlled access points (gates). Similarly, the houses' boundaries with other houses are emphasised by solid shared walls and a single entry front door, whilst internal borders between domestic spaces are less defined. The old city, on the other hand, bound itself within surrounding fortified walls with tight urban routes and fenced gates, with internal boundaries having less fortification.

The mechanisms of activity, the practice of privacy and protection are similar at these different levels of Old Cairo. They depend on elaborating the sense of mystery regarding what is behind the walls, and what appears following the crossing of a threshold. The organisation of internal-external boundaries imposes a series of rituals on those crossing the thresholds of each sphere (house, harah, and city). This reflects the idea of home as a defended territory. It deploys the spatial order and characteristics to strengthen the experience of being at someone's home whilst still in the public sphere. The sense of openness, transparency and clear visual accessibility is undesired and therefore is entirely absent at all those levels. The borders are clearly defined throughout your movement. Permission for entry is always required when getting into the old city, the harah, and the house.

The experience of labyrinth and mystery is, as such, an essential element of the spatial order of the home. The rituals of crossing the borders, experiencing long but broken pathways with indifferent facades on both sides, add to the confusion of the stranger/visitor, and put him/her in the weakest position in contacting the powerful residents. The harah as a concept, hence, mediates these experiences and is an appropriate manifestation of the idea of home as a built-in identity and spatial organisation whose architectural form and character (internally or externally) are meaningful elements to everyone.

It is at this stage I need to look at the theoretical grounds on which the idea and practice of home is established. In the following chapter, I will be discussing the idea of home as a practice in everyday life and how it becomes a territory of investigation: a complicated undertaking wherein architecture, social sciences and anthropological studies meet. I will also look in detail at the way the harah of Old Cairo responds to this idea and practice of home. The next chapter will stress the idea that the home is principally architectural territory of investigation that is informed by existing socio-cultural processes and the rituals of everyday life.

6.5   Wekalat Bazara'ah, Haret al-Tambukshiyyah, al-Gamaliyyah

BARRIERS AND THRESHOLDS IN OLD CAIRO: ARCHITECTURE OF THE VEIL

A large number of activities take place in the harah passage which in other parts of Cairo, or even during different historic stages of the harah, would be restricted to the physical setting of the dwelling … The manner and form of familiarity with which various intimate activities are carried out in the harah passage make it evident that the alley is actually considered by both sexes to be a private domain. Members of both sexes in the harah treat each other with familiarity similar to that existing among members of the same family. Even outside the harah, any male resident is responsible for protecting any female member of his harah. He is further responsible for what she does, and he has the right to interfere in her activities if he finds them inappropriate.[43]

Janet Abu-Loghud noticed that while most of the hawari in Cairo resembled the idea of home for their inhabitants, frequent journeys between home and workplaces took place on a daily basis: *"men with families often commuted to homes in more exclusively residential zones"*. The boundary of each home would start at the harah's gate, which all families were required to guard as a part of the protection of their own individual houses. The harah's gate was closed and guarded by a *bawwab* (doorman) or other individuals on a daily basis from sunset till sunrise.[44] This threshold was an essential spatial component that defined social boundaries and maintained the ultimate safe environments. Protection and shared ownership were strategies used by the residents which recreated the sense of home in the broader environment of the harah.

This inherent meaning of the harah as a home was emphasised by gatherings, with residents sitting and eating in this public space as they used to do at home. Edward Lane described his Ramadhan breakfast meal in the front of the house of his host in a Cairene harah. The meal was taken in the harah's public space, while his host's sons served him.[45] They had turned public areas into private without a change of the users or any change to the physical characteristics of what was described by Fadwa al-Guindi as *"the distinctive quality of Islamic construction of space"*.[46] Extending the home to the hara's public space was an indeterminate action by the locals to exercise their control over this private zone. To enforce responsibility of maintenance and control, Muhammad Ali's legislation of the early nineteenth century gave local members shared ownership rights of the shared spaces of the hawari.[47]

Basic features of home, exemplified in secure enclosure, entrance guarded door, familiar environment, ownership rights and duties, combined private and public venues, and finally shared identity and culture, were exceptionally valid in most of the hawari of Old Cairo, as we can see in Haret al-Darb al-Asfar, al-Nahaseen and many others. Visual intercourse and interpersonal interaction within its central space was the prime source of identity. As such, social boundaries were prioritised over the physical ones which continued to change over time. We saw in medieval times that, while building structures were frequently changing, social organisation and traditions remained consistent. This offered more flexibility to the community to adapt its settings without reliance on the always changing physical settings.[48]

The idea of home and ideology of domesticity can be seen at best in the social barriers between different domains in the Cairene house. The quality provided by the well-known medieval mashrabiyya emphasises how architecture was, in part, developed to suit certain social patterns of relationships. Establishing social barriers and thresholds was essential to the regulation of social intercourse in a wide mix of overlapping activities and interpersonal communication. They aimed to define the spatial and temporal settings that constructed the limits and effective zones of particular social spheres and how one moved from one to another. Women, who were the most protected and endangered members, were central to the construction of these domains according to whether movement within a particular zone was free or restricted. Even within the same house (within extended families in particular), conflicts could emerge easily where there was a lack of clearly defined boundaries or protective barriers. We could see a correlation between the women's position in the community as *harem* (plural of *haram*, means prohibited) and the symbolism of harem in the morphology of homes.[49] This relation is seen in the inherent connection between the Cairene woman's *hidjab* (the veil) and the way that the hidjab was established.

The concept of the veil is based on the right of the woman to cover her face, although she still can communicate and interact with others. It is precisely to *"allow someone to see but not to be seen"*.[50] The hidjab is much more than a material object; it is a social institution that was developed from its Islamic roots, and which was imposed on the wives of the prophet then extended to all free Muslim women. *"The wearing of the veil marks the transition from childhood to puberty and from spinsterhood to marriage"*.[51] In this sense it marks temporal social transition, while spatially, it is a process of screening the sacred from the secular. In conclusion, the hidjab is a dynamic barrier which could be lifted or even destroyed,[52] and about which conflicts could emerge. While the hidjab conceals the detailed identity and features of a woman's face, it does not conceal her identity entirely. Women tended to reveal their identity by their body movements, voice, particular gestures and eye contact: a practice that attracted men's attention while adding mystery and encouraging curiosity.

Home replicated this concept of partial privacy through establishing layers of barriers and thresholds. Barriers of privacy were not intended to provide total isolation. Rather, they formed protection against intrusion into private areas. Accordingly, we can see those barriers in certain types of openings, windows and entrances. Such mutual relationships between the woman and the house in a Cairene context are reflected in the literature. Bechir Kenzari and Yasser El-Sheshtawy asserted that the architectural motifs of Cairene houses replicated the concept of the women's veil when they referred to the *Mashrabiyya, the dominant wooden lattice window in medieval Cairo, as "an architectural veil"*.[53] They suggested that it is a veil made of wood to provide the harem with privacy that encouraged, at the same time, the action of gazing.

Mashrabiyya, literally, means *"a place for drink"*;[54] a wooden lattice window projected from the wall to allow cool air through to the porous earthen vessels of water put behind it. Kenzari and El-Sheshtawy have investigated the *mashrabiyya* as an architectural motif rather than a *"socio-political idea in the Arab-Muslim culture"*. While the *mashrabiyya*, as a window, aims to enable transparency and its historical genesis and meaning are related, they postulate the use of weaving rather than glazing. For them, mashrabiyya is a woven screen made of wood, however, posing intrinsic social, political and mystical meanings (Figures 6.7 and 6.8). Accordingly, they believe that *"the mashrabiyya lends itself to being interpreted as an architectural apparatus that crystallizes the desire for visual communication"*.[55]

The focus on the gaze effect and game of seduction is overwhelming: leaving issues of social interaction, oral communication, and inside-out connectivity barely articulated. The harem mashrabiyya of opposite houses were vocally and socially connected at higher levels despite the ground floor interruption by the alleyway. The main darb's width varied from 3–4 metres, with mashrabiyyas projecting from both sides, leaving a 2–3-metre gap. This gap allowed the exchange of conversation: resembling the continuity between domestic spaces which cross over into the ground level public space. When one considers that Oral communication was and still is the prime tool for circulating information, news and social intercourse within Muslim societies,[56] then we can understand the paramount importance of this connectivity.

6.6   Veil of a Cairene woman in the late nineteenth century

Courtesy of Rare Books and Special Collection Library of the American University in Cairo.

In summary, the mashrabiyya windows resemble the hidjab role in very physical terms. It provides a social barrier against intrusion, while allowing social interaction and visual continuity in variable degrees. *"One property of the veil is its dynamic flexibility which allows for spontaneous manipulation and instant changing of form"*.[57] It provides its wearer with the power to change a situation by uncovering or pulling the veil up.[58] The mashrabiyya allowed women access to and supervision of activities taking place in the public domain without their privacy being breached. It worked as visual barrier in one direction, while allowing communication in both directions.

6.7 Mashribyyah, lattice screen window made of wood in Old Cairo

Spatially, barriers appeared differently in outdoor and indoor space and at ground and upper levels. At ground level, while external walls appeared totally solid with the entrance door as the only threshold to access, internal walls were largely open with large mashrabiyya and wide perforations, allowing different spaces of inner public venues to combine into one large public venue. When moving to higher levels, more privacy was required. So, the mashrabiyya was smaller in size with tighter perforations. Such tightness eliminated the outside-in visibility, giving more privacy control to women. This was, moreover, associated with the change in designation of the space it served from ultimate private to semi-private. While *harem qa'a* (room/quarter) as an ultimate private space had the former type, semi-private spaces such as living spaces, corridors, services and circulation would have the latter (Figure 6.8).

6.8 Internal
façade of Bayt
al-Suhaimy in
Cairo showing
different types
of mashrabiyya
Note: Size and perforation
at ground (public) and
upper levels (private).

NOTES

1   Rapoport, *The Meaning of the Built Environment*, p. 170.

2   Rapoport, *House Form and Culture*.

3   Abu-Loghud, *Islamic City*; Al-Messiri-Nadim, *The Concept of the Hara*.

4   Denis, *Cairo as Neoliberal Capital*, p. 64.

5   Huwaida Taha has emphasised the huge difference in social habits and cultural worlds
     between the residents of those communities and the poor outsiders. See Taha, *Al-Tafawut
     alTabaqi fi Masr*, p. 12.

6   Campo, *The Other Sides of Paradise*.

7   Bertram (1996).

8   Ibn Manzur, Lisan al-Arab.

9   Raymond, *Cairo*, p. 39.

10   Raymond, ibid.

11   Ibid., p. 40.

12   AlSayyad, *Virtual Cairo*, p. 94.

13   AlSayyad, *Cairo: Bayn al-Qasrayn: The Street between the Two Palaces*, pp. 73–4. For primary sources refer to A. Al-Baghdadi, *Al-Ifadah wa Al-I'tibar*, written 1204AD (Cairo, 1946) and M. Ibn Sa'id, *Kitab Al-Maghrib fi Hula Al-Maghrib*, written 1243AD (Cairo, 1950).

14   AlSayyad, ibid.

15   They demolished all their significant buildings, palaces and big structures, except mosques.

16   Abu-Loghud, *Cairo*, p. 61.

17   The investigation of archival records of home ownership/trade in Cairo showed a pattern of division of properties, acquisition of additional areas of adjacent buildings and sometimes acquisition of higher levels of neighbouring buildings, which suggests that this process of property management has been going on for a long time, and certain regulations have developed accordingly?

18   Rabat, *The Culture of Building*, p. 198.

19   Hourani and Stern, *The Islamic City*, pp. 10–12.

20   Abu-Loghud, ibid., p. 24.

21   Abu-Loghud, *Cairo: 1001 Years of City Victorious*, p. 67.

22   *Qasabat* means narrow street with many bends. Al-Muizz Ledin Ellah is the founder of Cairo. In general it is a street named after its founder.

23   Al-Siuofi, ibid.

24   Ibn Manzur, *Lisan al-Arab*.

25   Akbar, *Emaret al-Ardh fi al-Islam*; Abu-Loghoud, *Cairo*.

26   Ibn Manzur, *Lisan al-Arab*.

27   Rabat, *The Culture of Building*, p. 197.

28   Rossi, *The Architecture of the City*, p. 21.

29   See also Rabat, *The Culture of Building*, p. 199.

30   Jenkins, *Social Identity*, p. 109.

31   Jenkins, ibid.

32   Kern, *The Culture of Time and Space*, p. 139.

33   Akbar, *Imārat al-arḍ fī al-Islām*, p. 386.

34   This is what we were told by Mohammad Al-Baghdadi, as mentioned earlier.

35   Hanna, *Construction Work in Ottoman Cairo*. Hanna mentioned several cases where buildings were ordered to be demolished by the court when neighbours complained about privacy, noise or health concerns.

36   Rabat, ibid., p. 200.

37   Lane, *The Accounts of Modern Egyptians*, p. xx.

38   Kenzari and ElSheshtawy, *The Ambiguous Veil*, pp. 17–18.

39   Rabat, *The Culture of Building*, ibid., p. xx.

40   Creswell, *Islamic Architecture of Egypt*, ibid., p. xx.

41   Kubiak, *Al-Fustat*.

42   The residence of Ahmad Katkhuda al-Razzaz in the Bab al-Wazir, 1778, with remains from the fifteenth century.

43   Nawal Nadim, "The relationship between the sexes in a harah of Cairo", unpublished Doctoral dissertation, Indiana University (Bloomington, 1975). It was also quoted in Janet Abu-Loghud, "Islamic City: Historic Myth, Islamic Essence and Contemporary Relevance" (USA: *International Journal of Middle Eastern Studies*, vol. 19, 1987), p. 168. While this concept is historical and has its root in the social behaviour of the harah's residents, it remains the same today. Almost the same wording was used by my main informant in al-Darb al-Asfar in an interview in 2008, to explain how they use the harah's passage and the way they interact together inside and outside the harah.

44   Lane (1860), p. 4; Abu-Loghud, ibid., p. 64. Sophia Lane-Poole described how someone's brother funded the funeral of their harah's bawwab, who turned out not to have died, and used the money to live in different quarters. See Lane-Poole, English Woman in Cairo. Letter dated, August 1847.

45   Lane, ibid., p. 56.

46   For more details about the construction of space in Islamic societies, see Fadwa al-Guindi, *Veil: Modesty, Privacy and Resistance.*

47   This was among other legislation aimed at re-organising and cleansing the city. See Abu-Loghud, *Cairo*, pp. 64–6; Sayyid-Marsot, *Egypt in the Reign of Muhammad Ali.*

48   This was a common practice in this period and during the entire medieval period. For more details and specific cases, see Leonor Fernandes, "Istibdal: The game of exchange and its impact on the urbanization of Mamluk Cairo", and Sylvie Denoix, "A Mamluk institution for urbanization: The Waqf", both in Behrens-Abouseif, *The Cairo Heritage.*

49   Reference from Courtyard Houses.

50   Kenzari and El-Sheshtawy, *The Ambiguous Veil*, p. 22.

51   Ibid., p. 19.

52   Ibid., p. 20.

53   Ibid.

54   Lane, ibid., p. 8.

55   Kenzari and El-Sheshtawy, ibid., p. 24.

56   Janet Abu-Loghud recognised the acoustic factor as one of the main differences between Muslim quarters and Hindu quarters in India, a characteristic that she applied to other Muslim cities. For detailed discussion, see Janet Abu-Loghud, "Islamic city: Historic Myth, Islamic Essence and contemporary relevance" (USA: *International Journal of Middle Eastern Studies*, vol. 19, 1987), p. 161.

57   El-Guindi, F. (1999), *Veil: Modesty, Privacy and Resistance*, p. 97.

58   Ibid., p. 96.

7

Medieval Homes of Cairo in 1800AD

MEDIEVAL HOMES OF THE HAWARI IN 1800

Although Medieval Cairo has been extensively investigated on several grounds, mainly on its social structure, historical buildings, and urban fabric, we have known little about how these worked together on a daily basis. A thorough look at all volumes produced on Cairo and its medieval history would reveal the political top-down tone that was preoccupied with stories on the ruling elites and preconceptions about the lives of the individuals. In fact, many of the histories of the period were written with elites and scholarly audiences in mind, according to Mario Ruiz.[1] On one hand, this could be a result of the wealth of information about the elites and the ruling class, in the face of a scarcity on the lower class side. Similarly, the chronicles of Egyptian historians and foreign travellers looked mainly for stories about the elite as worth recording. This left us with small fragments of information to weave together to understand the way everyday life and use of space took place in Old Cairo. A precious source was the archival records of marriages, trade, real estate operations, and court rulings. These records were largely comprehensive to the extent they encompassed richer information than the operation itself. To collate a credible narrative of Cairo of 1800 these records were extensively interrogated along with Al-Jabarti's accounts of the events and incidents in the old city as well as the contemporary accounts of Edward William Lane, which appeared a few decades later.

SOCIO-POLITICAL STRUCTURE AND HIERARCHY

At the turn of the nineteenth century, the harah was still the primary political and administration unit as well as being the basic social system in Cairo.[2] Despite the introduction of the *Tumn*[3] (means one-eighth, pl. *Atman*) by the French, it was limited to the formal political hierarchy, while ordinary people maintained their medieval organisation. Each harah had a leader, *Shaykh al-Harah*, acting as a mediator between the ruler and local community. While *Shaykh al-tumn* was appointed by the French, *Shaykh al-Harah* was socially appointed by his own community's nobility and was honoured in public gatherings and received rulers, ambassadors and dignitaries.[4] During the first

7.1   Buildings of
1800 in al-Darb
al-Asfar located
on a later map
of 1937 Survey

quarter of the nineteenth century, Lane summarised Shaykh al-Harah's influence as
*"to maintain order; settle any trifling disputes among the inhabitants, and to expel those
who disturb the peace of their neighbours".*[5] In addition, he mentioned recruitment of
servants for small fees, searching for stolen goods and bringing offenders to justice,
as some of the tasks assigned to him.[6] This meant that, in each harah, *Shaykh al-Harah*
resembled the government and the state: a system, as Edward Lane argued, inherited
from the Ancient Egyptians.

While this character seemed authoritative and ultimately powerful, he had to
be accepted and facilitated by the local subjects. First, Shaykh al-Harah had to be
respected among his locality and serve their interests.[7] Parallel to this authority, there
were other state agents such as the *Muhtaseb*, who oversaw pricing in the markets and
the preservation of public morality, but did not interfere in residential areas.[8] The *Ulama*
(religious scholars) were other state-society mediators, who used to defend the interests
of the public against the unfair resolutions and taxation of the rulers.[9] The security of
each harah, however, was maintained by the organised youth and gangs (*futuwwat*),[10]
who were usually protectors of their locality, but troublemakers for other localities. By
the end of the century, each harah was identified as a territory for a particular futuwwa,
whose power was an indicator that measured the strength and power of the locality.[11]

It is difficult to differentiate the political system from the social organisation of the
hawari communities in 1800.[12] Leadership and order of seniority were associated with
nobility, seniority and social status, with community leaders having to be of a higher
rank and superior position and authority among fellow community members. Tradition
associated these roles with local religious scholars, shaykhs, Imams or guild leaders
whose opinions and decisions were largely respected, a structure inherited from early

Islamic states.[13] Trades, marriages and conflicts witnessed this structure and order of seniority at work. Furthermore, the way community leaders sat down with their hosts would follow that order, according to Lane's accounts. This system, along with the evolution of the city's material conditions, had served as the foundation of the then existing political community.[14] Hierarchy allowed the elders, having experience and wisdom, the authority of decision making, problem solving: and opposition to the elders was considered as bad manners. Leadership could, therefore, have followed the largest and most noble family among the harah's community.

In haret al-Darb al-Asfar, for example, and based on archival records, we could practically presume that Shaykh Abdel Wahab Al-Tablawy *(whose house was later named Bayt Al-Suhaimy)*, the harah's most noble man of that time, held a superior position in the community as a leader during most of the eighteenth century, the period prior to the studied time.[15] He was a noble religious scholar who lived in a house that accommodated other properties in the waqfiyyah, or Mubaya'a.[16] Among other notable residents were Mustafa Ja'afar, the famous eighteenth-century Syrian coffee merchant,[17] and Shaykh Zein-Eldin Mustafa al-Diasty, the al-Azhar scholar.[18] Being a religious scholar meant that you could easily get rich through easy access to high profile merchants and rulers. Such popular history of scholars as community leaders was mostly the case in relatively high profile residential hawari that lacked certain professional affiliation and were inhabited mainly by native Muslim Cairenes of average middle-class families and counterparts from lower levels of the social stratification.[19] The residents included a few rich and high profile scholars *(Al-Suhaimy, Al-Banani)*, trade merchants *(Mustafa Ja'afar)*, lower middle-class families *(Fakhr Ibn Ali Halabi)*, servants and slaves working in rich members' houses.[20] There was no evidence of the presence of foreign ruling class people *(Mamluks, Ottomans)* or the poor, who used to live in open collective hawsh.[21]

Archival records of daily activities during the first decade of the nineteenth century support the above social composition of a typical residential harah. Three archival documents were found, which were dated between 1791 and 1810AD, describing trading of parts of houses among local residents, with the affiliation and names of senior leaders included as witnesses. For example, document no. 637, Record 534 dated 1st *Shawwal* 1205h (3 June 1791) described a swap deal *(Estibdal)* of a half portion of a house with a similar portion of another house in al-Darb al-Asfar;[22] document no. 61, record 537 dated 11th *Muharam* 1218H (3 May 1803) and document no. 194, record 537, dated 7th Shawwal 1225H (15 November 1810AD) referred to the names and affiliation of religious scholars as witnesses. Typically, the documents included a brief history of the properties which were previously part of Awqaf[23] properties, referring to their *Waqfiyyah* (the *Waqf* document) as evidence of the acquisitions' history of ownership. Interestingly enough, these records indicate that religious sheikhs were descendants of other scholars, meaning that being an imam or sheikh, was a profession of inheritance within the family, and perhaps the community.

With a rate of mortality reaching 50 per cent in young people during the first half of the nineteenth century, *"it was not the norm to enjoy the company of both parents for a long period of time".*[24] This demographic stagnation had affected the stability of the family as a social structure at that time, which could deteriorate, change or decline at an unexpected rate.[25] This could explain the high number of servants and slaves employed in Cairene households who became an essential part of the family in the majority of social stratifications.[26] Edward Lane had described an upper middle-class family that had reached 20 members in a large, extended family household.[27] For example, he described the harem as follows:

> *The harem may consist, first, of a wife, or wives (To the number of four); secondly, of female slaves, some of whom, namely, white and Abyssinian slaves, are generally concubines, and others (the black slaves) kept merely for servile offices, as cooking, waiting upon the ladies, & c.; thirdly, of female free servants, who are, in no case, concubines, or not legitimately so. The male dependants may consist of white and black slaves, and free servants; but are mostly of the last-mentioned class ... it is seldom that two or more wives are kept in the same house: if they are, they generally have distinct apartments. Of male servants, the master of a family keeps, one or more to wait upon him and his male guests; another, who is called Sakka, or water carrier, but who is particularly a servant of the harem, and attends the ladies only when they go out; a bawwab, or door-keeper, who constantly sits at the door of the house; and a "sais" or groom, for the horse, mule, or ass.[28]*

It is understood that Lane's accounts could not be fully accurate or reliable due to his limited accessibility to certain parts of the house, especially the harem, whilst the lack of similar accounts from other Egyptian historians prevents credible verification. However, in my view and based on the space arrangement and physical evidence we have in hand, Lane's account is more logical for a house of the size of Bayt Al-Suhaimy in Haret al-Darb al-Asfar, that was not a typical form within the city at the time.

PHYSICAL FORM AND SPATIAL DEVELOPMENT

Old Cairo's lanes and alleyways underwent several changes to their spatial setting in the time between the foundation of the city and its evolution as a metropolitan centre during the eighteenth century, with the first clear and detailed map drawn by a French missionary in Egypt in 1800. A typical alleyway cut its way through a dense fabric of attached houses with a few branches to either side. While the main lane was the central Darb or Harah (here refers to the street), the branches were either atfa or Zuqaq, as explained before. In al-Darb AlAsfar, two branches, Atfet al-Darb al-Asfar to the north and Zuqaq al-Darb al-Asfar to the south branched off the main darb, both of them dead ends that did not lead to any exit. They were formed as cracks in a solid mass of houses with inner courtyards. The only entrance at the time was to the whole harah and was at Share'l al-Gammaliyyah to the western side, with no access on the eastern side towards the busy Al-Muizz thoroughfare, which was blocked at the time. Instead, the eastern side was the most exclusive and least open to local public activities.

The map of 1800, in fact, did not show any large open space in the alleys that could accommodate social gatherings and public events or activities.[29] However, a marginally larger pocket beside the harah's main entrance, resembling the Ouqda, was evident in most of the hawari.[30] Surrounding the narrow alleyways of the hawari, the buildings' façades were continuously solid and flat, with higher level wooden grated windows at ground level and projected upper floors dominated by mashrabiyya,[31] typical of medieval Islamic cities. The height of most buildings was around three to four storeys. Ground levels usually had built-in stone seats beside the houses' entrance doors and high-level flat windows.

At the time, public sphere and alleyway spaces were dominated by men's activities and interaction; women, especially of the lower order, used to appear in public for their work, shopping or service work for richer families. Similarly, they were not obliged to put on the traditional veil, which was normally mandatory and mostly associated with women of higher order.[32] The harah's men used to drink coffee in collective settings, mainly in the front of their local shops, or house doors. Coffee houses, on the other hand, were meant to serve lower-middle-class workers, and those peasants living in the old city or coming to work and visit. They were busy around the clock (day and night); however, they were adjacent to the harah's gate on the outside.[33] As their presence was extensive, with around 1,200 coffeehouses existing in the old city on the arrival of the French in 1798, coffeehouses were the main venue for doing business, discussing trading deals and making public announcements or sharing news.[34] Above all it was the local centre for entertainment, where public recitation of romances took place.[35] They appeared mainly beside hawari entrances but never inside them. In fact, both coffee houses and hawari gates acted as social markers of boundary zones at which social behaviour, and codes of public/private customs changed.

Daily movement to and from the *souks* (markets) along with beggars and pedlars' movement and noise kept the main alley busy and noisy most of the day. Cairenes had to salute (saying: *al-Salam-u-Alaykum*) when they saw or passed by others, even from a distance, as per Islamic customs and duty.[36] Beggars engaged in shopping communications with local women through their mashrabiyyas, and with men who spent their day time on the *Mastabah* (stone seat) alongside their home or shop entrance door. Inviting neighbours, passers-by and fellow workers to sit and join the host for a meal, a drink (coffee) or even smoke, was the main and frequent everyday interaction. Such public format was best captured during public festivities, which were mostly religious and associated with the Hijri calendar.[37] They were celebrated within the public space, where children moved and played safely.[38] It was a tradition for poor children to ask for alms, and men would be prepared with several coins for that purpose.[39] While the French imposed an obligation for every house to have an exterior lamp to keep the alley lit,[40] during these festivities the alleys were constantly lit with the noise of lively activities such as *Zikr* (recital of Islamic verses).

Celebratory occasions such as weddings, *tuhoor*,[41] and funerals were public gatherings that had their own obligations, customs and format (Figure 7.2). First, all members of the harah were obliged to take part in these events. Tuhoor celebrations involved a march around the quarter while the boy was carried out in a specially decorated chair with musicians leading the march. Similarly, wedding celebrations lasted for three days, during which two marches of the bride and the groom took place separately. The bride and her female friends toured the area on the back of horses or donkeys to the *hammam* (public bath) followed by musicians. While the return from the pilgrimage was an occasion of importance for the individual, it was also celebrated by the people of the harah. Part of this celebration was to paint the front door of the Hajj's house and in some cases paint his front façade, a few days before his arrival. This ritual was meant to distinguish and celebrate the man in question by marking his house as he returned to it. This meant that spatial settings correlated to social and cultural rituals and interaction, and in some instances this was articulated in a physical format.

7.2   Sketch drawing of a local marriage celebration in the streets of the city
Lane (1860).

FROM THE ALLEY TO THE HOUSE

According to Nelly Hanna, "*The middle class house is a family home, in which lived the extended family, including parents, children (married and unmarried), uncles, widowed or divorced aunts*".[42] This pattern of the extended family prevailed over the high and middle social strata and prevailed in most of the old city's history until as recently as the first half of the twentieth century, when spaces, houses and architectural expression of this way of life largely disappeared, giving way to a smaller, micro family system. As different classes had different structures in 1800, I shall highlight the shared characteristics and components among the upper class, higher middle and lower middle classes in Old Cairo at the time. The dominant pattern of family had one master, who managed the household, socially and economically, while the mother was the de-facto female leader and prominent character. Sons and daughters did work with their parents but did not earn money in return. Rather, the finances of the household and different sons' and daughters' expenses were also covered by the family's honoured master.[43] Younger members were divided according to age (adult/children), and gender (sons/daughters), creating different frames of references within the confines of a single home. The elder son retained his father's authority and responsibility in his absence, while the elder daughter helped her mother and replaced her wherever needed.[44]

Despite its clarity as a structure and way of life, everyday interactions proved more complicated. We have to put the praxis of gender division, hierarchy of authority and management into context. Women, despite ill-guided orientalist views that they were lazy and dedicated to pleasure, were in reality active producers: of materials, textiles, managing their domestic activities and needs, as well as leading businesses and owning a number of properties and waqfiyya[45] in a very similar way to the men. The house was basically divided into two parallel lines of production and social networks, and each worked within a distinct sphere of influence.[46] This gender segregation within the domestic sphere was highly influenced by the Ottoman rulers' conservative culture, first brought to Egypt in the sixteenth century.[47] To better understand how this family structure worked, see Figure 7.3 that shows a standard family structure based on descriptions by Nelly Hanna and Edward Lane, who represented the spatial division of homes that are very similar to those described in archival documents I surveyed of this period. Lane talks about the motives behind the spatial division of the house during the first quarter of the nineteenth century, saying:

> Though the women have a particular portion of the house allotted to them, the wives, in general, are not to be regarded as prisoners; for they are usually at liberty to go out and pay visits, as well as to receive female visitors, almost as often as they please. The slaves, indeed, being subservient to the wives, as well as to their master, or, if subject to the master only, being under an authority almost unlimited, have not that liberty. One of the chief objects of the master in appropriating a distinct suite of apartments to his women is to prevent their being seen by the male domestics and other men without being covered in the manner prescribed by their religion.[48]

For the sake of comparison, this analysis of the socio-spatial division of the household and the cultural rationale behind it is compared with the spatial order of domestic spaces described in some house sales documents during similar periods. For example, a court clerk writes:

> It [the house] includes an entrance built in stone with wooden door through which we enter a partially covered courtyard (hawsh). There is a door at the right of the entrance that leads to a staircase leading to a landing that leads to a Coptic meqa'ad with one iwan and durqa'a overlooking the front facade … adjacent to this meqa'ad a stair and door that lead to the door of the harem qa'a with its services, including a toilet and store …[49]

We find most of the descriptions define two distinct zones; the ground level public zones, including meqa'ad or mandharah with entrance and courtyard as male dominant zones, while the harem is always at a higher level and accessed through a particular staircase. Such division was based, chiefly, on gender and was separated by strong borders, which were crossed either by marriage or by parenthood.

7.3 Organisation of the extended family in 1800

The men's side was strictly structured, starting with the father, married son(s), and adult son(s), whilst the women's side was more flexible and included the mother at the top, the unmarried aunts (optional), sons' wives, adult daughters (unmarried), and children (boys and girls). Married daughters and married aunts moved in with their husband's family after marriage, signifying a transitional change in kinship, which would henceforth be through the husband, not the father/brother link. Servants and slaves were essential members of the middle-class family, with their number and designation widely differing from one household to another. Most middle-class families could not afford large numbers of servants and slaves; however, Lane described the presence of at least one slave/servant in most middle-class houses.[50]

Wealthy extended families in Cairo are thought to have reached 20–30 persons.[51] Upper middle-class houses' sizes and space distribution helped us construct a figure of 10–12 persons per family as an average, including three generations and with provision of 1–2 black slaves. Lower rank families were usually composed of two parents and 4–5 children, a total of 6–7 persons. The private spheres were structured in entirely different ways among those three levels of families due to the complexity and size of associated spaces.

The analysis of the socio-spatial organisation of homes in Old Cairo in 1800 depended on two types of evidence. The first comprised two surviving houses of families of the higher order, Bayt Abdel-Wahab al-Tablawy's (known as Bayt al-Suhaimy), and Bayt Mustafa Ja'afar. Both resemble different scales of this class of houses.

While the former is, arguably, the largest of its type in Cairo that was owned by a native Cairene, the latter represented an example of a merchant home that is smaller but elegant and rich in detail. The second piece of evidence comprised two lower-middle-class houses' documents (without drawings) describing either a house or part-house that no longer exist. So, I will rely on the description from the documents. While the first group of houses will be analysed in detail due to the availability of information, the second group analysis will be constructed by combining the discovered descriptive documents with plans and drawings of similar houses of that period in Old Cairo.

BAYT (HOUSE) ABDEL-WAHAB AL-TABLAWY (1648, 1699, 1730, 1796)

Located on the northern side of al-Darb al-Asfar main alleyway, with an area of 2,250.50 sq.m, the house is organised around five different levels, with the ground and first levels being the main inhabited floors. It has a 33.20m long southern façade overlooking Haret al-Darb al-Asfar[52] and two entrances. The main guest entrance was to the south east while the harem entrance was to the south west. The other three sides adjacent to neighbours were solid with variable heights. The eastern neighbours, in particular, are assumed to have been low rise (ground floor only) structures as windows at higher levels appeared on this side. Other neighbours had similar building heights as shown by the absence of windows on their sides. The house had four different interlocking wings arranged along its four sides and around a central courtyard (Figure 7.6). Each wing could function as an independent residential unit, according to the 1800 model, each serving a branch of the family and with a separate staircase. The four wings were organised within the complex organisation of a large house with relatively private zones which was developed through continuous expansion by different owners over two centuries.

The original south eastern part was built in 1648AD by *al-shaykh Abdul Wahhab al-Tablawi,* as mentioned in the decorative cornich scripts in the first floor Meqa'ad, with major additions built in 1699.[53] The ground level of this part[54] consisted of an entrance space (5.10x2.0m) with a stone seat for the *bawwab* (guard). The ground level was a collective guest reception venue which had two main *qa'as* (halls) called *mandharahs* (guest reception spaces), one on either side of the entrance, and service spaces at the rear. The first floor had a series of harem qa'a, and a *meqa'ad* (terraced living space) overlooking the courtyard, one harem qa'a and a small room on the first level. On the second level, there was another harem qa'a and the loft of the first level's qa'a, leaving the third and fourth upper levels for small rooms and open-to-sky roofs (Figure 7.4).

The form of the original house, therefore, appeared to be L-shaped with four storeys. From the way the secondary mandharah at ground level and the harem qa'a on the first level were oriented, with large windows overlooking the courtyard, we could confidently state that the courtyard was an essential component of this original house. By 1699AD, the house had been extended by *al-Hajj Ismail*, the son of *al-Hajj*[55] *Ismail Shalabi*, in both north-eastern and north-western directions, gaining an enormous area. Later, as the record showed, the building was owned by another Islamic scholar, called *al-shaykh Mohammed Imam al-Qasabi*, who bought the house legally.[56] Later the building was bought by *al-shaykh Ahmed al-al-Suhaimy*, the famous Islamic scholar of his time, who died during the second half of the eighteenth century.[57]

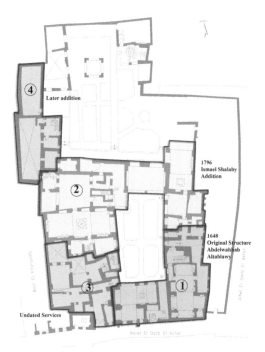

7.4 Historical development of Bayt al-Suhaimy

The house ownership was then transferred by inheritance to his two sons: Ahmed, who died in 1201*Hijri* (1786AD), and Mohammed.[58]

In 1800, the house was almost completely in the form in which it appears today in terms of major habitable spaces and their organisation. Having four mandharah(s) and four harem qa'as, each space had its own character. The south east mandharah was the winter reception hall, with its massive void and wooden panels decorating the walls and ceiling and no windows at ground level. In contrast, the north east mandharah was a summer space with a water fountain and full width and height mashrabiyya to the north, allowing cool air in. Two harem qa'as, similarly, have highly decorative walls (south west, no. 1, wooden panels; north east, no. 3, ceramic tiles) and were thought to be used by the mistress of the house (summer and winter). Harem halls were used for women's public duties and activities including hosting female guests, tradeswomen and sometimes tradesmen, and entertainment by showgirls.

The house was built of soft calcareous stone from the neighbouring mountain in Moqattam, which was a conventional building material of the time for the foundation wall and extended to the soffit of the first level.[59] This stone gave the ground level façade a light-yellowish hue which darkened with time.[60] The first floor projected over the ground floor's solid façade, which added approximately 60cm space to the house. This projection was supported on corbels and piers. All upper floors were built of plastered mud brick.[61] While the external façade was predominantly solid at the ground level, with two doors, upper floors were relatively open and transparent, providing more inside-out contact between the harem quarters and the outside world. Upper floors were projected approximately 60cm into the al-Darb al-Asfar void, with projection by some mashrabiyya of a further 50–60cm.

Windows were extensively used at upper levels to project into the main alley void. The façade was harmonised around two basic typologies of windows, which reflected the function of the rooms they served. While relatively large perforation patterns were used for ground level qa'as, services and circulation spaces, the projected and tightly perforated mashrabiyya were used for harem qa'as. This distinction emphasises social requirements and differences between spaces and utilities. Courtyard façades, however, were full of large windows that allowed dual visualisation and continuity of spaces from inside-out at ground level. This was mainly because this level was assigned to public activities as a collective venue, as mentioned earlier. Higher levels did not differ from the external façade and were dominated by tightly perforated and decorated mashrabiyya.

Typical of medieval organisation of large houses, this home could be read as multiple wings in an urban cluster surrounding a shared space (the courtyards). In that sense, it was not very different from the organisation of the Rab', which was occupied by people who were from different origins and not relatives. In Bayt al-Suhaimy, most occupied spaces were arranged around the central courtyard, with four ground floor mandharahs distributed along the four sides of the courtyard, each with a separate entrance to allow them to hold several simultaneous gatherings. In a similar order, higherlevel harems followed the same principles, each harem qa'a being accessible through a hierarchy of transitional and somewhat private corridors. Upper levels were vertical extensions to wings, providing some extra inner spaces rather than complete floors. The fourth level included a few rooms used for various purposes, while the rest of the floor was an open flat roof covered with a coat of plaster.[62]

In this prestigious house of the time, al-mandharah was the key and most valuable space to receive honoured guests and visitors, mainly notables. The literal meaning of the word mandharah is a "*scenery space*", which introduces an image of the owner's power and wealth to his guests. It was a three-part space, mostly rectangular, with variable heights; the central durqa'a in the middle with an entrance and exit, and two *iwans* (the main one was deep enough to host the Master and his honoured guest, and a shallow secondary one was for lower rank guests). The *durqa'a* (literally means the jewel of the space) was a semi-square shape at the centre, and the two *iwans* were of more rectangular form with their floors raised by approx. 25–30cm. While the *durqa'a's* height extended to the roof, cutting through four levels and ending with a small dome, the two *iwans* had a double height ceiling that extended only to the second floor. These three parts had their own social significance, as each part introduced particular social status to its occupants. The durqa'a, for example, although the most decorated and rich part of al-mandharah (as the entrance lobby and first area guests saw), also worked as the space of the servants and slaves, who were not allowed to sit or stand in either iwan. At the centre of its floor, was either specially designed marble, or in most cases, a finely sculpted water fountain called *fasqiyyah*, made of white marble, to cool down the space during the hot summer. Its ceiling was made of wooden ribs and decorated with different wooden inserts with conventional Islamic graphics, with windows appearing on intermediate levels overlooking the ground level. It, therefore, connected the building vertically.

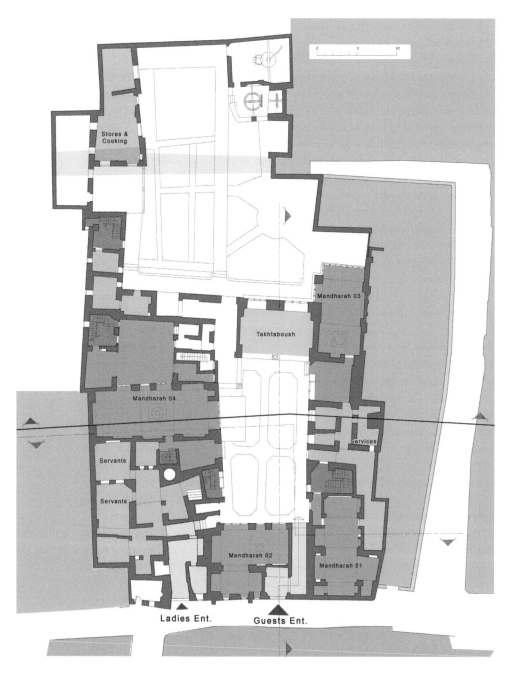

Stores & Cooking

Mandharah 03

Takhtaboush

Mandharah 04

Services

Servants

Servants

Mandharah 02

Mandharah 01

Ladies Ent.

Guests Ent.

0   5   10

7.5   Ground and first floor plans, sections and elevations of Bayt al-Suhaimy as they stand today

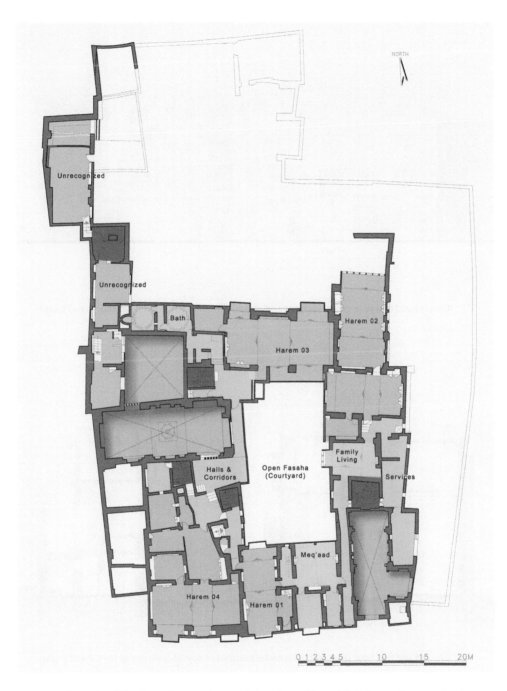

NORTH

Unrecognized

Unrecognized

Bath

Harem 02

Harem 03

Family
Living

Halls &
Corridors

Open Fasaha
(Courtyard)

Services

Meq'aad

Harem 04

Harem 01

0 1 2 3 4 5   10   15   20M

7.6   Ground and first floor plans, sections and elevations of Bayt al-Suhaimy as they stand today

Original Building

Extention 2

Main Facade
(Beit ElSehaimy)

Extention 2

Original Building

7.7   Ground and first floor plans, sections and elevations of Bayt al-Suhaimy as they stand today

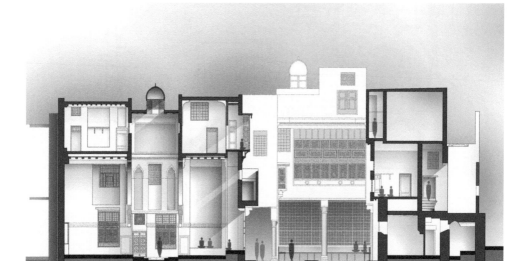

7.8   Ground and first floor plans, sections and elevations of Bayt al-Suhaimy as they stand today

The main *iwan* was raised from the durqa'a by 25cm. It was deeper and more comfortable than the secondary one, with raised-floor seats covered with shallow pillows and recessed into wooden wall panels. It was intended for those of the highest social status, and honoured dignitaries would take their seats there in a hierarchical order (similar to the seating in the public sphere). It had, in some cases, a secret escape door behind the master for escape in case of danger.[63] Although *Iwans* retained a higher social status, they did not have a decorated floor, as they used to be covered with rugs in winter or mats in summer. This was because they were used for collective praying, eating and in certain conditions sitting.[64] During those practices, the wearing of shoes on the floor plate was forbidden, as purity would have been damaged. Therefore, shoes had to be left on the durqa'a floor. To support and serve public activities, *al-Mandhara* would have been surrounded by service spaces such as an attached room, a corridor, and *kursi raha* (toilet), and in some cases a kitchen.[65]

The harem quarters' accessibility was under tight control and protection and required minimum interpersonal communication and involved single family members (husband, wife, and children) in an isolated and sacred space that was haram (prohibited).[66] However, harem wings had more variables than al-mandharah according to their relevance to the occupant couple, seniority and position. It is acknowledged that within the same family, difference in privacy measures and sub-cultures could be found, especially in terms of conservative mentality. Conservatives and more liberal personalities might have been present in the same house.Therefore, harem quarters differed largely, showing variable sizes and organisation of spaces. Two harem quarters (no. 4, no. 2) were of complex organisation and large size. They were a combination of associated spaces, starting with a lobby; introductory spaces functioning as family living space with attached rooms; the main harem qa'a with its three-part configuration; storage room and winter sleeping space. Others (no. 1, no. 3) were simpler in configuration, with the quarters being a combination of an introductory space and winter sleeping space with a small window and built-in recessed cupboards (it was either a three-part or two-part space). One of the essential elements of the harem qa'a was the winter sleeping room, or as I could call it, a *"pocket"*, due to its tight size when compared with the qa'a itself. In winter, Cairo became cold at night and large, open mashrabiyyas did not provide a warm environment for sleeping.[67]

With similar features to al-mandharah, the entrance plate, the durqa'a, was the lowest and highly decorated plate. It constituted the glory of the space and had the pleasing natural view of the water coming out of the fasqiyya. This plate was accessible to the mistress's servants and eunuchs and, as was the case with other durqa'as, was the place where shoes were left. With the aim of stressing the two different spheres of interaction within the same space, it was essential that the durqa'a had a raised ceiling, whatever the difference in height might be. In addition, the iwan(s) were raised by 25cm from the durqa'a, defining the ultimate private social sphere for the master and the lady. This part was the activity space for ladies: where eating, praying, and sleeping took place. We understand that, in keeping with the multi-functional use of space, it served as a reception area for female guests during daytime. The *iwan* was surrounded by cupboards in all solid sides and mashrabiyya windows wherever ventilation and daylight were desired. In most of the harem qa'as there were two iwans rather than one; however, they weren't socially equal. The main iwan (larger and deeper) was for the mistress of the quarter and her guests, while the secondary iwan was of lower order[68] and was, moreover, open to a variety of activities such as accommodation of children, female visitors and in some cases female dancers and artists.[69]

7.9 Socio-spatial practices in al-mandharah, early nineteenth century
Lane, *An Account*.

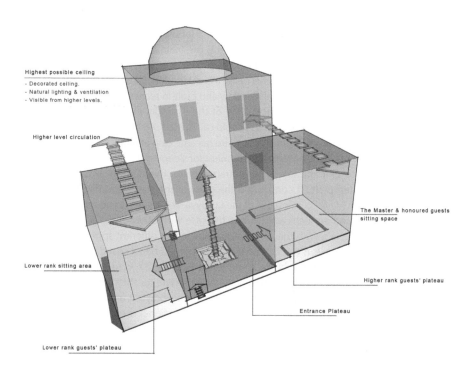

Highest possible ceiling
- Decorated ceiling.
- Natural lighting & ventilation
- Visible from higher levels.

Higher level circulation

The Master & honoured guests
sitting space

Lower rank sitting area

Higher rank guests' plateau

Entrance Plateau

Lower rank guests' plateau

Spatial Organization of the Semi-Public Sphere:
Al-Mandharah (exclusive manhood sphere)

7.10 The main Qa'a (al-mandharah) explained: socio-spatial organisation and hierarchy

7.11   Architectural motifs of al-mandharah

Notes: a) Durqa'a's fasqiyya, b) Durqa'a's high level dome decoration and side day lighting.

7.12   Sketch of a Cairene lady working on her Mansaj in the harem. First quarter of the nineteenth century
Lane, *An Account.*

There were a variety of activities women used to do in this space,[70] such as making their own clothes and other textile products using *al-Mansaj*[71] (Figure 7.12). Female retailers used to visit them during weekdays to buy or to sell products. In addition, it was found that ladies were interacting actively with their society, creating their own social networks and production units. However, this active environment used to change at night in the presence of the husband, when she had to be totally at his service.

The *hawsh* (the courtyard) was also an essential component of medieval houses as a central space around which different elements were arranged. It comprised public space within the private sphere; however, it was disconnected from the outer public sphere by an indirect path (two turns), a door and its keeper.[72] It was an architectural solution for environmental conditions. It responded to the hot weather by allowing northern cool currents of air to get through the courtyard plantation towards southern façade spaces. On the organisational level, it compensated for the tight alley, which could not provide either natural lighting or proper ventilation for such huge houses. It was a gathering spot for different components and sub-units of the family. Male visitors were admitted through it to their comfortable sitting area, whether indoor mandharah or outdoor takhtabush,[73] while women crossed it eventually to access their upper level chambers. Ladies and servants used to observe public activities and guests in this court from high level mashrabiyya.[74] In the absence of outsiders, "the courtyard became a workplace for men and women".[75]

BAYT MUSTAFA JA'AFAR (1713AD)

The house was built in 1713 on the order of Mustafa Ja'afar, a famous Syrian coffee merchant who wanted a house overlooking the main Qasabah's trade route. His name was mentioned in the accounts of Al-Jabarti,[76] who described him as among the famous merchants of the city. His family included two wives, one of whom was a slave, and one son. After his death, the house was inherited by four people, his son and two wives, and a freed slave called Suleiman Bin Abdulla. The household, afterwards, was led by his son: which was believed to be the situation in 1800,[77] until it was taken over by the authorities in the late nineteenth century to be the National Office of Antiquities. This house was second in class within the haret al-Darb al-Asfar's social sphere, following *bayt al-Tablawy*, in terms of size and capacity. The internal decorations of al-mandharah demonstrate the taste and interest of the owner in a highly impressive and decorated house.

This house was built at the alley's then dead end, with its western façade overlooking the city's main thoroughfare, Shar'i al-Muizz, although not accessible from it. Its southern façade and only accessible entrance overlooked haret al-Darb al-Asfar, the main spine. The location, I argue, was selected to overlook his trade in Shar'i al-Muizz, while it maintained private and protected access within a gated and secure residential community. Its ground floor included (typically) two courtyards, the main and socially active one to the south while the northern one was associated with services and cooking activities. The ground level was chiefly arranged around a double height mandharah, separating the two courtyards and extending between the eastern and western edges. Double height harem qa'a occupied the second and third levels, directly on top of al-mandharah. Alongside these two major spaces, the first level was assigned to semi-private activities, *with* small harem qa'as for the other wife and the son's family. This spatial organisation divided the house into three horizontal layers (ground level for public, first level for semi-private and second level for ultimate private) and three vertical zones (the southern hawsh with surrounding public spaces, the principal qa'a in the middle and services zone to the north).

The ground floor (Figure 7.13) was a simple and smart distribution of public venues that was accessed from its only entrance to the south east corner through an arcaded corridor. Ground floor public venues included *al-takhtabush* (open sitting area), facing south with high level windows. To the western side were two small rooms, mainly for food storage.[78] This side of the ground level was blocked by al-mandharah to the north, which occupied the whole width of the building (7.80 x 19.80m). It was a double height space with extensive decoration and arabisc work over the wall and the ceiling panels. Al-mandharah had two doors on the opposite side opening onto its durqa'a and providing light and air flow from the north to the south. Higher level windows overlooked al-Muizz thoroughfare. Whereas the southern hawsh was an ultimately public venue, with al-mandharah and al-takhtabush to its north and south sides, the northern hawsh was surrounded by a covered arcade for cooking and service purposes.

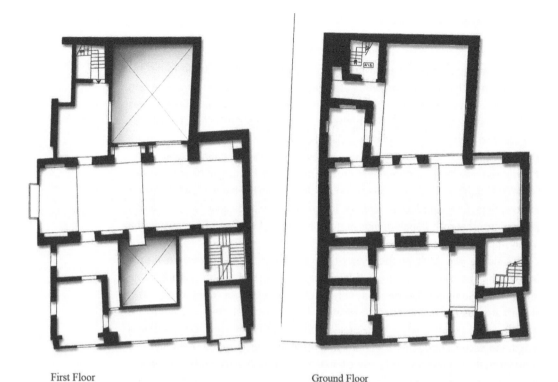

First Floor   Ground Floor

7.13 Ground (right) and first (left) floor plans of Bayt Mustafa Ja'afar

First level spaces were arranged along the perimeter of the house, with inner courtyards to provide daylight and ventilation. While al-Mandhara cut through that array of spaces, they had to be linked through it. Therefore, a wooden-fabricated bridge was introduced to link the northern and southern wings for the purpose of continuity and connectivity. In the meantime, this bridge overlooked the internal space of al-mandharah and was surrounded by high level wooden screens that allowed visibility from inside out only. The harem qa'a came to separate the two wings again on the second and third levels. The building had two façades, the southern and the western, with each one allied to entirely different contexts. They followed the same principles for openings mentioned above in bayt al-Tablawy, with one difference; inner courtyard facades were tight enough to have projecting mashrabiyya.

Main guest receptions took place at ground level between *al-takhtabush* and *al-mandharah* (outdoor and indoor spaces). While a double height mandharah occupied the full width of the house (east-west direction) and was central to the house layout, the main courtyard moved slightly to the south and was squeezed in size. In this arrangement, guests faced less confusion (compared with other houses) getting from the entrance through the courtyard, where both public venues were clearly visible. While south-side movement involved the approach of dignitaries and guests, the north side of al-mandharah (including service rooms and yard) was busy with movement of servants and slaves. The location of al-mandharah (between two courtyards) and its flat ceiling are probably the main spatial difference from al-Tablawy's house. Guests of social status were received mainly in this space, managed by the hierarchy of iwans as mentioned earlier. However, Mustafa Ja'afar's mandharah seems richer in its details and decoration than bayt al-Tablawy during the same period.

7.14   External façades of al-mandharah of Bayt Mustafa Ja'afar

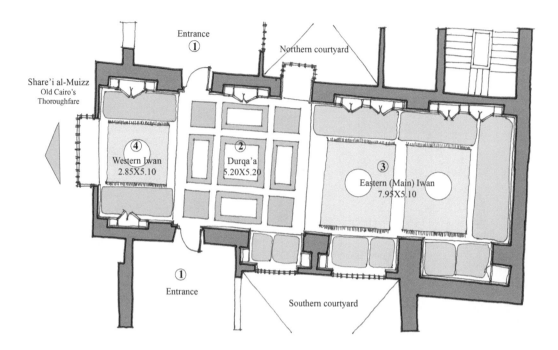

7.15 Furnishing
layout of the
third-level harem

This could be seen as a manifestation of wealth in smaller house configuration.
Food was served from the northern side's service and servants' quarter. The only
interruption to this void came from a wooden bridge introduced at the western side
of the qa'a on top of the west iwan to enable higher level circulation linking the north
and south wings, which otherwise were not connected. This bridge was used by
women to observe public activities and celebrations taking place in al-mandharah
without being seen. Therefore, two high wooden screens were placed on both edges.

There is no evidence that suggests that private activities in this house were different
from the norms and traditions explained in the previous house. As such, it is believed
that private and semi-private activities were typical traditional practices of the time.

HOMES OF THE LOWER ORDER

By lower order, here, I mean a lower middle class group who were professional workers,
ordinary tradesmen, or owners of retail stores. Nelly Hanna mentioned that this group
were the main buyers of houses during the eighteenth century.[79] These houses were
limited in size, with some of them lacking in decorative space. As expected, these houses
contained a limited number of rooms (3–5 rooms) and most of them did not have a
hawsh, and if it existed, it was used for functional purposes such as stables, livestock,
and food storage.[80] Three particular documents were found to describe houses being
sold in the same area as the previous two houses in al-Darb al-Asfar around 1800 (within
a decade either before or after),[81] with the other few in al-Gammaliyyah. The first was a
half portion of a house that was bought by Fakhr Bin Ali Halabi by 1205H (1791 AD), on a
plot inside al-Darb al-Asfar:

opposite to Baybars' school, which was the house of Amir Rajab Ouda Pasha,
including a door that gives access to a corridor ending with a door leading to a hawsh
that contains a water well and Coptic Meq'ad (takhtabush) and stairs leading to a
harem qa'a.[82]

This description shows a simpler organisation from houses of the higher order, but includes a hawsh, a takhtabush and a harem qa'a (Figure 7.16). In this organisation there is no mandharah, while it is apparent that the harem was a single space on top of al-takhtabush. This house was swapped, according to the same documents, with a similar half portion of a house located just to the west of Bayt al-Tablawi. This included a *"door that gives access to a lobby with two stores and stairs leading to another lobby leading to a ruwwaq (hall) with wall wardrobe. Then this leads to a kitchen, toilet and stairs leading to another room then to the roof".*[83] The third house had been bought by a religious scholar (from al-Azhar University) called Shaykh Zein-Eldin Mustafa al-Diasty by 1218H (1803AD). This deal was limited to a 2/24 portion of a house that was composed of *"two entrances, one leading to a stable (suitable for three horses) and the second leads to Durqa'a and Ruwaq that includes two iwans and Durqa'a, services and other rooms".*

From these examples we can construct a sketch of houses of the lower order limited to a door leading to hawsh, with takhtabush, in some cases, or small *qa'a* (multi-purpose space that could be a mandharah or living space) and stairs leading to a top floor harem qa'a. Some buildings did not have a hawsh (such as the third house) and usually the front door opened onto an open lobby (*fasaha*) and a ground floor multi-purpose hall (*qa'a*), with stairs leading to a top floor harem qa'a. The latter resembled vertical apartment arrangements that later would develop into contemporary apartment buildings.

While those archival documents did not have any attached drawings or diagrams, they were influential in drawing a diagram of space organisation of this type of house. This could be put into real configuration when read with some surviving examples outside al-Darb al-Asfar. Nelly Hanna displayed, in her book *Habitier au Caire*, some houses of the eighteenth century which comply with this organisation. Bayt Radwan Bey, for example, resembled a compact lower-order house in Haret al-Khayamiyyah that was composed of two principal floors, the ground and the first, and an intermediate level which included Meqa'ad in addition to a single qa'a on the roof. While the ground level was predominantly shops and stores alongside the entrance to the above apartment, the first level was an entirely residential apartment with two large Qa'as (one harem and the other was a multi-purpose harem/mandharah), one large living space (fasaha) and a toilet (kursi raha).

People of this lower order, unlike their counterparts of the higher order, used to separate their private and public lives. Due to limitations of wealth and available space, men used to meet in public spaces (whether at their stone seats, shops or at the nearest coffee houses). Women, on the other hand, used to visit each other, or to communicate through the Mashrabiayya if they were neighbours. Lane affirms *"there are, it is true, many women among the lower classes in this country who constantly appear in public with unveiled face; but they are almost constrained to do so by the want of burka' (face veil)".*[84] However, when an unexpected guest arrived, the main qa'a (usually nearest to the entrance) was reorganised into a mandharah's format temporarily to receive the guest. Edward Lane described a visit he made to one such home, where he was invited to a (harem-turned-mandharah) qa'a, and where the host's mother joined their discussion while sitting just outside the partly-opened door, a rare sign of liberation towards women's interaction in this type of space configuration:

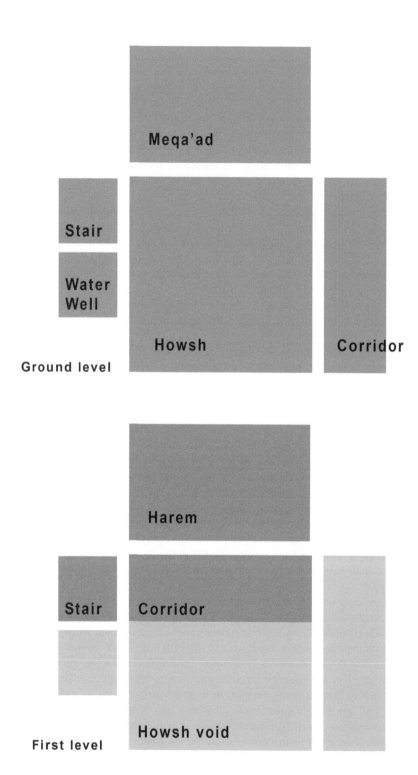

Meqa'ad

Stair

Water
Well

Howsh

Corridor

Ground level

Harem

Stair

Corridor

Howsh void

First level

7.16   Space organisation of houses of the lower order as described in Doc. 637

*... many small houses in Cairo have no apartment on the ground-floor for the
reception of male visitors, who therefore ascend to an upper room [harem]; but as they
go upstairs, they exclaim several times "destoor" ("permission") ... in order to warn any
woman who may happen to be in the way, to retire or to veil herself ... Yet there are,
among the Egyptians, a few persons who are much less particular in this respect: such
is one of my Muslim friends here, who generally allows me to see his mother when I
call upon him. She is a widow, of about fifty years of age; but, being very fat, and not
looking so old, she calls herself forty. She usually comes to the door of the apartment
of the harem in which I am received (there being no lower apartment in the house for
male visitors), and sits there upon the floor, but will never enter the room.*[85]

By the end of its medieval history, the alleyway's communities appeared to reflect
the local integration of its social structure and hierarchy in its spatial order and the
distribution of activities. The public sphere extended into the houses of high profile
residents, forming collective gathering venues spatially ordered to respond to
community order of seniority. Its spatial configuration and decoration had to reflect the
wealth and power of the master. On the other hand, houses of the lower order lacked
such emphasis, and the shared public sphere worked as a private sphere used, secured
and protected by the whole community. Meals and private talks frequently took place
in front of houses and shops. This inclusiveness of the public sphere was combined by
lengthy and confusing thresholds between public and private spheres within houses.
The harem quarters, mainly in wealthy houses, were separated from any public spheres
and activities by a series of tight, dark and non straight corridors, spaces and sometimes
private stairs and were guarded by servants and eunuchs. Such separation did not
mean total isolation, though. The mashrabiyya of the higher level harem ensured that
women could communicate with their fellow female neighbours of adjacent houses
and maintained continuous observation of public activities taking place in alley.

It was apparent that strong conservative ideology of Ottoman rulers influenced this
strict space organisation and spatial order of prestigious family houses (usually linked to
the ruler's class). Strong social and political order, manifested in the community hierarchy
and leadership, ensured that all classes were represented and shared spaces were
managed efficiently by local leaders whose wealth and resources were at their disposal.
The detailed analysis of 1,800 hawari, in particular that of al-Darb al-Asfar, was surprising
in terms of denying the long rooted notion of exclusiveness of the medieval house as a
strict private sphere, at whose front doors the threshold between the private and the
public is strongly manifested. The presence of multiple venues for public gatherings
around the central courtyard suggested that at certain occasions and festivities the
ground floor merged with outdoor shared space to form one extended public sphere.

NOTES

1   Ruiz, M.M., 2009, "Orientalist and Revisionist Histories of 'Abd al-Rahman al-Jabarti'". *Middle
     East Critique*, 18(3), pp. 261–84.

2   Nawal Al-Messiri Nadim, *The Concept of the Hara*, p. 314.

3   Each Tumn was composed of several larger units called Shiyakhat, which combined several
     hawari, under one unit, which later became the basis for the contemporary administrative
     organisation of this area. See Nawal Al-Messiri Nadim, ibid., p. 316; and Janet Abu-Lughod,
     Cairo, p. 25.

4   Lapidus, 1967, p. 92; Al-Messiri Nadim, ibid.; Staffa, 1979, p. xx. While his official rule was basically political, handling the requirements of the rulers, especially taxation management, overseeing rules of apprenticeship as well as maintaining order, he enjoyed a higher social status among the community. For more details see: Baer, G., 1982, pp. 149–60; Ismail, 2000, p. 369.

5   Lane, ibid., p. 124.

6   Ibid., p. 125.

7   There is no evidence in archival records or the accounts of al-Jabarti and Lane referring to cases where Shaykh al-Hara was imposed on or was not respected by the community.

8   Salwa Ismail, "The popular movement dimensions of contemporary militant Islamic: socio-spatial determinants in the Cairo urban setting", in *Society for Comparative Study of Society and History*, 2000, p. 369.

9   Al-Jabarti described them as powerful voices, which were feared by the rulers of the time. For examples, see Al-Jabarti, *Aja'ib al-Athar fi al-Tarajim wa al-Akhbar*, Part 1, pp. 12, 41–3.

10   Ismail, ibid.; Al-Messiri Nadim, *The Concept of the Hara*, pp. 39–42.

11   Sawsan Al-Messiri, *The Role of the Futuwwa in theSsocial Structure of the City of Cairo* (paper for the conference on Changing Forms of Patronage, 1974), p. 25, quoted in Al-Messiri Nadim, ibid., p. 316.

12   Janet Abu-Loghud, ibid.

13   While Islam as an Ideology emphasised that all people are equal, it distinguished the persons who had religious significance though their knowledge, fellowship or education. Shaykh was a usual honorary title, which literally refers to a person with religious knowledge; however, in nineteenth-century Cairo, every household lord had the title of Shaykh. For more details (refer to Lane 1860: 132–3).

14   Safran, 1961, p. 3.

15   Al-Jabarti and Ali Mubarak refer to al-shaykh al-Suhaimy every time they mention Haret al-Darb al-Asfar in their books.

16   All ownership documents of the harah's properties were described in relation to bayt al-Suhaimy (in front of, beside, or on the corner of, … etc.) Refer to Document no. 637, Record 534.

17   AlJabarti, Aja'ib al-Athar.

18   Doc. 61, Record 537, p. 30.

19   Fargues, 2003, p. 31.

20   Such composition is interpreted from affiliation of residents seen in archival records, historical accounts in addition to the surviving archaeological evidence.

21   Researchers and historians agreed that both classes in this period resided in spaces of the outskirts of the city, and while the ruling class resided in the area around the citadel and the lakes, the poor lived outside the city gates. See Susan Staffa, (1977); Janet Abu-Loghud (1971); Andre Raymond (2000).

22   This document described the swapped portions of each house that included "a door leading to a corridor leading to a hawsh (courtyard) with a water well and a Coptic Meqa'ad and stairs leading to the upstairs harem" of the first house, swapped for a part that included a "door leading to fasaha (large space) leading to a riwaq (room) above leading to more stairs or the roof" and another portion of shop outside the harah.

23   Property endowed to a certain purpose, mainly to support a religious institution. Sylvie
      Denoix defined two types of waqfs; the Khayri waqfs endowments, considered to be of true,
      pious foundation, and the Ahli waqf, a device to secure private property within a family. See
      Denoix, *A Mamluk Institution for Urbanization*, p. 192.

24   Fargues, *Family and Household in Mid-nineteenth Century*, p. 23.

25   Clot Bey, 1860.

26   In Fargues's study, the average number of slaves and servants per family reached 0.268 per
      house, and as high as 0.542 for households of Waqf managers and neighbourhood Shaykhs,
      who represented some members of al-Darb al-Asfar. See Fargues, ibid., p. 32.

27   This is an interpretation of Lane's description of the organisation of extended family. See
      Lane, ibid., pp. 132–4.

28   Lane, *An Account*, p. 133.

29   Nasser Rabat, *The Culture of Building*.

30   The absence of the heart space in that map was due to the lack of detailed survey by the
      French: whose main concern was to provide a map for their military and planning purposes
      rather than addressing spatial settings of each unit.

31   Lane, ibid., p. 6.

32   Ibid., p. 178.

33   Both Doris Behrens-Abu Seif and Janet Abu-Loghud referred to the French Expedition and
      confirmed the presence of 1,200 coffee shops in Cairo in 1800. Accordingly, with simple
      calculations, we comfortably state that at least one coffee shop was present in each harah,
      and it was thus for al-Darb al-Asfar.

34   Lane described many incidents and stories that were being told in the coffeehouses, the
      main venue of communication at that time.

35   Lane, ibid., p. 391. Lane differentiated between several types of storytellers of Cairo, the
      Sho'ara (sing. *Sha'er*, and means Poet) based on the type of stories they told. For more
      details, see Lane, ibid., pp. 391–425.

36   Lane, ibid., p. 198.

37   Hijri calendar is the Muslim calendar which started with the Prophet's migration from Mecca
      to al-Madinahin al-Hijaz. It is a moon-based calendar where every Hijri month follows the
      moon's patterns.

38   Lane, ibid. Children would not have to be from poor families to do this. It was a part of the
      festival that stipulated the solidarity of the community, when one gives to children of others
      as he does with his own children.

39   Lane, ibid.

40   Abu-Loghud, *Cairo*, ibid., pp. 64–5.

41   Introducing boys into manhood. For the detailed description of those celebrations, see
      Lane, *An Account*.

42   Hanna, *Habitier au Caire*, ibid., p. 299.

43   Hanna, ibid. The business was a sort of family-led enterprise.

44   Lane, ibid.; Hanna, ibid.

45   Endowed properties and charities.

46   Sedky, 2001, p. 2.

47   Ismail, *The Popular Movement Dimensions of Contemporary Militant Islamism*, pp. 368–71. Previous times, such as Mamluk's era, were marked by the free movement of women and their participation in public events. Sedky quoted Van Ghistele, in describing women's freedom during the Mamluk era, saying "One sees women coming and going and paying visits to their folk". He, furthermore, traced the origin of the women's isolated quarter, the Haremlik, and Mashrabiyya lattice window to the Ottoman culture, imported with their army in the sixteenth century. See Sedky, ibid.

48   Lane, *An Account*, p. 175.

49   Archival record no. 95, Record 500, pp. 194–5 (Appendix B).

50   Black slaves were the cheapest. See Lane, ibid., p. 133.

51   This typology was confirmed partially by several studies and historical analysis: among which, see Nelly Hanna, *Habitier au Caire*, and Clot-Bey, *Apercu general sur L'Egypte*.

52   This information was recorded in the official survey of the building in 1931 upon the formal ownership transfer from the private owner to the Egyptian Council of Antiquity (Formal report for Monument no. 339).

53   The decorative scripts stated that: "This blessed place was built as a gift from God, with his blessing, by the poor slaves, who hope for God's forgiveness; the wise and active Shaykh Abdel-Wahhab al-Tablawi in the year 1058(H)" (1648AD).

54   I defined the limits of this part using the archaeological survey of construction and architectural configuration. Having two columns adjacent to each other on the southern façade was for no reason but the extension of the building, which required an edge support, trying to copy the details and decorative elements of the original building. On the north eastern side, the difference in level marked the original building limits, especially that which came at the doorway of one of the harem qa'as.

55   Al-hajj means the man who has performed pilgrimage to Mecca. It is, to this day in the Cairene culture, a title given to every individual Muslim who has performed the pilgrimage. It is used as a reference for wise people.

56   Formal report for Monument, no. 339, p. 3.

57   His death date was mentioned differently in three historical resources: the first by al-Jabarty as 1178 hijri (1764AD) and by Ali-Mubarak (1888) as 1188 hijri (1774), while al-Qal'awy (died in 1814) mentioned it as 1177 hijri (1763AD).

58   Formal report, ibid., p. 3.

59   Lane, *An Account*, p. 6.

60   Even though the harah's name, meaning "the yellow darb", suggests that most buildings were left with their natural yellowish stone colour, we have no evidence to support this hypothesis. This name was given to the harah centuries earlier, and could not control the external finish of new buildings for such a long period.

61   Lane described it as burnt mud bricks "and of a dull red colour. The mortar is generally composed of mud in the proportion of one-half, with a fourth part of lime, and the remaining part of the ashes of straw and rubbish". See Lane, ibid., p. 6.

62   Lane., ibid.

63   Lane, p. 20.

64   Lane, ibid., p. 12.

65   Hanna, *Habitier au Caire*, p. 58.

66   The term harem was derived from haram, to refer to ladies, the leading personnel of this space. The term haram refers to prohibition imposed by the holy book, the *Quran*, which related to married women in that they could not be alone with strange men. This interaction is haram, from which, the wife was referred to as *hurma* (prohibited). Strange men should not approach them, as they are hurma. The harem was, accordingly, developed to mean the private space of women.

67   Therefore, there was a need for a small isolated pocket of low height and minimum size to provide a warm sleeping area. Every harem wing had a small sleeping pocket, which was either accessible from the main qa'a (North West harem, no. 3, and South west harem, no. 4), from the introductory room (north east harem, no. 2) or through inner stairs to another loft level (south harem, no. 1) where space was limited in the original building.

68   Sophia Lane Poole described how other wives of the master were left to sit in that lower order iwan, while the mistress of the house joined him in the main iwan. See Lane-Poole, S., 1846, *English Women in Egypt*.

69   Such activities were mentioned by different accounts and research. For example, see Lane, 1860, and Keddie, 2006.

70   Keddie, ibid.

71   Lane, ibid., pp. 187–8.

72   Lane, 1860, p. 9; Hanna, ibid.

73   Campo, 1991, p. 80.

74   Ibid., p. 80.

75   Campo, ibid., p. 80.

76   Al-Jabariti, ibid.

77   Report of Monuments no. 471, National Council of Antiquity, Nadim, Documentation, restoration, conservation and development of Bayt al-Suhaimy area.

78   Official monument report, no. 471, pp. 1–3.

79   Hanna, *Habitier au Caire*, p. 80.

80   Hanna, ibid., pp. 84–5.

81   The prices paid for those houses were relatively small and did not exceed 100 Riyals (each riay, 90 Nisf Fadhah; half silver). This equalled 9,000 Nisf which equals approx. 14,300 Bara: according to Nelly Hannah, average prices for middle-class houses ranged from 6,000–26,999 Bara.

82   Doc. 637, Record 534, p. 308.

83   This house was located (at that time) at the western corner between Zuqaq al-Darb al-Asfar and the main Darb.

84   Lane, *An Account*, p. 177.

85   Ibid., p. 178.

8

The Changing City 1880s–1930s

Cairo, therefore, will no longer be an Arab city, and will no longer possess
those peculiarities which render it so picturesque and attractive.[1]

Sophie Lane-Poole's reaction to
Muhammad Ali's plan (1805–1849) to modernise Cairo

FROM MEDIEVALITY TO EARLY MODERNITY

In the first paragraph of his book, *The Story of Cairo*, Stanley Lane-Poole described the city of 1902 by saying: "*There are two Cairos, distinct in character, though but slenderly in site. There is a European Cairo, and there is an Egyptian Cairo. The last was one El-Kahira, the victorious, ... it is now so little conquering, indeed has become so subdued ... in truth European Cairo knows little of its medieval sister*".[2] This grasp of the explicit differences summarises the city's development throughout the nineteenth century, the period that witnessed the setback for the hawari from being dominant to being marginal urban communities. Muhammad Ali (1805–49),[3] intentionally, had banned all traces of medieval architecture and their stylistic peculiarities for different reasons. Banning the use of mashrabiyya in new buildings was made under the *pretext of health and safety*;[4] as it was made of wood, it could cause a fire.[5] The rationale behind this ban could be understood if we consider the larger plan behind changing the image of the medieval.[6] It was a decisive diversion from the medieval past through, for example, widening the roads, creating squares and gardens, painting the houses in white, and adopting new housing models. In short, he was putting a modern mask on the old city fabric and soon realised it was not possible.

In contrast to Muhammad Ali's approach, Khedive Ismail (1863–79)[7] abandoned the hope of improving the hawari and focused on producing organised, planned western homes in the suburb of the old city. To achieve this goal, a raft of legislation was produced to control building activities and set standards for building, modifying or maintaining homes. This legislation was motivated by the dream of a modern city, regardless of any potential effects on the form and practice of home in the city. Here, I discuss that dream of modernity influenced the idea of home both socially and spatially. It is apparent that the transitional period around the turn of the twentieth century, in particular, was influential in transforming homes and their environments from significance to marginality, from a centre of power into a vicinity of the poor.

Over the past two centuries, the hawari have been slowly changing from their medieval image into a more complex and modern character. Change was a consequence of a combined set of factors. From one side, social structure of local areas was significantly changing, on the back of emergence of western-styled and civilised centres outside the old city, which attracted rich merchants to relocate, leaving old hawari deprived of their rich inhabitants and resources. From the other side, the government started to implement a heavy handed policy to control the built environment and urban space with the aim of enforcing a European-style image (a European nation in the East).[8] The embryonic form and image of planned Cairo were unmistakeable at the turn of the twentieth century, while the harah's position as a fundamental urban unit had forever been compromised by the well-established and maintained *Ismailia* (European) Cairo.[9]

Throughout the nineteenth century, Cairo was moving towards an industrial economy, requiring many peasants to join the newly-established factories and to earn regular wages. Muhammad Ali's (1805–49) project for modern Cairo commenced with cleaning the hawari and painting buildings' facades in white, demolishing ruined houses and regulating street lighting, and façade designs.[10] However, his most important contribution to the changing urban setting was the focus on large size industrial projects such as cotton and oil factories, which required large numbers of regular workers. Those came from the countryside, the troubled and unstable countryside villages, searching for secure jobs in Cairo and looking for cheap accommodation.[11] This situation applied heavy pressure on the hawari: an already congested space with poor and low quality services, to accommodate the waves of new migrant workers. On the other hand, late nineteenth-century developments made by European real-estate companies attracted the rich merchants due to their European flavour, paved and lit boulevards[12] (Figure 8.1). The place of such influential players in local communities of the old city was filled by those poor peasant migrants who came to join the new waged working class. They occupied roofs of houses and vacant plots, creating the *Ishash*[13] and *Ahwash*.[14] This moved the hawari down in the social ladder, with their physical structure, the buildings, falling into disrepair due to lack of maintenance and financial resources.

As a result, the hawari of the early twentieth century were no longer the preferred sites for the merchants' homes. Large courtyard houses were, consequently, replaced by compact multi-storey houses. Introverted organisation of homes was turned inside-out, with large openings on the central lane. For example, we could not recognise any new courtyard houses that were built within the confines of Old Cairo during the first quarter of the twentieth century.[15] The last recorded courtyard house was *bayt al-Kharazi* (known as *al-Kharazati*) in al-Darb al-Asfar (1881), which was taken over by poor migrant families by the middle of the century.[16] The dominant house type, then, was a three-four level, load-bearing, compact building, which resembled the modern form of the extended family house of the nineteenth century. Each floor in this new typology was connected to the harah more that it was with other floors, with every floor turning out to be a separate apartment.

The implementation of the model houses of Muhammad Ali appeared late in the nineteenth century after failing to catch on during his lifetime (Figure 8.2). The hard boundary and tightening urban area of Old Cairo, surrounded by new developments of Ismaili Cairo, made inevitable the decline of the historically-dominant model of courtyard houses. The extended family structure, the core of community, had to find alternative ways to reside in smaller plots with lower affordability. Studying Old Cairo over that extended period of time has to consider such social and economic change: how does it take place, and how does it lead to the form and culture of the contemporary situation? Without understanding such forces of change, we would never understand the rationale behind the contemporary form of houses, or even the idea of home. This chapter aims to conceptually frame and document the socio-cultural development during late nineteenth and early twentieth-century Cairo. It, however, focuses on the turn of the twentieth century as the temporal agent for such a significant change in the image as well as in the lifestyle.

Early forces of change emerged when western-educated Egyptian elites, on returning home from their educational missions in Europe (mainly France), were appointed to lead national governmental and cultural institutions.[17] Their objective was to emulate the European model of knowledge and philosophy within the confines of their national territory and with respect to their local culture and tradition. There is no more evidence on this fact other than the writings of Gamal El-Din El-Afghani, Muhammad Abdu, Rifa'a Al-Tahtawi and Qasim Amin.[18] While the first three figures were religious scholars with progressive views on Islam, the fourth was of civilian background. They used media outlets and institutions available to them to boost anticipated ideological reform, the main objective of their missions, especially through contesting radical culture and restrictions on freedom of thought, practice, and knowledge as well as the discriminative gender segregation of the time.

8.1   Place de l'Opera, Cairo at the turn of the twentieth century

Courtesy of Rare Books and Special Collection Library of the American University in Cairo.

8.2 Model Houses of Muhammad Ali: a) lower middle class houses;
b) upper middle class houses (mostly outside Old Cairo). (After Magda Ikram Ebaid)

Some would claim that such intellectual and ideological reform is irrelevant to the practice of the ordinary people living in the hawari, suggesting that such forces of change were apparent only within the confines of elitist domains.[19] However, we can see relevant evidence in the relaxation of restrictions over women's activities in the public sphere early in the twentieth century, the changing of housing from introvert to extrovert arrangement, and more women going out to work. Such changes in attitude would have never taken place, in such a medieval context, without a significant change in people's beliefs and morals as a consequence of a gradual influx of the reformers' ideologies and principles of reform through local shaykhs and educated scholars.[20] The division of home into separate and isolated wings was replaced with a more interconnected organisation and women were freely moving throughout their homes (the house and the harah). The harem wing was reduced to a room among others, and the family would gather in a shared living space.[21] The large service backyard was abandoned and compact houses seemed to dominate the built environment. In short, the idea of home was changing in the mentality of local people. While the home remained the women's castle and a sacred place, they were increasingly welcomed in the public sphere and outdoor environments. Homes became more fluid and flexible in terms of practising everyday life and in terms of integrating indoor and outdoor activities.

* * * *

The very nature of the hawari lies in the interwoven aspects of its structural mechanisms. Any change in one aspect leads to a chain of other changes; social change affects economic activities and both contribute heavily to the alteration of the dominant culture. Once culture changes, architecture steps up to articulate and shape that change into a spatial elaboration. The shift in social class from the predominance in 1800 of the high middle income group led by religious scholars and trade merchants to that of middle class tradesmen and retail-shop workers and owners by the end of the century had proliferating influence on the lifestyle in such communities. Property sales records (old contracts) of houses in the 1880s in al-Darb al-Asfar, for example, listed five residents who worked in spice trading, two owned shoe-shops and a few were working as estate agents. These represented an average middle-class population whose resources were rather limited.[22] The few scholars who remained among the residents retained relatively large houses. Among them were *al-Shaykh al-Mansoury* and his descendants:[23] who have retained their inherited house and land until today. One new foreign merchant, a Russian, Mohamed Fadl Al-Kharazy, resided in the harah from 1881 after building his large house between *Bayt Mustafa Gaa'far* and *Bayt al-Suhaimy*.[24]

With the departure of rich merchants and high profile scholars to the new quarters of European Cairo and the absence of wealthy families, the hawari had to find alternative means of income generation. Shops, foodstores and workshops started to appear in the main darb as early as the first decade of the twentieth century. Mohamed Al-Sioufi's record of the economic activities in the harah during that period shows that several retail stores and three workshops: mainly carpenters and upholsters, appeared on both ends of the main alley.[25] The workshops were basically services supplying the local society rather than commerce or trade on their own merits.

Al-Darb al-Asfar (Percentage of residents)

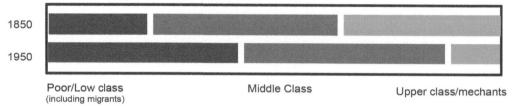

Al-Darb al-Asfar (Percentage of occupied areas)

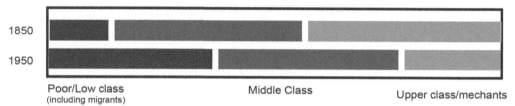

8.3 Relative changes in social representation in Al-Darb al-Asfar 1850–1950

Notes: Above: relative percentages of social levels of buildings' residents. Below: relative percentages of areas occupied by each social class.

A local carpenter was manufacturing timber windows, doors and furniture for the nearby localities, according to one of the oldest residents.[26] The increment in commercial and industrial occupation gained momentum during the first half of the twentieth century, occupying the vast majority of ground level spaces by the 1950s–60s.[27] The old carpenter was the grandfather of a contemporary resident, who reported that the workshop was very successful and busy with mashrabiyya and Arabisc work until the late 1940s, when such products became old fashioned. Glass windows replaced the traditional timber ones, and new peasant migrants had established their own informal businesses and lived off the proceeds.[28]

Families and descendants of many old families remained in Old Cairo, while those of the higher order had left the old city in search of a new modern life by the turn of the century. Meanwhile, the emerging heavy presence of migrants and their relative financial and spatial autonomy allowed them a private domain of independent practice and sub-culture.[29] They had their own leadership and local political structure. Such duality in culture and occupancy gave rise to frequent disputes. Migrant groups in Old Cairo had displayed very interesting patterns, with each harah or community having waves of migrations linked to a particular place of origin in the countryside. We found that migrants from al-Fayoum's villages had formed coherent groups of migrants in some of Al-Gammaliyyah's hawari. Similar groups from Asyuit and Sohag villages were found in other places. Mostly, they came to Cairo in search of waged and settled jobs, as land cultivation provided neither a secure job nor a proper income.[30] They filled empty spaces and abandoned buildings with tents and ishash. They occupied shared rooms, built temporary shelters above roofs, and assembled tents on empty land. We do not have a detailed history of their presences or narratives of their practices of living style. However, we have fragments of information that suggest they established their own activities.[31] Men worked in unskilled jobs, basically in construction; the women's job was child rearing, while some had established their own food industry.[32]

Such extensive migration had a significant impact on the hawari's social structure, and lifestyle. It caused imbalance in its organisation and moved the community down several steps in the social stratification. At the back of this change, without maintenance, investment, or even interest in keeping the aesthetical values of its inherited structures Old Cairo declined into ruins, and lost some of its charm of medievality, while picking up problems of modernity and industrial activities, for example, the roofs of historical and traditional houses turned into waste zones.

Following this long-term transformation, Old Cairo reached mid-century with three categories of residents:

- Old families with a long history in the hawari communities. Their early twentieth-century buildings were organised for extended family occupation.
- Working class nuclear families, mainly newcomers, who came to the harah for its cheap accommodation and short distance from the city.
- Migrant villagers from Upper Egypt, who resided in properties throughout communities, Rab's and somtimes transformed old houses into collective mass-housing units, in what was a hawsh-like camp. Unlike the other two categories, they were uneducated, unskilled and lived in relatively bad conditions.

From Figure 8.3, that shows the relative change that took place during the transitional period, we could understand that the presence of the middle class remained relatively constant, while the increasing number of migrant villagers and the poor had filled the space left by the wealthy, upper-class merchants. This explains the ruined condition of Old Cairo during much of the twentieth century as the devastating consequences of the absence of local financial power, shared spaces and services that had lost their resident sponsor.

On the level of legislation and order of law, Cairo had suffered from a lack of formal legislative structure and administration of its urban units. There was a need to replace the local socially-based system with a modern and formal administrative structure. In the 50-year period around the turn of the century, many legislative and administrative structures were created and developed. The Khedive of Egypt (Tawfiq 1879–92; then Habbas Helmi 1892–1914) issued several laws and decrees to organise ordinary people's activities such as building, trading, crime, and taxation. With the growing number of people living within each *tumn*, administrative tasks became too complex to be managed by shaykh al-harah.[33] During 1886, a tax council was formed for every tumn, comprising elected local building owners. On 10th June, tumn al-Gammaliyyah had elected their own tax council, which for years to follow was responsible for taxing occupied buildings.[34] Their meetings were weekly or monthly and took place in the area concerned among the then 148 Hawari listed in al-Gammaliyyah's structural organisation.[35] This council worked in tandem with each shaykh al-harah, which was no longer responsible for taxation, and *Tanzim*[36] officers, who monitored the building processes.

The governmental perspective of Cairo was a total transformation towards a modern capital whose reform entailed the reform of its governmental and municipal structure. However, building processes took place on the basis of private and limited interests and were not in line with the new thoughts on modern image.[37] In 1847, for example, and in a break with the old tradition of naming hawari after professional affiliation,

Cairene hawari, zuqaqs, and streets were named formally on plates and all houses and buildings were numbered on their front doors, in an effort to regulate and organise the city in modern terms.[38] Tanzim department was the leading institution attempting to handle the confusing structure of the hawari in an effort to implement an urban order.[39] Tanzim Officers were, hence, a mix of Egyptian and French engineers with deep interest in European urban ideals. Substantial actions of reform included issuing the alignment acts, supplying new road maps, between 1885 and 1899, and implementing standard road widths called *Khutut Tanzim* (lit. lines of organisation). These were supported by a law and decrees with mechanisms of implementation in the long term.

From 1883 onwards, new building activities and restorations were not allowed without prior permission from the *Tanzim* department, based on approved drawings made by "architect/engineer". A decree issued on 8th September 1883 from the Ministry of Public Works[40] said: *"Clause 1: Any building activities within cities should obtain a formal permission before the work commences"*. Clause 2 of the same decree insisted that the application for building permission should be accompanied by *"drawings of the plot, roads and neighbours drawn on a scale 1:200"*. The new legislation required every new building to follow the defined road alignment maps (*Khutut Tanzim*).[41] The same decree set the minimum dimensions of a room to be 4(L) x 3(W) x 3(H) metres. Any non-approved activities were considered illegal and were subject to a penalty or demolition.[42] The extent of these restrictions was described by Mahboub during the 1930s: *"They* [power of rules] *are wide to the extent that by the application of Tanzim Alignment laws of 1881, 1887 and 1889, building lines can be decreed for the widening or modifying of any public street or road. No new constructions can encroach on these lines, and, moreover, heightening or any forms of maintenance, including even plastering of such portions of existing buildings as are cut by these lines, are forbidden".*[43]

In a retreat from the traditional notion of the shared space being a private property of the community, the newly established governmental institutions exercised ultimate control over non-documented private properties, declaring that any "non-owned" building was a "public/state" property. In addition, permission was required to undertake any outdoor activity/ building, even for the benefit of the locality or in the public interest. In 1896, a decree was issued to regulate the usage of public spaces like hawari and Durub as well as main streets and roads.[44] It allowed the usage of areas not more than half-width of the road for parties, festivities and weddings, subject to an approved permission and payment of fees. Funerals, on the contrary, respected the area's limitations but did not require permission or fees. Such permissions were to be issued by the Ministry of the Interior and Public Works.

Those restrictions and complex processes of permission-seeking and fees reduced the margin of power of local leaders, depriving them of the authority to manage, control or even support shared activities in their local public sphere. The new laws and power of authorities left little for local members to control in their socio-cultural context. In a reflection on how harsh the power of the state was, the eastern dead-end wall backing retail stores of the al-Darb al-Asfar was ordered to be demolished, opening new, easy access for foreign tourists to the historical buildings of al-Suhaimy and Mustafa Ja'afar houses.[45] While those decisions completely changed everyday lives of the residents, local leaders did not have any power to protest or change them. Accordingly, residents perceived such local authority to be limited to internal conflicts and struggles.

ARCHITECTURE AND HOMES IN TRANSITION

The master builder, the central character in building activities in the Cairo of 1800, had almost disappeared, at least legally, during the 1880s, following new building regulations and decrees, and was replaced by a complex technical process of planning, drawings, application and construction. The new system required qualified professionals to implement the state's vision, and only architects with experience in producing scaled drawings for formal review and permission purposes. Furthermore, the construction work had to be supervised by the architect/engineer and Tanzim officers as per a legislative decree issued on 8th September 1883.[46] The decree defined several conditions and recommended certain forms and spatial orders, in which a courtyard in an introverted house was a preferable solution. It set several hygienic regimes and spatial requirements such as minimum internal room dimensions, natural ventilation for all rooms, and orienting openings to the north, with provision for a few to the south (for cross ventilation). Moreover, it required a toilet for every closed apartment and shared for individual rooms.

In what could be largely perceived as detailed building specifications of the present day, the decree specified details for recommended construction materials and finishes, external and internal as well as inspection procedures. It stated that walls should be made of *"limestone or bricks with lime mortar"*.[47] *"Internal walls should be plastered and receive one coat of oil paint"*. All work should be inspected frequently by the Ministry of Public Works' engineers.[48] The decree was amended later to add further specifications regarding exposed terraces and mashrabiyya on the main façade.[49] The revised decree, issued on 9th January 1899, was inclusive and decisive about these issues.[50] It specified the depth of the exposed balcony should be at a height of no less than 4.50m from entrance level and should not be longer than 1.00m for wide roads (more than 6 metres) and 0.50m for narrow roads (less than 6 metres). Exposed balconies should be at a distance no less than 1.00 metre from the neighbouring edge. Those regulations, although general and intended for implementation across the whole city, were introduced especially to organise the new quarters of the city which were still under construction and development.

The legislation might appear to have been necessary for organisation of the growing metropolitan Cairo. However, the influence of the Europeans, especially the French engineers, was apparent. During the last quarter of the nineteenth century, French-educated Egyptians controlled most governmental and cultural institutions, in participation with western (mainly French) experts. Tanzim Department in 1889, for example, employed four Egyptian and six French engineers and all formal correspondence and communications were written in French.[51] Under this organisation, the traditional master builder's role was subdivided among several institutions: the architect, for design and production of accepted scaled drawings; the Ministry of Public Works, for design reviews, permissions and inspection; and finally, the new builder/contractor who had to construct the building exactly as per the drawings and specifications.[52]

But, we need to know how this disappearance of such fundamental character influenced the local idea of home. The master builder (or *Mi'mar*) was an expert by tradition who had comprehensive understanding and knowledge of local culture as well as social and ritual norms. His planning principles encompassed a deep understanding of inherent socio-cultural structure that was different from one community/quarter to another. Design and build was a combined site and context-driven operation, in which instant decisions were in relation to neighbouring buildings; constraints were present and response was immediate. In the absence of such direct contact with site and culture, buildings became isolated entities that were irrelevant to each other or to the communities that accommodated them. This could perhaps, explain, some incoherence in decision making in such aspects as floor levels and height, building materials and decoration.

Homes of Old Cairo had gone through a process of transformation during this transitional period, replacing the introverted, large plot size houses with extroverted high rise (3–4 storeys) compact houses, in which the horizontally-stretched Harem turned into vertically organised wings. Transformation of socio-cultural and economic conditions definitely took its toll on the built fabric in every aspect. As we learnt earlier, several properties in 1800 were divided into part-shares by inheritance; some shares had been sold separately or swapped with others.[53] The vast majority of selling records in al-Bab al-Ali's court around the turn of the century followed the same convention.[54] The result was small plots for reduced, compact, vertically-extended houses that soon developed into joint-household units, in which family relatives used to occupy separate apartments within the same building. Kenneth Cuno reported that the joint-household system predominated urban housing in Egypt at the turn of the twentieth century.[55] In his study, Cuno found that the majority of houses in Cairo consisted of joint-household units of seven members (58 per cent) rather than the single houses of the second half of the nineteenth century.[56] At the turn of the twentieth century, only a few houses of courtyard introverted organisation were still in actual use.

Emerging domestic typology had to accommodate families within parts of previous houses, with new socio-spatial logic that was organised vertically. The basic unit of the family, the master of the family, would occupy the first two levels: the ground level had an entrance, main hawsh, for stable and food storage, a Majlis male and family reception space, a kitchen with oven and a few services [R3.1.07].[57] If more space were available, the harem wing could be part of this level. Upper levels were arranged to include a variety of sleeping (harem) wings with small service areas, each of which housed a sub-family branch, mainly elder sons and their families (Figure 8.4). A second stage of this development came when each family residence was divided into separate apartments whose occupants were not related to each other. This was taking place, chiefly during the 1940s–60s, parallel to the modern movement in Europe and the rise of apartment buildings in cosmopolitan Cairo and promoted by the state.[58] The vacant multi-storey houses were then filled with immediate occupants from different backgrounds. For the first time, every unit of the building had to work independently and include all services. The shared kitchen on the ground floor started to disappear and small kitchen units were implanted at every level. While surveying some of these houses in al-Gammaliyyah, tight kitchens and bathrooms were found to be common features (Figure 8.5). This suggests that kitchen spaces replaced parts of the previous bathroom space. At least this was the case in two buildings in Zuqaq al-Darb al-Asfar (no. 2, no. 3). In buildings, which were built before the 1950s, kitchen spaces, unlike the bathrooms, seemed not to be an original part of the design of every level.[59]

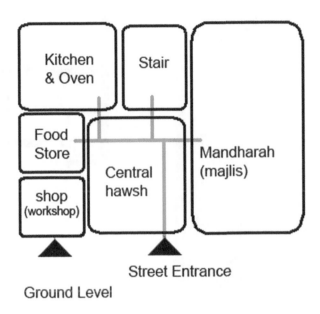

Ground Level

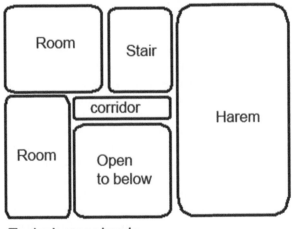

Typical upper levels

8.4   Typical zoning of medium-sized houses around the turn of the century as interpreted from archival documents. Number of rooms/floors could be multiplied as per the need and family size

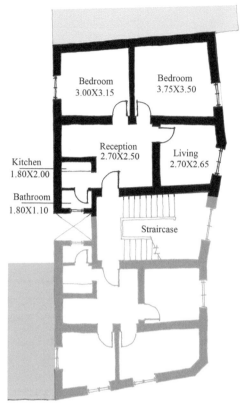

**8.5   House #1
Zuqaq al-Darb
al-Asfar**

Note: Typical plan and
façade of early twentieth-
century houses.

This transformation had profound implications for the social cohesion and physical infrastructure of these buildings. Vertical extensions meant reducing areas of direct communication between houses from one side and the public sphere from the other, both visually and vocally, in a break from the local tradition that relied on the continuous flow of activities between private and public spheres and where zones of transition occupied and transcended the boundaries between indoor and outdoor spaces. Under the new order, there was a need for alternative means of communication between high levels and the alley space. Mashrabiyya was not a solution, being inward-directed and considered an old fashioned and outdated element. Operable windows, based on the modern *Rumi* style,[60] and terraces emerged to dominate façades, being specially recommended by new building regulations. Balconies became the transitional spaces providing outward overlapping spaces with the alley for those living in high level apartments to stay in contact with events taking place on the ground. Similarly, pushing apartments towards higher levels facilitated the increase in the number of retail shops and workshops intruding into ground floor areas opening to the harah space.

The house roof was then turned into densely covered floors occupied by Ishshash of the poor migrants, or used for livestock such as chickens and ducks. However, this practice was not particularly widespread before the middle of the century. It was not a predominant phenomenon of Old Cairo. On the other hand, the tightness of plots and the limited area at every level for openings left architects with no option but to seek repetition of opening order on various levels. With, typically, 6–8-metre-wide façades and two overlooking rooms, there were no options but to provide one opening for each space and at least one balcony on every level. The regulations on balconies limited the depth and length of such exposure. As a result, buildings constructed under such conditions looked similar and variety of opening sizes and style (types of mashrabiyya and other openings) was replaced by a vertical modular order.

THE TRANSITIONAL ARCHITECTURE OF THE HOME:
A SEARCH FOR MODERN ORDER

The transformation towards vertically-extended houses was a direct result of the change in the social structure and affordability for residents, as well as the natural division of land by inheritance. It was rare to find a single heir of a house. Archival records of buying and selling houses at the turn of the century were (similarly to 1800) based on part-houses. For example, a house in *al-Darb al-Asfar* was sold in 1882 and was composed of *"a courtyard (Hawsh) that is partially covered and includes a water well, manual laundry corner, it leads to stairs leading to a Meqa'ad (works as mandharah) with a durqa'a on the main facade and more stairs lead to a Harem qa'a with a toilet at a higher level and other spaces and rooms with toilets"*.[61] This configuration represented a simpler form of the traditional house, in which the mandharah disappeared and was replaced by a multi-purpose open *meqa'ad* (Figure 8.6). The traditional harem quarter, on the other hand, was reduced to a closed room and a toilet as mentioned in the above text. Other family spaces were moved to higher levels. Other sale documents of houses of that time in this context were no different.[62] However, later houses were even smaller and the hawsh became tighter and provided for livestock in the form of a small stable.

On the other hand, there were some large houses which had been built during the 1880s and followed the medieval organisation of the pre-1800 houses (i.e. central courtyard with surrounding living spaces). Bayt al-Kharazi, was built in 1881 following the typology of old medieval houses, but of much simpler organisation: a central court, with two opposite halls, one for men (mandharah) and the other for women (harem) (Figure 8.7). As a Russian merchant who wanted a secure house and access to the trade centre of the city (al-Mui'zz Street), location and form of the house responded to al-Kharazi's requirements. The house had two entrances; main entrance on al-Darb al-Asfar, secondary one on al-Mui'zz Street on the north-western corner. The main entrance led to an arcaded lobby with a store room to the east of the central hawsh. The house had its own water well and a cooking quarter (to the north) that was open to the sky through small shafts. Upper levels included several rooms of different sizes.

This was perhaps a good representation of the transitional organisation of housing development between the medieval courtyard-centre form with intertwining levels and the contemporary, compact and multi-storey apartment buildings in this particular context. Main halls were simple and rectangular: without significant decoration or differences in floor levels. The façade had a level array of windows that were modular and of standard size (Figure 8.8). This could be the consequence of the restrictions on using medieval ornaments and decorative elements (such as mashrabiyya) under government laws, in addition to the adoption of classical European style which was preferred by the ruling officials of the time.[63]

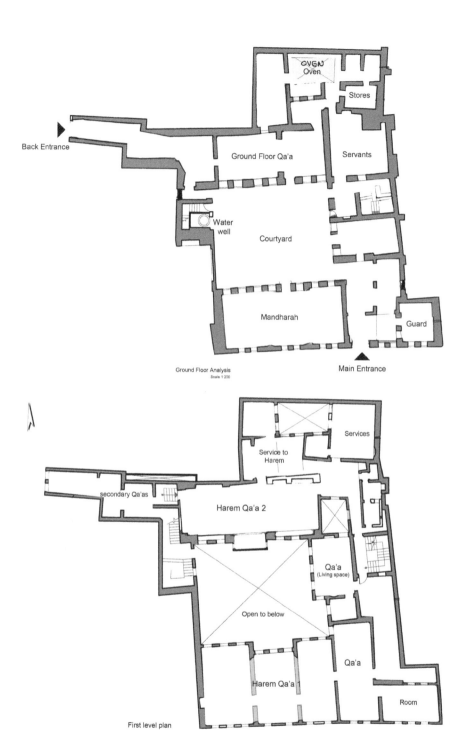

Oven

Stores

Back Entrance

Ground Floor Qa'a

Servants

Water well

Courtyard

Mandharah

Guard

Ground Floor Analysis
Scale 1:200

Main Entrance

Services

Service to Harem

secondary Qa'as

Harem Qa'a 2

Qa'a
(Living space)

Open to below

Qa'a

Harem Qa'a 1

Room

First level plan

8.6–8.7   Floor plans of Bayt al-Kharazi

Note: Ground floor (top) and first floor (bottom).

8.8 House interiors reflecting the modular pattern of openings

In addition to those two common models, other types did exist in al-Darb al-Asfar during that period, but could hardly be considered a common typology. In the map of 1937 (Figure 8.3) we could trace three relatively large houses (nos. 6, 9, 12, 15), which were different in form and type but were alien to the common typology. In the first three houses, the main buildings did not overlook the main darb; rather, they were at the rear edge of the plot, leaving the front as an open but fenced space. Al-Mansoury family's residence (no. 12, main darb),[64] for example, retained a unique organisation that did not overlook the harah directly. The building was in the middle of the plot, with a large garden at both front and back. It resembled the modern villa type houses. The territory of the house was clearly defined with a boundary and included its own private, semi-private and semi-public spheres. However, we could not place too much emphasis on that type due to its scarcity in that context.[65]

Semi-public spheres of homes had limited space in which to remain part of the indoor space in new compact houses. While al-mandharah of the 1800s was an essential space in every house, regardless of its size, it was a rare space during the first half of the twentieth century. This is obvious in property documents: with al-mandharah as a designated space disappearing from the description of 1880s houses. This is supported by the survey of early twentieth century house layouts. Along with the reduction in plot sizes, the vertical arrangement of the residence turned al-mandharah into a small room that was separated from the kitchen and other services. In a few examples, it was decorated for receiving guests.[66] Houses of that period were planned to be small units of family apartments, with a shared family space at ground level (#6 and #1 Zuqaq al-Darb al-Asfar). The conservative culture of 1800 was then beyond the affordability of the residents and the capacity of available plots. Ground level living spaces started to receive male guests, while female guests were allowed house visits and used to access their hosts' upper floors [R3.1.07].

INTRODUCTION TO MODERNITY

The departure of the rich merchants and elite from Old Cairo, while contributing to the physical and financial deterioration, did not result in a similar decline of shared social activities. Despite the physical deterioration, the community continued its rich and active social context during the first half of the century. Literature of the political history of Cairo during that period proved that the hawari was an essential base for resisting British occupation. Coffee houses during the 1920s, according to Naguib Mahfouz,[67] were the sites for meetings, funding and arranging revolts and attacks against the British soldiers. They were also local public venues full of people and entertainers every night until the morning. In *Palace Walk*,[68] Mahfouz described such a continuously active environment: *"There was no clue by which to judge the time. The street noise outside her room would continue until dawn. She [a wife awaiting her husband's return] could hear the babble of voices from the coffee houses and bars, whether it was early evening, midnight, or just before daybreak".*[69] Such a lively picture was affirmed by both Stanley Lane-Poole at the turn of the century and by Ahmed Mahfouz[70] in *Mystery of Cairo*, a rare book about Cairene life at the turn of the century. Festivities and parties at that time were almost identical in keeping with the traditions and procession of their counterparts of 1800, especially wedding parties (the separate tours of the bride and the groom with their friends).[71] Similarly, social practices and cultural norms continued to play a crucial role in the hawari's daily life, however, at minimum cost [I5.1.09; R1.2.08]. In contrast with the secure living conditions of 1800, the people of lower status had to work out their own living resources by establishing home-based food industries, opening workshops, grocery stores or renting any ground floor spaces for retail or storage purposes.[72]

This active public life was challenged by a series of laws aiming to change the physical features and organisation of those spaces. Between 1880 and 1910, various laws were issued to put the public sphere into a shape and order that suited the ruling regime's desired image. The clearest attempt appeared in the decree issued by the Ministry of Public Works on 22nd February 1882, whereby Article 12 was intended to clear the roads of the stone seats spread across the old city. It says: *"All structures intruding from the building into the road, such as stone seats, stairs, should be removed with the exception of historic, religious or artistically valued buildings until their facades are refurbished on the alignment line".*[73] Article 13 of the same decree instructed that *"all existing arches on public roads should be demolished once they are damaged or attached structures are to be restored [refurbished]"*. It continued: *"The construction of those arches on public roads is prohibited from now on".*[74]

While the stone seats were removed and re-built several times,[75] it is presumed that legislations relating to permission and fees for public festivities were not implemented, at least in Old Cairo.[76] It was apparent that the government wanted to deal with Old Cairo and its hawari as it did with Ismaili Cairo with its paved and wide boulevards. The regulators and officials did not understand that the public sphere of the hawari, including its physical features (stone seats, stairs) and social events (festivities and mawlids), were essential parts of local culture and daily life. They housed essential economic activities, serving food to visitors for mawlids that flourished throughout the year [R4.2.08]. Removing the seats and getting permission for using public spaces could have eliminated most gatherings and basic socio-cultural patterns of life.

If public and private spheres were changing in form and organisation, this change would essentially reflect on zones of transition as overlapping domains between the two. During this challenging period and subsequent socio-spatial changes, transitional zones were the most affected domains, spaces and environments. The alley, which was a traditionally private and secure space, had to host activities associated with the expanding ultimate public sphere. The city's public sphere and its busy life and traffic extended into the harah's area when the former dead end was opened to public traffic from the busy Shar'i al-Muizz. It had become an open urban space to serve tourists and other commercial activities. The old dead end, which was a transitional sphere between the house and the harah, became a transitional zone between the harah and the city at large. The new open space appeared on later maps of 1937, made according to alignment line number 34(b) and 36(b). The resulting space was a *saha* (plaza) of size 8.00 x 20.00 metres,[77] the largest and most exposed space, exactly where the alley became a dead end (Figure 8.9).

Such an open junction with the old city's thoroughfare had affected the public sphere in two directions. First, it encouraged commercial activities and retailers to extend to the inner spine of the harah. Second, it reduced the sense of local security and privacy in the *darb* and its branches, while increasing the number of entrances and allowing a continuous flow of passers-by. Accordingly, this change had altered the edge and territory of the local public sphere in spatial and socio-cultural terms. The open *saha* turned out to be a reserve for Al-Muizz thoroughfare, and therefore was excluded, at least socially, from al-Darb al-Asfar. Most shops that opened in the saha's space were oriented towards non-resident customers. This situation was emphasised by the behaviour of people whose homes were in that zone. Residents of buildings overlooking that open space experienced minimum if any interaction with the rest of Harah's people. They considered themselves irrelevant to the harah's community and represented a higher class [R20.1.09].

On the other edge, the western entrance continued to work as a strong boundary and secure entrance due to its spatial configuration with a relatively tight entrance. But, opening the spine from both sides allowed short-cut movements by strangers or local passers-by who wanted to cross from al-Muizz to al-Gammaliyyah streets. Security measures in such semi-private zones were more vulnerable and had to be altered. Instead of monitoring strangers whose destination used to be inside the harah, residents had to monitor the flow of passers-by to make sure they exited the main alley and did not harm or stalk any of the local residents. At night, every resident had to be alert and active social interaction of the locals tended to be mainly in the evening, with a reasonable level of active security and protection.

The *Saha* was characterised by two opposite façades; early twentieth-century houses to the south (tight plot, four levels high, modular design) and historical houses of 1800 to the north (large size, low height, traditional in style and decoration).

8.9 Open
junction with
al-Muizz St

Note: It has become the
biggest open space in
the area.

The development of the architectural character of homes was represented in the differences between those two sides. Differences in style and urban pattern extended to openings and decoration, in which there were stark contrasts between old and new houses. On the south side, Bayt Habb al-Rumman was built in 1934 and building no. 2 was built during the first quarter of the century.[78] Opposite to them, Bayt al-Kharazi was built 40 years earlier. New houses of the 1930s–40s, in general, were built on small plots and the harah became a tighter, more vertically extended space than the original of the medieval ages.[79]

The public sphere became divided into two domains: the business and industrial domain at ground level and the residential homes at upper levels. The communication between the two domains was increasingly limited to high level terraces/balconies. The increasing presence of workers turned the main spine of the harah into an industrial space. According to the survey of commercial activities of that period, industrial activities become dominant only by the mid-twentieth century. Most of those trades/shops were owned by the harah's residents.[80] Shop owners, on their side, used every metre of the open space to serve their business; they used building entrances, dead-end space and ground floor rooms as store rooms or industrial labs/workshops. The services of the harah could not withstand such heavy pressure, and its physical structure suffered a serious decline.

EVERYDAY LIFE AND THE SOCIAL SPHERES

In studying everyday life of that period, I referred to the stories of old residents, some of the documents and writings of the time in addition to the resourceful writings of Naguib Mahfouz, who lived in the area during the first quarter of the century, that depicted traces of life in al-Gammaliyyah quarter and its hawari. Mahfouz's writings were basically a description rich with analysis of everyday life and associated socio-cultural norms. According to Mahfouz, daily routine of local families resembled two parallel patterns, males' and females'. The father led the males at work, while the mother led the females (daughters and wives) in performing domestic duties as a team. Servants had disappeared from the domestic organisation and spaces.[81] Women had to wake up early and hurry to the kitchen on the ground floor to prepare family breakfast; the only and main collective meal for family members. Males ate together, while females ate later. According to such routines, males used to return home at different times and therefore eat separately. Lunch and dinner were most probably taken in their own rooms with their wives.

STREET NEAR BAB-EL-KHARK

293

8.10 Security measures in the alleyway: man sitting on his doorstep in the late nineteenth century
Lane-Poole (1902).

One senior resident who lived in al-Darb al-Asfar during the 1940s declared that his family's original house of the 1920s–40s (no. 6, Zuqaq al-Darb al-Asfar) included a family living on the ground floor with the master's bedroom, living space, a kitchen, and a *hawsh* for raising animals and livestock.[82] Every upper level included a wing for one son along with his wife and children. Cooking and eating had to be performed in a collective setting in a ground floor kitchen that overlooked the *hawsh* for ventilation and supplies. The father was the only financial power and his sons worked with him without financial independence. *"Once the father died"*, the resident said, *"every level was separated into independent units and later modifications turned the ground level apartment into stores and shops"* [R4.1.07].

Even though many families were experiencing changes in their lifestyles and their members were becoming more educated, the traditions and social norms remained in firm continuation until mid-century. In the private sphere, the master was the dominant character along with his wife. Sons were followers of the father, regardless of their age or wealth. Naguib Mahfouz gave us a pleasing picture of the breakfast meal.[83] It used to take place in *"the dining room on the top floor adjacent to the parent's bedroom. On this storey were also located a sitting room and a fourth chamber".*[84] Only father and sons shared the meal, and sat in a constant order around a low table. *"No one dared look directly at their father's face. When they were in his presence they would not even look at each other, for fear of being overcome by a smile. The guilty party would expose himself to a dreadful scolding".*[85] This description of life patterns would not be complete without stressing the indoor-out continuity. This appeared in ladies' communication through mashrabiyya and balconies. Male friends would meet at each other's shops or in the coffee houses. No visits were paid at home unless for a formal visit for a formal event like an engagement, wedding, or even conflict negotiation. Festivities and partying practices, reported by senior members, were similar to those of 1800 recorded by Edward Lane. However, the practices were simpler in form and at a cost that suited the lower income group.

* * * *

The turn of the twentieth century was found to be the most challenging period for the survival of the hawari of Cairo, with lack of support, intensive migrations and destruction to its infrastructure and physical systems. Old Cairo was a battle ground for the emerging intellectual reform movement and its position on women's participation in society and the public sphere, and where radical culture was deeply rooted and most powerful. However, change in the popular mood towards women was facilitated and encouraged by the emergence of the modern quarters of European Cairo, rulers' pressure to drive the wealthy merchants to reside in newly developed zones, while formal institutions took forceful action to reform the built environment. Medieval houses and harawi were the first to be affected by this process of change, fundamentally and structurally. Courtyard-centred houses became an unnecessary luxury for the emerging low class community. This form of organisation was replaced by the more affordable, compact and extroverted multi-storey family houses on smaller plots of land.

In summary, large houses were divided into smaller plots due to inheritances and exchange laws, and later every portion was developed as an independent unit/ building. The notion of having a house to exclusively accommodate one extended family was in decline and giving way to more divided and independent apartments. In the meantime, the state's formal institutions exercised powerful control over activities in the public sphere, manifested in the opening of the western dead end to through traffic, declaring two houses as historical sites and requesting formal permits prior to any public festivities and celebrations in public spaces. The sovereignty of the local social group was, hence, compromised and their control over the public sphere was challenged by emerging institutions. Similarly, opening the dead end resulted in social and spatial division of the previously united public sphere into two culturally defined domains, the popular to the east, with a village-style culture, and the classy eastern side, with heirs of historically-rooted families who continued to live, despite their financial and professional success, in their inherited houses/apartments. Moreover, commercial activities evolved (benefiting from open accessibility) as profitable resources for many ground level spaces, resulting in transformation of ground level apartments/houses into workshop spaces.

NOTES

1   Sophia Lane-Poole describing her criticism of Muhammad Ali's policies and plans to modernise Egypt and abandon the old city, during her stay in Cairo. She was the sister of Edward William Lane, and frequently visited high profile women in their harem quarters. Lane-Poole, *The Englishwoman in Egypt*.

2   Lane-Poole, *The Story of Cairo*, p. 1.

3   Muhammad Ali was the ruler of Egypt from 1805–47, and is considered the founder of modern Egypt, where he built a modern army and navy. He dreamed of Egypt as a powerful and advanced state and established the first industrial economy in Egypt. For more details, see: Sayyed-Marsot, *Egypt in the Reign of Muhammad Ali*.

4   Lane-Poole, *The Englishwoman in Egypt*, p. 49. In her letter of July 1843, she stated, "a proclamation has been issued by the Pasha for extensive alteration and repairs throughout the city. The houses are to be white washed within and without; those who inhabit ruined houses are to repair or sell them; and uninhabited dwellings are to be pulled down for the purpose of forming squares and gardens; mashrabiyya are forbidden, and mastabahs are to be removed". This was quoted, as well, in Abu-Loghud, *Cairo*, pp. 93–4.

5   Kenzari and El-Sheshtawy, *The Ambiguous Veil*, p. 22.

6   Abu Loghud, *Cairo*.

7   He was the grandson of Muhammad Ali and was educated in France. He shared the dreams of his grandfather and was instrumental in making the plans a reality. During his time, the Ismailiyya quarter was planned and constructed with several boulevards, opera house, and palaces. All were ready to receive European royalty who visited Egypt for the opening ceremony of the Suez Canal.

8   The force was apparent in several decrees issued during the 1880s to apply constraints on new buildings in the form of specific standards such as room sizes, materials and external and interior finishes. These all were driven from Ali Mubarak's vision (as quoted at the beginning of the chapter). See for example, decree issued on 1st September 1882, imposing the first building regulations (shown in page 178).

9   Researchers are divided as to when such change started to take place or become apparent. Janet Abu-Logud and Andre Raymond considered the last quarter of the century as the significant period.

10 Abu-Loghud, *Cairo*. She called this era "Cleansing the stables". He prohibited the use of mashrabiyya claiming that they cause fire hazards. However, most historians referred to his desire to eliminate traces of a medieval city, whose major feature is the Mashrabiyya.

11 Sayyid-Marsot, ibid.; Beinin and Lockman see the main reason for that as the sweeping transformation of agrarian relations in the nineteenth century and concentration of land into the hands of a new class of indigenous large landowners. See Beinin and Zachary, *Workers on the Nile*, p. 24; Beinin, *Formation of the Egyptian Working Class*, p. 15.

12 Jean-Luc Arnaud argued that moving those merchants to new quarters was a principal plan of the government. Merchants living at the edge of the old city, in particular, were given large plots in new quarters to build new homes. They were the first landowners in the new quarters, while "keeping their businesses (shops, stores) in the old city, which should remain the central site for trade in Cairo in the long term". This was planned to feed the new Ismailia quarter with active movement and lively environment. See Arnaud, *Le Caire*, p. 153.

13 *Ishshash* (pl., single: *ishshah*) means a temporary structure, mainly of timber.

14   *Ahwash* (pl., single *hawsh*) means gated open land filled with tents. However, this phenomenon was not present at all hawari. Some of them did not experience such migrant waves (at least not extensively) as Al-Darb al-Ahmar. However, Al-Gammaliyyah, our case site, was one of the attractive sites for those migrant waves.

15   As per my investigated records of the time: using archival documents and archaeological analysis. But this does not mean it is a fact. However, this means that this type was either abandoned or became rare at that time.

16   Al-Darb al-Asfar restoration project has recognised 33 families who were living in his house during the last quarter of the twentieth century. Those families are believed to have been living there for a long time. Interview with the project director: Dr. Asaad Nadim, who is also one of the current business owners in the harah.

17   Such as Ministry of Education, Ministry of Public Works and others.

18   Those names were listed by Albert Hourani, in his seminal work, Arab thoughts in the liberal age as leading reformers in Egypt, during the period 1850–1940. See Hourani, A.A. 1962. *Arabic Thought in the Liberal Age, 1798–1939*. London: Oxford University Press.

19   Studies that handled the reform movement focused on the intellectual debates and the figures who were involved in it. See for example: Muhammad Emara's *Qasim Amin*, a rare work, handling the reflection of this issue on the life of ordinary people.

20   During that period, the reform thoughts relieved the pressure to obtaining large houses in keeping with the organisation of isolated quarters. The reformers insisted: women should not be isolated and Islam says so. Therefore, there was no need for such a complex system of spaces. It is important to state that such debates never stop in Islamic societies. Educated scholars, as usual in Egypt, mediate in communications between the elites and ordinary people.

21   The harem quarters started to disappear from the descriptions of houses during the 1880s, and by the end of the century, the *Ghurfit Noom* (Bedroom) became familiar in descriptions of houses.

22   Archival document no. 194, record 500 dated 16th Shaaban 1281H (1865AD).

23   Archival document no. 54, record 25 dated 1307H (1889AD).

24   Archival document no. 64, record 24 dated 1307H (1889AD). Mohamed Fad al-Kharazi was one of the few exceptions, as a rich foreign merchant to arrive and reside in the harah at such a late date in the nineteenth century.

25   Al-Sioufi, *A Fatimid Harah*, pp. 24–39.

26   Interview with a resident coded [R3.3.07].

27   Al-Sioufi, ibid. Several senior residents confirmed that situation and the survey of al-Sioufi.

28   Interviews with residents and professionals in Old Cairo. Codes [I2.1.07; R1.1.07]. Al-Darb Al-Asfar project had moved many families of this group outside the area in order to restore these sites. Those working in such businesses were given money to start their business elsewhere.

29   This subdivision of the hawari during the twentieth century was not a special feature of Old Cairo; the same phenomenon appeared in different districts such as Bulaq al-Dakrour, and Manshiyat Nasser. See the work of Singerman, *Engaging Informality*; El-Kholy, *Defiance and Compliance*; Ghannam, *Remaking the Modern*.

30   This was, as per Joel Beinin, the major reason for all internal migration waves to Cairo of that time.

31 These fragments are collected from different resources such as early twentieth-century, commercial surveys of Mohamed El-Siuofi; *Demographics surveys of al-Gammliyyah*; GOPP reports on al-Darb al-Asfar. Finally, information was developed with reference to al-Gammaliyya quarter in general, of which several studies were made.

32 Such activities and narratives were similar to other cases analysed by sociological scientists. For example, see Al-Messirri Nadim, *The Concept of the Hara*.

33 In 1881, the Minister of Public Works requested the employment of four additional engineers and an office in each quarter to follow up building activities. See the formal report for restructuring of the Ministry of Public Works, 17th July 1881, Ministry of Public Works. Archival Document, file number 6/2/A Public Works (Archival Code: 0075–035966). Cairo: The National Centre for Archival Documents.

34 The formal election schedule was announced in the formal newspaper. See *Al-Waqa'i El-Masriyyah*, issue 72, dated 9th June 1886, p. 639.

35 The list of al-Gammaliyyah's hawari was published in *Al-Waqa'i El-Masriyyah*, issue 77, dated 7th July 1886, p. 639.

36 Tanzim, literally means organisation, and it was the name of the civil works department in the Ministry of Public Works until the 1940s.

37 Mahboub, *Cairo: Some Notes on its History, Character and Town Plan*, p. 289.

38 *Al-Waqa'I al-Munammarah*, no. 64. The decree ordained that the numbers were to be written in black ink on white background and framed in black. Every side had to be numbered with either even or odd numbers. See Abdel-Wahhab, *Takhtit al-Qahira wa Tanzimaha* (Planning and organising Cairo-since foundation), pp. 23–4.

39 See several decrees for new alignments of Cairene hawari issued during 1889–1910, in which some plots were either added or excluded from the public road network. For example: decrees dated 28th November 1910 (File 6/3/D, Public Works, Code: 0075–036184); 9th November 1908 (File 6/3/D, Public Works, Code: 0075–036183). Cairo: The National Centre for Archival Documents.

40 Formal Decree for the condition of building houses within big cities, 8th September 1883. Ministry of Public Works. Archival Document, file number 6/2/A Public Works (Archival Code: 0075–035972). Cairo: The National Centre for Archival Documents.

41 Due to the lack of professionals at this period and their high cost, I expect that many building owners left their buildings un-restored and limited the opportunities for low-class people to build new houses. Ruined buildings were continuously occupied without maintenance. In the popular *Al-Moqattam* newspaper, a French Architect called Korsicous published daily advertisements for building houses and receiving his fees in delayed installments. See *Al-Moqattam* Newspaper, issues 3279, 80, 81 dated 8th, 11th, 15th January 1900. Cairo: Sarrouf and Co.

42 All these decisions were taken by the *Tanzim* department and Ministry of Public Works under the Alignment Act of 1881, and the Law of Expropriation of 1906. Mahboub, ibid., p. 289.

43 Mahboub, ibid., p. 292.

44 Decree on the usage of Public spaces, 23rd April 1896. Ministry of Public Works. Archival Document, file number 6/2/A Public Works (Archival Code: 0075–036012). Cairo: The National Centre for Archival Documents.

45 Despite the wide search for the decree of this modification, I could not find it. Either the relevant documents were missing, or the British authority made such an addition without formal decree from the *Tanzim* department.

46   Formal Decree for the conditions on building houses within big cities, 8th September 1883, Ministry of Public Works. Archival Document, file number 6/2/A Public Works (Archival Code: 0075–035972). Cairo: The National Centre for Archival Documents.

47   Ibid.

48   Ibid.

49   See decrees for amending the decree of 1883. For example: decrees dated 9th May 1889 (File 6/2/A, Public Works, Code: 0075–035989); 16th June 1895 (File 6/2/B, Public Works, Code: 0075–035990); 4th June 1896 (File 6/2/B, Public Works, Code: 0075–035991). Cairo: The National Centre for Archival Documents.

50   Formal Decree for the conditions on building houses within big cities, 9th January 1899, Ministry of Public Works. Archival Document, file number 6/2/B Public Works (Archival Code: 0075036014). Cairo: The National Centre for Archival Documents.

51   The Cabinet had requested translation of those correspondences to discuss them. Some letters on this request were found among the archival records of the Ministry of Public Works during the 1880s.

52   In the daily records of the Ministry of Public Works, several projects were announced for bidding on construction works. For such bidding, detailed specifications and drawings were specified by the owner or the institution which owned the building. See for example: record 5/2/1M (Architectural issued documents), no. 355 – buildings, for the year 1877.

53   Refer to many archival decrees and documents mentioned in the previous chapter.

54   They were similar even in terms of gender of landlords/owners: who were mainly females (widows, or freed slaves).

55   Cuno, K. 1995. Joint family households and rural notables in 19th-century Egypt. *International Journal of Middle East Studies*, 27(4), pp. 485–502.

56   Cuno, ibid.

57   Case reference from archival records.

58   Naguib Mahfouz, described his family's move from al-Gammaliyyah to al-Abbasiyyah, the new district of the city, as early as the 1930s.

59   This was confirmed by three senior residents of the harah who witnessed the old and the modified organisation. Two of them were building owners.

60   Rumi-style windows are timber windows developed during Muhammad Ali's era to simulate European windows that are relatively tall and divided into three vertical parts.

61   Archival document no. 194, record 500 dated 16th Shaaban 1281H. Al-Bab al-Aali Court Records, The National Centre for Archival Documents.

62   The two examples of al-Darb al-Asfar from the first quarter of the century confirmed this organisation, which informed the research of the typology of the harah's houses of the time.

63   Such encouragement and the adoption of modern European-style windows and façades appeared clearly in other parts of Old Cairo (such as al-Darb al-Ahmar).

64   An extended family of high profile, with a deep rooted history in the harah extending back to the mid-nineteenth century.

65   Unfortunately, those houses were not available for survey.

66   In some cases, the owner could afford additional decorations to provide higher quality space to receive guests. However, such rooms were used most of the time for family gatherings (R3.1.07).

67   Naguib Mahfouz is a Nobel Prize Winner in Literature. In his Trilogy Novels, which were set in nearby al-Darb al-Asfar, "he used the streets and alleys of historic Cairo at the turn of the century as the backdrop of his socialist-realist novels", that appeared as documentation of the social history of Old Cairo at that time. See Williams, *Reconstructing Islamic Cairo*, p. 274.

68   This novel, was first published in 1956, and describes life in the context of al-Darb al-Asfar from 1910–20, where Mahfouz lived out his childhood. Therefore his novels about Old Cairo are considered as documentation of real life of the time, especially in Gammaliyyah where al-Darb al-Asfar lies.

69   Mahfouz, *Palace Walk*, p. 1.

70   Mahfouz, *Khabaya al-Qahira* (Mysteries of Cairo).

71   The similarity was found between the description of Naguib Mahfouz, *Old Resident of the Harah* [R4.1.07] and those of Edward Lane.

72   Al-Siuofi, ibid., p. x.

73   A Decree issued by the Cabinet concerning the conditions and the structure of the *Tanzim* Department and its responsibilities, 22nd February 1882. Ministry of Public Works. Archival Document, file number 6/2/A Public Works (Archival Code: 0075–035967). Cairo: The National Centre for Archival Documents.

74   Ibid.

75   References to be added.

76   As described by elderly members (R1 and R4), during the 1930s and 1940s, people were arranging festivities and parties in the public spaces freely and without intervention of the authorities. The brief records of Stanley Lane-Poole about the life in the hawari of old city confirm our proposition that social activities like weddings, festivities and funerals were actively taking place on a daily basis. It is hard to believe that every one of those activities took place after getting permission. Our suggestion is that such legislation was strictly applied on main roads and thoroughfares rather than on closed hawari and atfat, which were rarely visited by the officials.

77   As measured from al-Darb al-Asfar survey map (number 345) of the year 1937 and verified by the year 1958. The map shows land plots and open spaces of the harah during the defined transitional period.

78   The former date was mentioned by a family member of the original owner, while the second is based on the archaeological evidence.

79   Data collected about selling houses in al-Gammaliyyah at the end of the nineteenth century along with a map of 1937, proved that many land plots were divided into smaller pieces of land, which were later built up as multi-storey houses (4–5 levels). This height did not work with the traditional fabric: which was naturally developed to suit low height structures.

80   The archival documents informed us that many of the harah's residents were tradesmen by the end of the nineteenth century, which was in contrast with its nature in 1800.

81   This was partially due to the laws that prohibited slave trade during the 1870s, as well as the lack of local affordability; a consequence of the social class downgrading of the community.

82   Traces of some of these spaces still exist today, mainly at lower levels, but most of the upper floor was totally transformed.

83   Due to different working or studying times, everyone came home at different times, and accordingly took his meal separately. "When they [sons] came home in the afternoon, he [the father] would already have left for his shop after taking his lunch and a nap". This was a common pattern of life at that time. See Mahfouz, N., *Palace Walk*, p. 19.

84   Mahfouz, N., ibid., p. 19.

85   Mahfouz, ibid.

9

Contested Territories of Modernity

THE HARAH AND THE DISCOURSE OF MODERNITY

Many modernist theorists link modernity to mobility, advancement of transportation, and information technology. For them, mobility is an essential prerequisite to modernity. Zygmunt Bauman, for example, draws similarity between "fluidity" and "modernity",[1] in that both are metaphors that *"grasp the nature of the present"*. The search for modernity in a traditional context is problematic due to the apparent ideological and spatial oppositional relationships; modernity versus tradition, accessibility versus enclosure, equality versus hierarchy, hybridity versus authenticity. These differences, however, do prevent inhabitants of the medieval cities from carving their pathway to practise modern lifestyles in their own capacity, culture and language, with ordinary residents having to come to terms with the requirements of modernity and mobility (such as transportation, television, satellites, air conditioning, and many others).

In the discourse of modernity, recent social studies demonstrated evidences that modernity does not contradict authenticity.[2] Quarters of Old Cairo have always been part of the modern and well-connected metropolis of Cairo through means of communication and transportation. Its hawari are not excluded from either such a global system or the influence of its culture.[3] Residents move out of their traditional context in daily trips to schools, jobs, markets and the shopping centres. Simultaneously, outsiders move in for work, business, and visits. Satellite dishes and air-conditioning machines are spread on the roofs and façades of old houses as an unmistakable sign of modernity. The heavy presence of computers, domestic appliances and mobile phones has put them among modern users of global technology.[4] In fact, the authentic Gammaliyyah quarter has become a by-product component of modern and cosmopolitan Cairo.[5] Its socio-historical roots provide its residents with a set of "rules of social life", which Giddens considers as *"techniques or generalizable procedures applied in the enactment/ reproduction of social practices"*.[6] Although Old Cairo emerged after the transitional period as a home of manual craft, business and touristic destination, these rules seem to have acted as an agency of local social configuration and traditional values. The power of the local social contract was superior to the influence of modernity.

9.1   Contemporary
buildings in
al-Darb al-Asfar
(1950s–present)

Surprisingly, semi-skilled and skilled workers/employees, who approach the hawari's public sphere every morning for work, had to align with the local social norm. In the daytime, users of the harah's space are, to a certain extent, different from those of night time, yet the organisation is maintained.

The harah is an essential urban unit that has been *"under many development pressures and internal destruction forces, symbolizing the interaction between continuity and change, deconstruction and reconstruction, as well as heterogeneity and homogeneity"*.[7] Excluding the harah from the discourse of modernity disregards the current mechanism of local community and undervalues local socio-cultural practices and economic activities. Farha Ghannam[8] studied the diversity found in the modernity in Cairo's traditional quarters, emphasising that modernity is displayed more in everyday life of the traditional Cairene quarters than in the self-proclaimed modernity of the state.[9] Fashion, music and shopping are among many fields of activities that related the local people to the modern world rather than isolating them. Western characters and music found in historical hawari are, hence, evidences of modernity, the *inherent western product*.[10]

The recognition of everyday socio-spatial practice of modernity of the Cairene harah reflects on and contributes to the construction of modernity itself. Hastrup and Olwig argued that conceptualising cultures as separate entities of localities was a way of imposing an artificial order on a disordered world.[11] They dismissed the view that a local context could be read in separation from the global world. Even though Giddens does not dismiss the power of organisations in local contexts, he stresses that the individuals, no matter how local or traditional their specific contexts of actions, *"contribute to and promote social influences that are global in their consequences and implications"*.[12]

Giddens' thesis locates such actors as the harah's individuals as active players in creating and initiating modernity rather than being subjected to external modernising influences.

On the other hand, Alsayyad and Roy suggested "*Medieval modernity*" as a term that questions the progression of urban development in the modern world. For them "*medieval*" is invoked not as an historical period, but rather as a *trans-historical analytical category* that interrogates modern urbanism at *this moment of liberal empire*.[13] They argued that the spatial settings and geographies of the gated enclave, regulated squatter settlements, and the camp are medieval concepts of urbanisation that are recurring in the increasingly fragmented and divided modern urban landscape. This draws "*attention to the enduring paradoxes of urban life and form ... suggesting that medieval forms of organization and community can lurk at the heart of the modern*".[14] Thus, the analysis of the contemporary harah out of the context of the modern world and global culture seems misleading and insufficient to address everyday interactions, forces and influences.

Under forces of modernity and its hybridity, the notion of the harah as an integrated social group with a shared professional or religious identity does not stand today.[15] Each harah entered the twenty-first century as a predefined symbol of identity, whose morals, values and subcultures define the local character with a distinguishing mark. The harah's name becomes an adjective, a custom inherited from one generation to another. The name itself becomes a signifier that defines an intrinsic value system and social contract that appreciate interaction, communication, solidarity, friendships and shared history. Even within the same locality, sub-value systems are found. In Atfet al-Darb al-Asfar, the dominance of the al-Fayoum-origin group constitutes a subculture and value system distinguishable from those of the rest of the inhabitants. As a result, it is popularly known as *atfet al-Fahayma* (the people of al-Fayoum): a single name for a set of values that represents a complex character of fellow countrymen, of lower order, living in exclusive groups, hard-mannered, and maintaining minimum interaction with other inhabitants. In short, this created a micro-scale enclave in such a locality, bringing in issues of segregation, differences and sometimes conflict of interests.

Architecturally, the same name provides yet another system of perception that registers buildings as components of this value system: especially when they are of monumental nature, or have a significant impact on the everyday life of the residents, such as the touristic destination of the monumental complex of al-Suhaimy-Al-Khazarati-Mustafa Ja'afar houses. The continuous and consistent harmony within alleys of the old city today presents a distinguished style of buildings that "*is identifiable because it manifests a certain relationship*"[16] with its place and occupants. The modern harah is becoming a home for hybrid low-income groups who reside in apartment buildings reaching up to five and six storeys. It is interesting how such vertical buildings took their place comfortably beside the old structures on a piece-by-piece basis (most were built during the 1930s–50s and a few were built in recent decades). The contemporary harah, as such, became more vertical and denser, whilst displaying its own image of modernity: with satellite dishes on roofs and air-conditioning units stuck to a few façades (Figure 9.2). Following on from the transitional period, the mixed-use (commercial, industrial and residential) activities of the ground floor had sustained their superiority with some input from technology and global business.[17]

9.2 Satellite
dishes dominate
Old Cairo's skyline

Note: Unmistakable mark
of modern life in al-
Gammaliyyah quarter.
A view from outside the
old city walls.

The popular view of the hawari's environment among the general public in Egypt is of informal and deteriorating urban neighbourhoods of limited accessibility: full of people of low order who lack basic services and decent living standards. On the one hand, to call somebody *ibn hawari* (the son of the hawari) is an insult denoting a deficiency of good manners and appropriate behaviour. This became a class designation that was developed into terms such as Baladi (villager's way of life) and *Ibn al-Balad*[18] (villager), *Sha'bi*[19] (popular and indigenous) and *Bee'a*[20] (low-standard, literally means environment and refers to ill-mannered context). On the other hand, it is viewed as a sign of strong community: a safe place full of support and solidarity. Egyptian drama and TV productions played an essential role in establishing this dual perspective. The well-known *Trilogy of Naguib Mahfouz* during the 1950s and the *Layali El-Hilmiyyah*[21] series of Usama Anwar Ukasha highlighted these contradictions which work as magnetic forces juxtaposing the modern-westernised culture with the native residential hawari. In those examples, the main characters, simulating the city, are split between two worlds; their local side reflects the native culture with rich heritage while their modern side represents more advanced living standards but with materialistic and cold relationships.

While the harah today constitutes a smaller and less influential unit than in 1800, the alley and its street hierarchy continue to form a spatial organisation that represents a defined social space with sets of rules that regulate the social life of its inhabitants.[22] The "lived space of the harah", Salwa Ismail argues, is characterised by the spatial configuration and modes of everyday sociability, which *"give rise to particular understandings and experiences of privacy and the boundaries between the public and the private"*.[23] The proximity of the homes and the sharing of harah space bring neighbouring families into close contact in everyday interaction, with women, children and teenagers spending extended time in their local alleyway.

9.3 Provisions for maintaining privacy in al-Darb al-Asfar: operable external curtain

Privacy could not be strictly practised in such a context. The harah, today, is seen as a combined environment of private/public spheres where the boundaries between the two are blurred and indeterminate. However, particular features such as movable curtains and operable louver windows help local residents to practise control over the boundaries, based on their temporal needs (Figure 9.3).

Here the veil is becoming a strategy rather than a fixed typology of windows, more dynamic, subjective and temporal. In contrast with the traditional mashrabiyya, the modern veil re-uses its genesis, the woven textile, allowing a clearer interpretation of its function as a dynamic element that is used by individuals to draw their boundaries.[24] The appearance of a man in his high level window in the morning requires a response from women in overlooked homes: either to pull the curtain down or to close the wooden window. Conversely, when women from these homes chat together, windows of both houses are opened wide: inviting others to join in. Such action involves spatial adjustment to the indoor and outdoor spaces. This shows that gender segregation still plays a significant role in the spatial practices, represents strong social norms, and requires visual barriers and physical boundaries between spheres.

The modern institutional structure of Cairo is called *Muhafazat al-Qahira* (Cairo governorate) and is divided into different districts, each of which has sub-units of quarters (in Old Cairo). For example, there are 19 *shiyakhat* of *al-Gammliyyah* quarter in *Ha'I wasat* (Middle District).[25] These *shiyakhat* largely retain their nineteenth-century boundaries as they appeared in the population census of the 1880s.[26] The executive council is responsible for "*following up the activities of the institutions involved in the management of part of the city within the boundaries of the local unit and evaluating their performance*".[27] The quarter (*Shiyakhah*) is the lower-tier of government and includes a quarter secretary, financial auditor, and heads of technical services under the direct management of the head of the quarter.[28] This hierarchy, however, is for daily management purposes; important decisions about any locality are taken at a higher level. Development plans and the rehabilitation projects of Old Cairo are managed by governmental institutions such as the Ministry of Culture, Ministry of Housing and Governorate of Cairo in cooperation with international funding agencies such as the United Nations Development Programme (UNDP).[29]

Input of local people in this decision-making process, if it even exists, is very limited in the state-centred political model of contemporary Egypt: where the state defines and solves social problems.[30] The government looks at Old Cairo as one holistic historical site that can be a source for huge touristic revenues and whose ultimate development should be targeted towards this goal.[31] Within the complex context of historical fabric and extensive presence of monumental buildings, planning bodies faced fewer struggles in negotiating with the state. Until recently, there were no special requirements regarding building codes for areas of special nature such as Old Cairo, and the building process in this sensitive fabric followed the country's standard building regulations.[32] During the twentieth century, relatively high-rise apartment buildings appeared over small plots in the hawari, which produced a disordered and chaotic lack of coherence and integrity with the surrounding context.[33]

The first formal consideration of the peculiarity of the hawari, in Old Cairo, came in 2007 under the Prime Minister's decision no. 2,003 for the year 2007. Yet, no specific regulations, design criteria or specifications were developed other than controlling building heights in the historical zone. The decision rules that *"it is not permitted to give building permission in the adjacent zone or opposite to a monumental site, unless permission was given by a special committee formed by Cairo's Governor, including the Supreme Council of Antiquities, and following the general terms of this decision"*.[34] The decision, further, limits building heights in Fatimid Cairo zone to *"one and a half times street width, but not more than 14 meters high (from street level to the roof's parapet), unless it is adjacent to a historical building or monumental site, in which case the new building is limited to the height of the existing monument"*. Special consideration is given to building in the tight alleyways of *"Harah, Zuqaq, Darb, and Atfah"*: which follows the above conditions but with height limited to 11 metres.[35]

As far as Old Cairo is concerned, those regulations and rules are not expected to contribute to its urban fabric in the short term. This means that the current urban fabric is expected to change gradually during the next few decades with the replacement of the existing old structures. That would only apply if the law did not change and the power of law remained in control, which unfortunately, proved not to be the case in the troubling three years after the Egyptian revolution in 2011. The form and image of a Cairene harah could change within a time span of 50–100 years, based on the current rate of new building additions/replacement over the past century. According to the new regulations, new-build buildings in the harah will be no more than two storeys high (½ times the width of 3–5 metre-wide alleyways). It is believed that the purpose of this decision is to limit building heights rather than to regulate building processes. However, such decisions require in-depth analysis and discussion which is beyond the scope of this study.

MODERN HOMES AND THE INTERCHANGING PRIVATE SPHERE

The vibrancy of the public sphere comes from its complementary addition to the insufficient private spaces of the house that are no longer inclusive of any semi-public activities, chiefly guest receptions. The organisation of home today is more complex and fragmented than it was in its historical settings due to two main reasons; firstly, the increase in mixed-use properties and the spread of workshops and stores throughout the harah's ground floor. This deprived the house of its natural extension through spatial connectivity with the alleyway space. As a result, private spheres have become more limited by their physical boundaries, and the interior's bond to the exterior has been broken. The second reason is the reduction in the size of apartments and the emergence of smaller typologies (e.g. one-room apartments, bachelor accommodation). Today, three typologies of family structures are identified: first, the descendants of the extended families who continue to occupy their family properties, whom I will refer to as the "**Settled**" group (**S**). Usually, they are a group of nuclear families with cross-family blood relations living together in the inherited building. They enjoy relatively higher economic viability than the rest of the community, which is reflected in the quality of their houses internally and externally. This is seen as a modern adaptation of the extended family model, depending on the autonomy of every branch, while maintaining the basic requirements of the organisation.[36]

The second type is the nuclear family which occupies rented apartments as tenants. The group is referred to as the "**Tenants**" group (**T**). They mostly migrated/moved during the 1960s–70s and have continued to occupy the same unit until today: benefiting from the rigid rent rules which kept the rent fixed for an unlimited period [R1.2.08, R3.1.07]. The residents under this type are mostly educated and occupy stable jobs with fixed salaries; some work in governmental and public sectors or services and they are keen to register their children for basic education. This type resembles the majority of families currently inhabiting the harah.

The third group are those basically in a transitional situation and not permanent or settled residents. They are mainly bachelors, or small migrant families (mainly from Upper Egypt) who live in shared accommodation. Under this group, I shall include the recently arrived migrant families who occupy single rooms in a complex residential building (represented mainly in buildings no. 3 and no. 15 in al-Darb al-Asfar). This group is tagged in this study as "**Bachelor**" group (**B**). I combined the two groups as they both live in similar settings, and in most cases, they share services like a kitchen or bathroom (for example, the loft in the same building).

This typology spatially follows a particular order of shared identity, dividing the harah into three parts: east, west, and north. The eastern section, starting from the old gate to the edge of *Bay al-Suhaimy*, is the popular zone, with a diversity of family types, including high concentration of the **T** group along with the **B** group. This part adopts what Juan Campo[37] called "*rifi* (countryside) *subculture*" that is constructed on "*attachments to local customs, family honour and solidarity, and the land*". The western section included a higher concentration of the **S** group: with two large buildings occupied by the heirs of old families (houses nos. 1 and 2 in Zuqaq al-Darb al-Asfar). This part resembles the modern concept of the home featuring hard boundaries between private and public spaces, strict access and maintaining separation from the surrounding context.[38]

The northern section, on the other hand, has been dominated by migrants from the al-Fayoum region of Upper Egypt, who changed the atfa's name to *Haret al-Fahayma* (People of al-Fayoum) [R1.1.07; R2.1.07, R5.1.07]. They are distinct in their customs, lifestyle and spoken dialect. They used to live in poor conditions and gather in informal collective housing called *ishash* (tents). However, this practice was eliminated during the conservation project, and the remaining families have rented apartments. While the natives call themselves *Ahl al-Hara* (people of the hara), or *Awlad al-Hara* (hara's sons), the migrants are usually called according to their place of origin (old tradition). It usually takes the migrant family two generations to climb up the social ladder: especially from the status of a migrant to a native (and considered Ahl al-hara).[39] This duality of social structure within one quarter was not a property exclusive to Old Cairo. Farha Ghannam[40] reported that relatively new neighbourhoods constructed on the outskirts of Cairo (e.g. *El-Zawya al-Hamra*), have similar distinctions between native population and migrants from rural areas, where each group lives in a distinct part of the same locality. However, migrants found no open space to fill, so they filled deserted houses (*bayt al-khazarati*), and small rooms in complex buildings, as mentioned above.

Family sizes in the harah correspond to the national average, with no dominant family size or structure. However, there are relatively few large families, while medium-size ones are in the majority. In the survey of 1991, the majority of families were composed of 5–6 individuals (33.3 per cent), the second largest group was composed of 3–4 individuals (31.9 per cent), followed by 7–8 individuals (14.1 per cent).[41] At the time of this survey, there were 33 families of the migrant group living in bayt al-Khazarati only.[42] This group was relocated outside the area in 2000 during the conservation project. This majority of medium-sized families were guaranteed decent living conditions that suited the average income of LE 250 per month, with the same survey concluding that 67.1 per cent (74 families, out of 134) of the population had incomes of just below LE 300/month.[43] From the consensus of 1991, we could ascertain that the majority of families comprised between 3–6 persons (66.2 per cent, 88 families, out of 135), and the majority of units were a size of 40–80 sq.m (62.9 per cent: 83 families, out of 132). The similarity between the figures suggests that the average family of 4–5 persons lives in a unit with an average size of 75sq.m.[44]

As the aim is to understand the organisation of domestic space and how the space organisation relates to social practice, we have surveyed houses and interviewed their occupants. The preliminary findings of this survey conclude that while the apartment building is the dominant model, its size and internal division varies and is not essentially related to family size, but rather to the available plot size. The houses of **S** type are relatively large while the occupying family is not so large. The **T** group houses are basically smaller in size while the families are larger. The bachelor/migrant family group occupies large buildings which have huge diversity of internal layout and do not share particular features or typology. Here again, Al-Darb al-Asfar is used as a sample that exemplifies how a unique community in Old Cairo could be deeply divided, yet work in harmony for the collective good.

a) Settled Group Homes

Current **S** group houses are built on previous properties, usually owned by the same family. Those houses were basically built around the 1950s and 1960s or earlier to host different members of the same family. They were, originally, designed to accommodate the son's family in upper floor apartments. The master of the family used to live on the ground level with a large living space, in which the whole family gathered [R4.1.07, R1.2.08].

9.4   Enhanced and decorated interiors of the settled group buildings:
rich Arabic furniture collection in a well-decorated corner for guests

9.5   Settled group apartment building no. 2 Zuqaq al-Darb al-Asfar: a) typical floor plan and apartment unit; b) the building appears on the left side of the Zuqaq

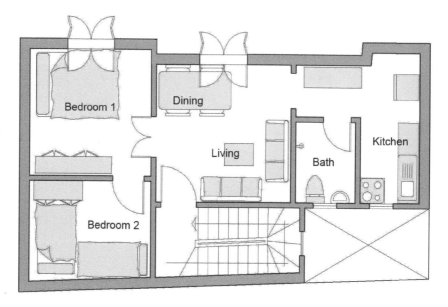

However, following the death of the founding father, the ground levels were turned into retail shops and stores for let.[45] Upper apartments became more autonomous units and were passed on from one generation to another.

While this type does not enjoy any distinct external visual advantage of style, they are relatively large and regularly maintained with, mostly, a locked entrance door. Examples of such buildings are Houses nos. 1 and 2 Zuqaq al-Darb al-Asfar, and no. 18 the main darb. The building is organised around a central entrance and staircase. Its ground level is used partly as a retail outlet and partly as a small residence. Upper floors typically contain two apartments per floor. The ground level is raised 10 steps to maintain the privacy of the ground floor residence and its windows follow the traditional practice. Each of the apartments usually consists of three closed rooms, one shared living space, a bathroom and a kitchen (Figure 9.5). The first floor and above are projected approx. 1 metre outside the first floor, adding more space to one room and giving room for a terrace (balcony).[46]

b) Tenant Group Homes

The majority of the hawari's apartments are occupied by tenants and vary in size, structure and spatial arrangements. Each apartment accommodates either an independent nuclear family or small extended families (such as married daughter/son living with their elderly parents).[47] Every building of this type consists of one or two shops/workshops and an entrance leading to the staircase at the ground level. The entrance and the staircase are situated at one side of the building, whose façade does not exceed, in most cases, 6–8 metres in width. Each level above ground includes one apartment for a nuclear family. The façade is generally divided into two modular units, each of which represents the space behind, with a front window or balcony. Upper floors are projected approx. 1 metre to the anterior of the ground level. Those buildings are made primarily from load bearing mud-burned bricks and plastered from outside.

9.6   Typical tenant group house buildings (no. 2, main darb)

Even though some of these buildings were deteriorating internally, their façades did not reflect such conditions[48] [I7.1.07].

Most of those buildings were split from previous larger plots which were subject to division or part-selling following inheritance by a multi-branch extended family. Typically, they are surrounded by neighbours on three sides, with a single tight façade. While there is no space for an inner courtyard, small shafts are used to ventilate the inner-oriented spaces. To permit fresh air through the house's open-topped stair shaft, all the doors and windows, including the entrance door, are left open, creating a modern form of the traditional malqaf.

c) Bachelor Housing

Bachelor housing is alien to the harah's heritage and social structure.[49] The migration of the last decades of the twentieth century culminated in the city being stretched beyond its capacity and some parts turning into urban-villages.[50] Three buildings of al-Darb al-Asfar were transformed during a few decades of the twentieth century to be the local sites for bachelor accommodation within the harah. Those are the hawsh in Bayt al-Kharazi, a poor hostel-type apartment building (no. 23, main darb); and scattered shared rooms of workers on top of aluminium workshops in atfat al-Darb al-Asfar (no. 15 main darb, and 3–7 Atfet). The first, however, was cleared of its residents by the end of the century and was turned into a monumental site.

Those buildings were designed and built, initially, to house families on a tenant basis. The average-sized unit consists of two bedrooms, a living space, a kitchen and a bathroom. Building no. 1 was built during the 1960s and had a long tradition of being a hall of residence for the nearby al-Azhar University: the vast majority of whose students came from the countryside and rural areas. In reality, most apartments are divided into independent rooms; each accommodates between 6–9 people in a military camp-like arrangement. No space is left for social activities or for gatherings and, when I was invited for a cup of tea, I had to sit on someone's bed. Occupants have to limit the time they spend at home to times for relaxing and sleeping. Such overcrowding and informal tenancy arrangements (bed-tenants are not formal) affect the quality of life and the building's safety. However, this did not concern the building's owner, who was interested only in getting the rent at the end of the month: "*the rent being paid according to this arrangement is greater than for any similar sized apartments in the harah*" [R1.2.08]. He declared that the units are rented to individuals and he is not involved in who lives inside the unit [R2.1.07]. Some of the units were customised to accommodate one nuclear family per room.

Bachelors living in this building reflect a broader spectrum of single people working in Cairo in terms of diverse qualifications, interests and age. I was invited for a cup of tea by a group of five people living in one room. The group included a professional painter, a university graduate, a tea maker (father of five and living there for 20 years), a young boy and an elderly man. Their space is not suitable for any type of activity, scarcely for a cup of tea, and sometimes a shared meal at the weekend. The second of the remaining types, in ateft al-Darb al-Asfar, are informal vertical extensions of scattered rooms on top of the metal workshops. However, they are not ordered, formally designed, or officially permitted.

DYNAMICS OF MODERNITY IN THE HARAH

What anyone would need in order to comprehend a particular social phenomenon is the specific social data relevant to the context under study. Such specificity should be able to capture the subtler features of differentiation within a particular community or neighbourhood.[51] Bearing in mind that communities are dynamic constructs, their pattern of development involves "social circulation", which involves the change of occupants of existing positions, and "social change", which refers to change of the positions to be occupied in the community.[52] This change is essential to the sustainability of the organisation as it provides the flexibility to absorb different pressures and changes in changing situations, such as change of economic activity, change of character, or the change of people resulting from the emergence of new generations. The harah of the twenty-first century is distinguished from its past by the higher proportion of educated individuals and their diverse occupations: which are no longer associated with a single profession or affiliation.

The public sphere today hosts more diverse activities than it did during the previous two centuries, while its connectivity to the outside world has been consolidated at a faster pace. Commercial and industrial activities are becoming dominant at the ground level in both indoor and outdoor spaces. Most buildings have retail stores and workshops in their frontages, while backyard spaces are reserved for storage and certain workshops (textile). Out of 50 buildings in al-Darb al-Asfar, 45 have commercial or industrial activities at ground level. The traditionally social spaces of a residential community have become active industrial and commercial venues, with trolleys touring the alley throughout most of the day. In essence, the private local spaces experience more frequent public activities during the daytime, while at night that space is rearranged to support private activities. Despite such diversity of actors and activities, al-Darb al-Asfar is still regarded as *"the community that relates local inhabitants to one another".*[53] The temporal ethos of managing the public space is becoming an essential part of the daily conscious behaviour of local residents who, given their limited indoor space, need to utilise it as an extension of their private arena.

The lack of intervention from the authorities in resolving local conflicts led to the establishment of informal local institutions that intervene at times of conflicts as historically inherited in the socially based Urf:[54] a popular but un-written convention that organises and informally regulates the principles of social interaction. It outlines the basis on which conflicts are to be treated, with senior members of the group responsible for implementing its orders. The "Urf here is that we resolve our local problems internally", a senior member asserts [R1.1.07; R3.1.07]. Similar to the Egyptian culture, reporting incidents within the harah (similar to the family) to the police is not a welcomed act and is considered disrespectful to the senior members of the harah [R1.1.07]. This system of local management, however, is limited to resolution of local tensions and disputes that do not involve legal or juridical consequences.

In alley communities, the continuous social circulation and change do not affect the stability of organisation that is, arguably, bound to the place more than to the individuals. The social surveys made by Mohamed El-Sioufi in 1981 marked a major change in the inherent structure of the community.[55] His survey denoted that the extended family typology has been vanishing since the 1950s/1960s, giving way to the dominance of a new typology in the form of nuclear family units.[56]

9.7 Commercial activities at the junction with share'i al-Muizz

This change was associated with significant economic change, which is represented clearly in the large-scale introduction of industrial workshops which was discussed earlier.[57] The gradual increase in commercial activities and strangers in ground level spaces did not change the urf and the order of seniority continued to be filled by new generations. This social circulation can be illustrated by the example of two members: Hajj Arabi and Hajj Muhammad Sa'ad, the eldest senior members in al-Darb al-Asfar today. They have retained this position in local social and political hierarchies since the 1980s–90s, having previously been ordinary members since the 1940s.

There are two major social surveys relating to al-Darb al-Asfar: The above mentioned El-Sioufi included physical, social and economic surveys, and that conducted by the General Organization for Physical Planning (GOPP) in 1991 focused on the built environment and social structure.[58] El-Sioufi's statistics of the 1980s recorded that 71 per cent of the surveyed sample (35 families, 501 individuals, in 42 buildings) had lived in the harah for less than 25 years, 22 per cent for 25–50 years, and 8 per cent for more than 50 years. 44 per cent of the residents were white-collar workers, reflecting the improved levels of education, and 68 per cent of the employed workforce worked outside the harah.[59] In the GOPP survey of 1991, it was found that two-thirds of the families were composed of 3–6 individuals, with a general average of 5 persons. In 60 per cent of the households, monthly income was less than 200 LE (USD 36).[60]

After the 1950s and 1960s, inhabitants became less involved in the functional/ business activities in the harah, which had to import outsiders in order to operate. Even though commercial and industrial activities were emerging at a greater pace, the majority of residents were increasingly becoming white-collar employees with office-based jobs.[61] The harah's public space started to grasp the essence of modernity exemplified in mobility and dual movements, or what we, arguably, call an exchange of population on a daily basis. In the early morning, residents leave their territory heading to work, while outsiders move in.[62] The public sphere belongs to the outsiders in the morning and to the inhabitants in the evening. Accordingly, the public sphere has to attend to the requirements of entirely different types of occupiers every day, whose requirements and social activities change significantly. The result is a hybrid culture of lower-class with noisy functional space during the day and middle-class with quieter social activities at night.

****

If we can classify the practice within the private spheres, two patterns could be recognised: the modern and the popular (*Sha'bi* – the common people). The modern pattern involves drawing strong boundaries between the private and the public, restricting social practice within the unit, having less involvement in the public sphere but relating mainly to the outside world. Families of the higher order used to pay more attention to their domestic space, decorating it and providing different internal venues for social activities. The size of the inner space is a positive asset to support this practice. Two interviewed women of this group had adorned their domestic space with decorative portraits, arabesque furniture. Houses in this part are similar to those in other modern quarters in the city.

People of the popular part, in contrast, make heavy use the harah's public sphere due to the tightness of enclosed domestic spaces. Such a culture of informality is celebrated as a social system with active interpersonal communication, ease of boundaries and familiarity with public spaces. In this part, most of the people find their basic needs met through social practice within limited distance of their homes. In contrast with the previous pattern (the modern), the popular practices involve blurred boundaries and ease of access from the private to the public and vice versa in an uninterrupted flow of movement. In several cases, community leaders accompanied me to visit houses without the presence of fellow house masters. They concluded *"we are one family and there is no fear once we (local males) are here, no strangers among us"* [R1.2.08; R3.1.07]. In the company of a senior member, families invited us to their house spaces and talked about their lives. They freely disclosed their private spheres and their arrangement of furniture. Within the home no boundaries exist, bedrooms host fridges, living spaces have movable tables, while dining chairs are hung over the wall to save space for other activities (Figure 9.8).

At the centre of this pattern is the notion of utility space which could be used for different purposes during different temporal sittings. The living space could be used for watching TV in the evenings, a dining space during meals, or its old sofas could be used as beds at night. This pattern is less celebrative and more functional, while receiving guests is not an indoor activity in this pattern and takes place in the public sphere. At home, the house is predominantly managed by women. Men come home to eat, sleep and for social gathering with the family. However, male gatherings with friends and neighbours are arranged in the harah's public spaces, either at the coffee shop or at a friend's shop doorstep. While men usually meet outdoors, ladies visit each other at home in the morning while the men are at work. Chatting over cooking and paying visits in the early morning are common practices.

SPATIAL REALITY AND EVERYDAY TRANSFORMATION

Max Weber, the leading social theorist, asserted that it is in the praxis (acts, courses of action and interaction) that we can trace the essence of a community, group or a society.[63] These praxes represent the very nature of real production and consumption of space in everyday life, which in the view of the local actors hold no significance of any sort.

9.8   Different strategies for space management within tight private spheres

However, they reveal the power, strength and social division/cohesion of a group of people in specific spatial settings. Michel De Certeau, on the other hand, drew a distinction between the settings and people's responses. In *The Practice of Everyday Life*, he tended to read people's everyday activity and movement and compare them with the actual spatial settings made by institutions and structures of power (governments/ corporations), or perhaps between *tactics* of individuals and *strategy* of power.[64] While this distinction was clear in the concept city (a city planned and designed by strategic decisions of institutions) as opposed to the way walkers map their trajectories of movement based on immediate and unconscious response to the quality of real space, the situation is more complex in the hawari.

The hawari is an inherited spatial setting and local public sphere, with deeply-rooted rituals, traditions and social rules. Hence, tracing people's actions and reactions would help to situate everyday tactics and determine how they have changed from previous periods or how they were affected by emerging forces of modernity. By investigating simple everyday spatial practices (as people's movement, activities and behaviour and the way they correspond to space and place) and the way the space is organised to accommodate them, we can trace the way the public space is utilised to suit basic social characteristics and the essence of the community in the contemporary situation.

Eating meals, drinking coffee and smoking *sheisha* (smoking pipe) are typical activities performed on a daily basis in the harah. Most of the interviewees (men) take their meals, mainly breakfast and sometimes dinner, in front of their shops, workshops or houses. A movable dining/drinking table set, previously stored away, is arranged without interruption of public movements. The outdoor space emerges as a sufficient alternative to the missing indoor social spaces, which were formerly used to host such activities (sharing meals, inviting visitors). Therefore, the intrinsic qualities of the private atmosphere are present in the open outdoor space. All interviews with male residents were made in the hara's public space. On several occasions, people volunteered to participate in both discussions and interviews without invitation. They believed in their right of intervention, once it was taking place in their shared public space.

To accommodate commercial activities, the harah required open accessibility and smooth movements from outside-in (workers/customer/raw materials) and from inside-out (products/transport). In this respect, the saha played an important role throughout the twentieth century, creating a new spatial experience with the most recent built-in terrazzo seats and surrounding restaurants. It has become a major gathering place and forms a decorative approach to the monumental houses in the harah, although mainly attracting outsiders. This large space is basically decorative and serves the monumental image of the historical buildings and associated festivities. It has been abandoned by local residents who feel exposed to outsiders and passers-by, while they appear regularly and actively in the other open (but tight) space on the eastern side.[65] It belongs to the major public and touristic thoroughfare of al-Mu'izz more than it belongs to its locality. This confusion of identity caused an implicit rejection by the locality whose boundaries are lost and whose borders have disappeared. As an indeterminate reaction, the boundary is pulled back: creating an internal boundary at the edge of the historical buildings and giving them up to another public sphere. The new spatial reality immediately affected the social sphere, causing the local public sphere to shrink, and allowing the ultimate public sphere of the city to take new ground.

In contrast with this imposition, the natural growth of the built-scape was getting tighter and going higher. The houses are attached to each other and are as high as four storeys on average. The main darb and Zuqaq al-darb al-Asfar are congested, with high apartment buildings. Atfet al-Darb al-Asfar, however, enjoyed more open spaces with less density of housing: benefiting from being an industrially dominated zone and the presence of Bayt al-Suhaimy as a preserved site on its western edge. The harah's main path followed its traditional roots, by preserving its distinctive visual characteristics: exemplified by the shady effect (resulting from the height of buildings in proportion to alleyway width) and the restriction of through visibility.[66] The increase in height of buildings has, however, increased the sense of closure and, accordingly, intimacy, creating more layers of social interaction.

Traditional practice, however, retained its viability and existence through different tools, one of which is to re-invent itself in new forms. For example, the disappearance of the water carrier (*sakka*) and the closure of the water store (*sabeel*) did not mean the passers-by did not need to drink water. Local shop workers, in response, reinvented the sabeel's place in the form of a mobile rack or display of traditional water bottles, placed on the sidewalk of the hawari. Figure 9.8 shows the positioning of these water bottles opposite the former sabeel Qitas beside haret al-Darb al-Asfar's gate.

a) The Daytime Workplace: Outsiders in the Harah

The local public sphere in the morning is overwhelmed by industrial activities and their workers. Every vacant corner or relatively large unused space is used as a store yard by workshops, especially aluminium workshops. The impact is to reduce and limit the effective area for social activities in the morning. This situation demonstrates the power politics during the day, when workshop owners have the upper hand. This dominance could be comprehended through the noise and pollution of machinery, communication in front of the shops and the flow of product storage and transportation. Parallel to this group, some retired residents do spend their daytime with friends and neighbours in the main darb. The host is usually the grocer or the shop owner and they usually sit in a shaded corner to chat, drink tea or smoke the sheisha. In all situations, the nearest ahwa delivers hot drinks to the whole harah, wherever requested, approximately every hour.

Paving and transforming the harah into a path for pedestrians reduced the pedlars and vendors who used to tour the hawari to delivering their products to the women at home. Children took on the delivery role, purchasing the basic domestic needs from the local grocer or shops adjacent to the hara's entrances. The dominance of commercial activities, however, is broken during holidays and weekends. In particular, arrangement is made for special occasions such as marriage or funerals, for which the community is mobilised to prepare the public sphere.

b) Night-times for the Locals: Spaces for Socialising

The dominance of workers in the morning is compromised and challenged at night, when most of the local residents turn up and take control of the public space. Even though working hours for most workshops usually extend into the evening, their occupation of the space is no longer exclusive. It becomes the venue for interpersonal communications and negotiation. Male residents tend to spend much time in the public space at night.

9.9   Commercial and industrial activities and gatherings during the day 2009: friends and neighbours gather in front of their shops

Frequently, during my interviews, I heard passers-by telling my interviewee "*I will come to you after salat al-'Isha*" (the 'Isha prayer, which marked night-time). Again the prayer times are used as basic temporal transitions, according to which events and meetings are organised. Young men, teenagers and children make the best use of outdoor space at night, inviting their visitors and friends to join them. The nature of these night activities is best experienced in Ramadhan, after al-Maghreb (sunset), when workshops have ceased to operate and the harah is fully lit and decorated.

Festivities, such as *E'ids* (feasts), marriages, funerals and *Sobou'I* (celebration of the seventh day after a baby's birth), are typically celebrated outdoors. Outdoor spaces are arranged accordingly to accommodate each one. Funerals and *Sobou'I*, in particular, require welding of the private space to the public one.[67] Doors are left open, similar arrays of chairs are arranged in the house and in the harah in front of the house. Microphones are employed to radiate joy (songs for weddings) or sadness (Quran for funerals). Such use of outdoor space for these events is associated with the condition of the hara as a poor area, whereas in upper-middle class segments, these events take place in specially designated but costly indoor spaces (hotels, community centres or social clubs).[68]

c) The Farah in the Public Sphere

The *farah* (wedding) is one of the most prominent activities, during which social interaction is manifested at its best. This is not limited to the wedding party itself as its rituals and occupation of the public space continue for days. The preparation of the bride's furniture, storage of materials/products takes place in the public space for a few days before the wedding. The whole alleyway becomes the central location of the continuous and consecutive ceremonies. The tight indoor space is insufficient to accommodate all guests, and does not work as a venue for the relevant rituals. Outdoor space works, therefore, as a spine of celebration that links all buildings together and merges outdoor with indoor spaces. It flips the external building façades into interior decorative panels of the stage; colourfully lit and beautifully illuminating. Live events span the time from day to evening, and the space from inside-out, with women dominating the public as well the private space. A larger part of the open space is used for furniture arrangement.

During my field work in haret al-Darb al-Asfar, I witnessed the local community's preparation of furniture for a bride; it was obvious that such functional activities were being undertaken as ceremonial rituals. Ladies were singing and young girls were dancing in the harah throughout the day while *al-Menaggdeen* (mattress workers) were preparing the mattresses. The absence of any families from such rituals and occasions is considered disrespectable to the celebrating family and may provoke tensions [R4.1.07]. Participation is a duty, especially for women and *"duties had no social distinctions"*, an informant says. *"Everyone should be present at any personal occasion"*, if someone does not show up, then *"it is a big problem that should be solved"* [R.1.1.08; R.3.1.07].

d) The Museum in the Harah

In contrast with the lively venues of the eastern part mentioned above, the western part of the harah is relatively quiet and lacks dynamic and lively local social activities while, in reality, it acts as an extension of the historical museum of Islamic Cairo. It resembles a formal modern city quarter which differs from the popular eastern part of the harah, socially, culturally and spatially.

It is strongly argued that the imposition of touristic activities and restrictions of occupancy over historical context created invisible and psychological barriers between the popular and the formal. Local people don't feel that the space belongs to them. It is not surrounded by their homes and is always filled by strangers (tourists). This has, indirectly, affected the social practice of the people living in this part of the harah, whose social engagement does not take place within the harah. While the monumental complex of Bayt al-Suhaimy hosts musical and cultural activities and celebrations, with gates open to local residents, most audiences do not experience the harah or its environment. Rather, their presence, sometimes, becomes a problem and separates the building from its context. One of my informants, whose shop is in the western part, told me that extensive formal parties take place in Bayt al-Suhaimy and *"associated security measures affected local businesses"*. The police, during some parties, such as the opening ceremony attended by the First Lady in 2000, closed the harah for a week, restricting residents' movements and creating psychological barriers between them and the monument. It turned *"the presence of these valued monuments into a disadvantage for the locals"* [R4.2.07, R5.2.08].

* * * *

The change in Old Cairo's social structure and the predominance of population of the lower order in the second half of the twentieth century have altered the priorities of local spatial configurations. Emerging housing types, such as those for tenant/bachelor groups, were a response to the needs of the new and dominant social order along with the decline of the extended family structures. The traditional arrangement of the extended family multi-storey houses became inappropriate for the new tenant groups who sought independent housing units with a certain level of privacy. Old buildings had to be reorganised spatially to develop these independent units within the previously interdependent arrangement and spatial order. This resulted in some awkward arrangements, with some services implanted in strange locations (a bathroom outside an apartment, or a tiny pocket kitchen partitioned off from a bathroom). Such arrangements could have breached conventional codes of architectural theory and practice; nevertheless, they provided adequate and urgently needed responses to individual family needs. When modernity forces its way into such a historical context, blurring of the well-established order and socio-spatial organisation is to be expected, especially in the absence of clear or well developed solutions. Obviously, the harah could not be flattened to receive the new order. New build apartment buildings, on the other hand, offered sensible adjustments to old models and provided enhanced solutions even in small plots. Houses built since the 1950s have designated spaces for services (kitchen and bathrooms) in each individual apartment and comply with basic architectural conventions, while seeming to be inspired by the spatial enclosure of traditional harem quarters.

We could argue, hence, that modernity forced adjustments to traditional spatial settings by imposing internal boundaries on what were originally interdependent arrangements. However, the response of the local builder-contractors was initially unsatisfactory in its makeshift solutions and blurred boundaries. It took them a few decades to align with new requirements, and the production of the second half of the century proved responsive and satisfactory within existing constraints (finance, resources, and available plot size).

NOTES

1   Bauman, *Liquid Modernity*, p. 2.

2   See Diane Singerman, *Avenues of Participation*; Farha Ghannam, *Remaking the Modern*.

3   Ghannam, *Remaking the Modern*, p. 18.

4   Ibid., pp. 17–22.

5   Ghannam, ibid., p. 5.

6   Giddens, *The Constitution of Society*, p. 21.

7   Abada, *Heterogeneity within Homogeneity*, p. 3.

8   Ghannam, *Remaking the Modern*, pp. 19–20.

9   Cole, *People, the State, and the Global in Cairo*, p. 794.

10   Ghannam, ibid., p. 21.

11   Hastrup and Olwig, *Siting Culture*, p. 2.

12   Giddens, *Modernity and Self Identity*, pp. 2 and 33–5.

13   Alsayyad and Roy, *Medieval Modernity*, p. 1.

14   Alsayyad and Roy, *Medieval Modernity*, p. 5. They considered similarities between the
      medieval and modern cities as both are imagined to be central nodes of international trade,
      and consequently its fragmented patterns, p. 2.

15   After the deterioration of traditional markets, crafts and guild organisations, the
      contemporary harah's residents had little in common as individuals except for their shared
      identity of being residents in a specific harah.

16   Smith, *The Dynamics of Urbanism*, p. 73.

17   Al-Sioufi, *A Fatimid Harah*. The statistics clearly identified the shift in local activities in
      al-Darb al-Asfar to mixed use during the 1950s and 1960s.

18   Ghannam, ibid.

19   Elkholy, *Defiance and Compliance*.

20   Abaza, *Changing Consumer Cultures in Egypt*.

21   *Ukasha, Layali El-Hilmiyyah*, the most viewed Egyptian TV series during the 1990s.

22   Ismail, *Political Life in Cairo's New Urban Quarters*, p. 13.

23   Ismail, ibid., p. 14.

24   Refers to the genesis of the veil as analysed by Kenzari and Sheshtawi, *The Ambiguous Veil*,
      pp. 17–18.

25   The governorate structure changed during 2008 following the introduction of two new
      governorates, Helwan and 6th October, which have been formed in parts of Cairo. The
      above information is the latest available as of February 2009 from Cairo governorate official
      website.

26   1880s census.

27   Habitat (1993) *Metropolitan Planning and Management in the Developing World: Spatial
      Decentralization Policy in Bombay and Cairo*, p. 116.

28   Ibid., p. 118.

29   UNDP and SCA, *Rehabilitation of Historic Cairo*. Official report.

30   Ahmed Sedky (ibid.) quoting Grindle and Thomas, *Policy Makers, Policy Sciences*, p. 22.

31   This view was implicit in both the UNDP report and the state's tactics of implementing
      rehabilitation strategies which did not consider case-by-case study or focus on individual
      groups' problems and needs.

32   Prime Minister Decision no. 2,003 for the year 2007 was the first to consider particular
      requirements for Old Cairo, even though it remained very general in its terms and
      conditions and lacked technical precision.

33   Abada, *Heterogeneity within Homogeneity*, p. 8.

34   Prime Minister Decision no. 2,003 for the year 2007, pp. 203–4.

35 Explanatory Report issued by Cairo's Governor dated 3rd September 2007, issued as an amendment to the Prime Minister decision no. 2,003 for the year 2007, p. 206. This decision is based on a report submitted by the Cairo governorate, which recommends a set of specific regulations and standards for Old Cairo should be put together by "a committee of scientists, experts, and architects to give design guidelines for the construction materials and architectural style, including façades, projections, doors, shading devices".

36 For the forms and needs of the extended family system in Arab societies see Nydell, *Understanding Arabs: A Guide for Modern Times*, pp. 72–4.

37 Campo, *The Other Sides of Paradise*, p. 90.

38 One senior resident of this group (in house 1 Zuqaq al-Darb al-Asfar) identified this particular feature as important to his privacy and security while living in this area. This, however, did not affect his well-being as part of the community. But his particular practice of privacy was respected by others. Due to such restrictions, I was not allowed to take photos in this building, but allowed to draw sketches.

39 Resident (R1) is the son of an immigrant family. His father migrated from Upper Egypt and resided in a room in Wekalet Bazara'a (a multi-family residential complex building) in the neighbouring hara of al-Tambukshiyya in the first half of the twentieth century. By 1954, his financial status developed and he was able to purchase a small plot of land in al-Darb al-Asfar to build his own multi-storey family house [R1.1.07].

40 Ghannam, *Remaking the Modern*, pp. 5–6.

41 GOPP, *Al-Darb al-Asfar Rehabilitation Project*, p. 53.

42 Nadim, *The Documentation and Rehabilitation of al-Darb al-Asfar*.

43 GOPP, ibid., p. 53.

44 Ibid.

45 In the Al-Siuofi study, it was apparent that several shops and workshops appeared for the first time during the 1950s and 1960s, which confirms our stand that those shops replaced ground floor living spaces. See Al-Siuofi, *A Fatimid Harah*, ibid.

46 The use of projected areas was in accordance with general building regulations for houses in Egypt during most of the twentieth century, which stated that higher level projections should be used as at least 50 per cent open terrace and the rest (50 per cent) closed space.

47 In most interviews with tenants, there was an older family master who was the principal tenant and a married son or daughter living with him/her. This, for example, was the case in buildings nos 6 and 10, in al-Darb al-Asfar alley, no. 13 Atfet al-Darb al-Asfar.

48 This is due to renovation of external plaster covering all the harah's façades during the conservation project which was completed by 2000.

49 Even though in its recent history the harah was used to accommodate bachelors, they were mainly students studying at al-Azhar University, which was in line with the harah's character of being the traditional destination of religious scholars.

50 Hoodfar, *Between Marriage and the Market*, p. 23. During the 1960s, the population of Cairo was growing at an unprecedented rate that saw its population double between 1947 and 1967. In 1976, the city was growing by 200,000 residents per year; most of these immigrants were fellaheen (villagers) who came to join the industrialisation revolution; see, Beattie, *Cairo: A Cultural History*, pp. 211–12. See also, Andre Raymond, *Cairo*.

51 El-Kholy, ibid., p. 45.

52   Wilson, *An Analysis of Social Change based on Observations in Central Africa*, pp. 58–9.

53   Al-Siuofi, ibid., p. 8.

54   It means the popular but unwritten conventions among members of a particular community. The 'urf change from one district to another. They are not likely to differ within the hawari of the same quarter.

55   El-Sioufi, ibid.

56   Ibid.

57   Ibid.

58   GOPP, *Documentation of al-Darb al-Asfar*. Three volumes.

59   El-Sioufi, ibid., pp. 26–7.

60   GOPP, ibid., p. 37.

61   El-Sioufi, ibid.

62   During my field work, I had to wait until afternoon to meet some local male residents.

63   Eldridge, *Max Weber: The Interpretation of Social Reality*, p. 26.

64   De Certeau defines "strategy" in his terms as: "the calculus of force-relationships with becomes possible when a subject of will and power (a proprietor, an enterprise, a city, a scientific institution) can be isolated from an environment" and plan a proper place. "Tactic", on the other hand, is: "a calculus which cannot count on a proper (a spatial or institutional localization)". I.e., it creates its own and different way of practicing the planned place. For detailed definitions and analysis see: De Certeau, *The Practice of Everyday Life*, p. xix.

65   During our field work, 2007–08, no local social activities were detected in this space. Only two local persons used to sit in this space occasionally.

66   Nasser Rabbat Developing Kevin Lynch's principles on formation of a distinct visual perception. See Rabat, *The Culture of Building*, p. 200.

67   It is worthy of note that transforming the main darb into a pedestrian path helped the local residents to expand their social practices into that space without the interruption of continuous vehicle movements.

68   Elsheshtawy, *Urban Transformation*, p. 303.

10

Narratives of Spatial Transformation in Cairo

SOCIAL PRACTICE BETWEEN SPATIALITY AND TEMPORALITY

Only the end of an age makes it possible to say what made it live, as if it had to die in order to become a book.[1]

Central to studying historical narratives of architecture is to understand the way human needs were changing, and how architecture adressed this change and through what means. Nancy Stieber denotes, above, how deeply essential and central the investigation of cultural meaning and content have become to the processes of architectural production. A researcher has to study the narratives of a situation that is still in action, but is rooted in the past, while maintaining the essence of continuity and change. In other words, it is a situation in which the past lives on in the present and the present stems from its historical structures and organisation.

Over the studied period, the coherent social dynamics of Cairo had been giving way to a new system of social partitioning associated with spatial categorisation of diverse income groups. Housing stock of a medieval city with large courtyards was not adequate for the lifestyle and living needs of the evolving low-income groups who needed small affordable housing across the city. Production of homes, as a result, had gone through a gradual transition from its rich medievality and affordable modernity. Architecture had to respond to the growing need for smaller spaces,[2] while the idea of home had developed socially, culturally, and spatially in response to change in social structure. As a consequence, local communities had to cast aside some of their exclusive nature. For example, bachelor accommodation has become allowed and the secure dead-end of some alleyways was ordered to be demolished to become open and accessible for tourists and trade.

The determination of everyday spatial practices , hence, focused on the contextual settings within which home is produced and reproduced. The form of a living room can be understood only if we know the way family members use it in terms of their number, gender, order, and the distribution of sitting areas and windows. We need to search, as Nancy Stieber mentioned: *"the contingent, the temporary, and the dynamic, on processes rather than structure, on hybridity rather than consistency, on the quotidian as well as the extraordinary, on the periphery as well as the center, on reception as well as production".[3]* To classify architecture as a cultural production means it should be viewed as symbolic practice, a product of concrete expressions *"that can be mined to reveal their codes, making explicit both their agency and their contingency".[4]* We, therefore, need to understand the processes of living activities rather than the rigid physical structure.

If the community is organised around social life and behaviour (social contract), and worked as a normative tool by which to legitimise relationships and local political processes,[5] then everyday architecture resembles the development of activities within certain spatial configurations that keep changing on daily basis. Even though architecture is globally recognised as a field of physical production, the hawari of Old Cairo provided empirical evidence on the flexibility of such physical space: which has adapted as a spatial stage to the socio-cultural situation and the immediate needs of the occupants. It is entirely impractical for a single space in the harah to be restricted to one particular activity. Rather, space works as an active theatre that can be reorganised and ordered to host diverse shows or display a new social drama on a daily basis. Social spheres, thus, are more appropriate and comprehensible notions for consideration and analysis than spaces in such complex settings. The spatial order of the local public sphere, for example, is an active workplace in the morning and open coffee house in the evening. Several family houses had changed to mixed occupancy (commercial at ground level, and residential at upper levels), while another extended-family house of the nineteenth century had become the home of 33 individual families with their food businesses and pedlars' carts by the 1990s. What had once been a socially active *darb* became a storage deck for metal workshops and a site for business transactions. The temporal settings proved to be influential in determining certain activities either in private or public spheres.

In Cairo, the image has little to reveal about actual practices of daily activities and occupancy of their homes. For residents, the practice of home is structured around individual and mutual and collective social interaction, restrained habits and behaviours, historically rooted traditions and morals. By social change, I mean the changing patterns of daily activities, interaction and needs, as by-products of the change in local social structures and organisation during the studied period. I attempt to trace the development of spatial practices, social interaction and cultural processes by which the idea of home is shaped and manifested, chiefly by its occupants. I follow the flow of activities, temporal and spatial settings of the social spheres, while tracing the modes of change, the meaning of home as illustrated in the socio-spatial organisation at different periods.

THE AGENCY OF TIME AND CHRONOLOGY OF SOCIO-CULTURAL CHANGE

Social difference and diversity are essential substances of cities.[6] The distribution of such diversity distinguishes one particular urban structure from another. The medieval hawari resembled the principal social units upon which, in 1800, Cairo was structured. Those hawari have become, however, marginal quarters in the contemporary cosmopolitan city. Such change is a by-product of the redistribution of social groups within the evolving structure of the modern city. Growth in the hawari's population was basically a result of internal migration waves from the countryside to urban communities.[7]

Expanding suburbs of the early twentieth century, such as *al-Abbasiyyah*, *Ismailiyya* had become the territory of the priviliged people; the wealthy, educated and civilised people, while the hawari became the backyard domain of the traditional, country and working class families.[8]

Such a phenomenon resulted in a polarised society, in which social difference became spatially traceable and cultural diversity was physically situated. As a result, the absence of a particular group and the emergence of another within a particular community called for cultural practice to be altered towards the interests of the common group and their cultural background. While change is a basic dynamic process of urban societies, such change in the structure of the hawari was significant to the extent that it downgraded such areas from the pinnacle of the city to one in decline, un-serviced and marginalised zone within a very short period of time (1880–1930). Between 1882 and 1917, the population in the al-Gammaliyyah quarter more than doubled from 30,084 to 62,329 people: without a parallel increase in the physical structure or the occupied spaces.[9] The quarter became congested and the one-bedroom apartment was the predominant residential unit, accommodating 43.8 per cent of the population (5,876 families).[10]

Two notional changes have been detected. The first is the direct influence of governmental institutions and authorities, who were motivated by the rulers' westernised ideology and the dream to create a European city on the Nile.[11] This focused on moving rich merchants' homes to new quarters while keeping their trades and businesses in the old city. The intervention in urban development was empowered by series of legislation and restricted any attraction or potential of normal development in the old city.[12] The centrality of decision making was partnered with the state's desire to recruit a selected group to inhabit new quarters, denying the hawari their local resources. Timothy Mitchell included in such intervention the facilitation of migration of waves of workers required for large civil projects; a policy that required a demographic disposition and redistribution of population within the city.[13] This suggests that the internal migration waves were, as claimed by Joel Beinin, planned by the state and not just natural movements of people.[14]

The second change was the indirect influence on the existing local economy in the absence of trade merchants and its viability in servicing the living needs of resident families. The change of social structure, with waves of migrants of lower order, resulted in heterogeneous communities with emerging self-contained enclaves within the hawari with home based businesses. Indigenous residents had to turn their ground floors into income generating facilities either by establishing businesses or groceries or letting them as stores, shops or workshops in an attempt to benefit from the new open accessibility and traffic flow. Both had changed local culture and influenced a change of everyday life patterns: from exclusive residential area into a commercial and industrial route. Al-Darb al-Asfar, therefore, provides us with the evidence that a change of the social stratification results in an immediate change in the local economy and popular culture in Old Cairo.

Contemporary administration has become more centralised in the government's hands. As one building owner mentioned, "*we have no control over our properties. The government and the law undermine our rights. The government multiplied the bills several times while they haven't allowed us to increase the rent since the 1960s*" [R1.3.09].[15] Local residents have had no choice nor have they been consulted about their own community and building developments.[16] Decisions on building codes, regulations, and several projects for development of historic Cairo, involving significant change in local economies and land use and even relocation of industry and inhabitants, were made in the absence of residents' views.[17]

One of the significant features of development is the changing types of accommodation. Integrated structure of communities in 1800 provided a wide range of accommodation types which varied in size, but retained a similarity of organisation and minimum space requirements. Families who could not afford a house resided in the *Rab'*. Consistency of daily practices granted the harah an established order within a mix of diverse income groups who shared the same pattern of activities and extended family relationships. Houses used to have similar components, even though the sizes were entirely different. While al-mandharah, harem, kitchen, entrance and the courtyard were essential elements of large houses, small houses used to merge the harem and al-mandharah together in temporal order, while the courtyard was merged with the entrance to be a roofed hawsh that was used to host the livestock. Accommodation started to change significantly around the turn of the century. The occupancy survey of al-Gammaliyyah in 1917 informs us that 43.8 per cent of families were living in one bedroom, whilst 5,331 of these families were of limited size (1–5 members) and 547 were large families (6–16 members or above).[18] 27.9 per cent of the families lived in two-bedroom apartments, with 2,676 families being of limited size, whilst 1,058 were large families.

Changing types of accommodation could be revealed through comparing those censuses with the sample of al-Darb al-Asfar survey of 1981, in which the majority of families were of nuclear nature (approx. 68 per cent) with an average household size of 5.2 people.[19] The majority of families moved in around the 1950s (68 per cent) and 44 per cent were white collar workers. Average house size in 1991 was around 75 sq.m and average family size remained at around five people.[20] Therefore, while the accommodation was predominantly congested, with one/two bedrooms, around the turn of the century: when migration waves were extensive and every vacant space was used for living in, by the middle of the century it had become organised and socially stable. During the second half of the twentieth century, the migration waves minimised and educated families began to settle in the harah.[21] The housing situation has not changed significantly since the above surveys, while commercial accommodation has continued to develop year-on-year.

Commercial accommodation, on the other hand, did not appear in the estate trading of 1800, which suggests it was either minimal or non-existent. The community today has become predominantly commercial at street level; almost every building (43 buildings out of 53) has at least one shop or workshop at its ground level. Products are partnered with humans in using and residing in the shared public space. The central alleyway is no longer an exclusively residential territory of a home, rather it is a space of conflict between humans and products, in which people and products are powers in negotiation to gain more dominion and superiority over the other. The powerful commercial value of the products is beginning to gain ground on account of human activities. The territory is in dispute between the human domain and industrial domains. Struggle is becoming an everyday practice, where power politics involve direct confrontation between the residents and the stranger shopkeepers.

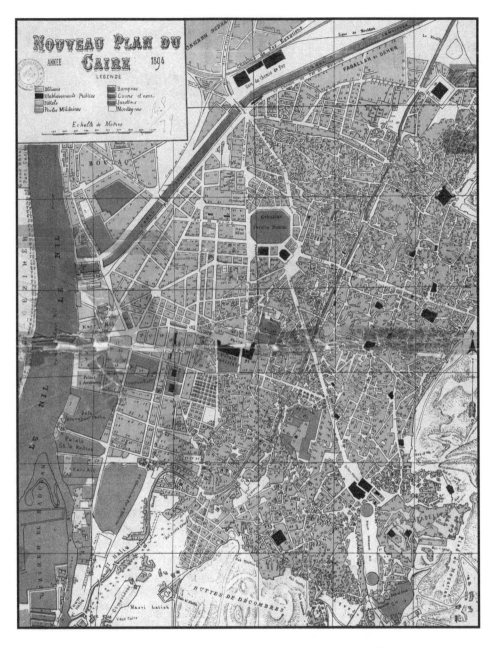

10.1   Map of Cairo 1894, showing the hawari areas against the
formally planned suburbs to the west of the old city
Courtesy of the Egyptian Geographic Society.

10.2 Local technology and business in the public sphere

Note: Outdoor computer games and drinks fridge; a local invention in the absence of municipal control.

The demographic change of the late nineteenth century influenced dominant culture and social levels within the old city. Other factors, such as the emergence of educated professionals and women as active members in the public sphere, required an alteration of the traditional system of management. The hawari had started to accommodate engineers, architects, accountants and teachers by the mid nineteenth century.[22] The local socio-cultural scene had shifted towards a liberal and secular society: with a demand for a coherent leadership system involving a greater number of educated members. Such demands were supported by the later exposure of women to the outside world, and the growth of bureaucratic institutions. Cairene women started to be involved more in the public sphere by the end of the century: during the national reform period in particular, claiming more rights and freedom for women.[23] Social barriers eased and cultural norms developed among the new groups of educated residents. The authority of *shaykh al-harah* came under the scrutiny of white collar professionals. By the end of the century, education became an authority in itself, to which many people referred in search for advice and help.

The division of the shared alley space between higher profile members' houses on one side and the migrant majority on the other resulted in two diverse practices of everyday life. The eastern end, however, was marked by people of the lower order with busy street life. While the eastern end retains similar settings and activities, the western end, because of its exposure to the outside world (*shar'i al-Muizz*), is considered part of the outside world and has lost its intimate character as part of the home.

The influence of this division in the character and identity is significant on the behaviour of residents on both sides. Native residents, in particular, distinguish themselves from the migrant groups, rarely participate in collective activities and believe they outclass them, culturally and socially [R25.1.09]. While collective meals are a usual public practice in the eastern side, in which several neighbours and friends gather, native families no longer eat in the street at all. Both sides of the same community draw their home differently. Migrant families seemed more flexible and inclusive of public spheres in their homes and extending their authority to such shared domains. To the contrary, the homes of native families were extremely limited by the physical boundaries of their ownership rights, with very limited expansion of activities to the public sphere (exchanging greetings, participating in collective occasions). Hence, the hypothesis that a collective home is a definite area for everyone living in the harah is quite deceiving and impractical. The idea of home is a unique vision and practice, which is transformed into a spatial configuration by its own actors. It is quite subjective and depends on everyone's practice within the socio-spatial environment as far as it extends.[24]

Collective meaning of home in Old Cairo, however, appears in the common code to be recognised by all inhabitants as their inclusive territory of home, regardless of individual interpretation of this territory. This code is an essential historical construct, developed through experiences and negotiating boundaries of the home. Relaxed communication between men and women, dress code and mutual cooperation in hard times are, for example, basic principles of this code as addressed by all interviewees. Non-compliance with this code invited hard responses and could result in collective exclusion of that member and his family. The common code of such a home, called by some scholars a social contract, resembled the conventions on which the home is formed and practised on a daily basis. That code is, in Bourdieu's terms, a *habitus* for its own people who utilise their historical production of the community that "*produces individual and collective practice, ensures the active presence of past experiences deposited in the form of schemes of perception, thought and action*".[25] The hawari, therefore, emerge as environments (spatially and socially) that synchronise past experiences, giving them disproportionate weight, in order to reproduce those experiences as a *practical hypothesis* that controls their responses to the immediate world of interaction.[26] Every resident is bound by that code within the territory of home and is not outside it.

In that sense, home becomes a defined territory, a spatial setting, in which the code is superior, even though its application is extended beyond that territory. Being spatially configured, the home could face a change in its practice in accordance with any spatial change of its structure. The hawari of Cairo are significant examples in such situations. Opening the western end of the harah had generated an immediate response representing the shrinkage of home inwards, excluding the significant and more open space of the harah. Dina Shehayyeb described another case in al-Darb al-Ahmar, where she witnessed a house on a dead-end street which had collapsed leaving a large open junction with a wider central street. In a few months, the vacant plot became a street junction, attracting many shops to open up there.[27] Soon, it became the main entry point to this lane and intimate practices withdrew inwards. Opening a coffee house at a harah's entrance gives rise to another example of the surrounding area becoming excluded from the shared public sphere. A coffee house is a public venue and in practice conflicts with the intimate nature and practice of home. It is not acceptable to let women move informally in front of a row of staring strange men. Imposing open accessibility or male-oriented public venues such as coffee houses in any locality invites immediate social response from the residents which could be translated spatially in the reorganisation of the territory and associated (collective or individual) practice.

The harah complies with Bourdieu's notion of the habitus' universe, even though it is neither an isolated nor self-contained world. The harah produces practices, which in turn reproduce regularities immanent to the objective conditions of the production of everyday life's generative process.[28] While some scholars suggested the harah was a self-contained self-sufficient community,[29] the hawari are actually interdependent and interwoven universes that work simultaneously within specific socio-political contexts and are constrained and informed by broader circumstances beyond the control of individuals or local communities.[30] They are influenced by national and regional cultures that exert a powerful influence on shaping local understandings of social reality, such as gender relations, education and order of seniority. People had to cross the harah's threshold on a daily basis to attend to their needs. Common people, for example, had to walk to the market to buy their own food ready-cooked on a daily basis as they could not afford the fuel costs.[31] Others had to cross the harah's gate in their daily trips to their workplaces.

Developing practices, therefore, become regularities in themselves, which define the transitional status from one social sphere to another. The demonstration of this transition appears in the way women and men are dressed (as for the above daily trips), their freedom of movement and interaction, and their representation in the public sphere. Dina Shehayyeb[32] observed that: *"within Old Cairo, the boundary of home (especially in al-Darb al-Ahmar) is determined by the way women, in particular, are dressed. They move freely with their home-style informal clothes, within the area they consider a home. Crossing this envisaged boundary requires the dress code and behavior to change. These boundaries, however, are not physically or spatially distinguished. Rather they are marked cognitively based on particular social reference such as a coffee house, where strangers monitor every passer-by, especially women".*[33]

Home boundaries are determined by Bourdieu's system of objective potentialities. That system is inscribed in the present, but, knowledge of the absolute possibility of people's reaction towards an action you might take controls your momentary decision. Therefore, home boundaries became those limits, established by inhabitants as defended territory, whose invasion in every manner (physical, social, behavioural, or even cultural) is not tolerated, but resisted. Such a "socially constituted system" of "cognitive realities and structures" actually controls what people do in successive social situations in their everyday practice of being at home.[34] A resident expressed the power and mechanism of such a local system of objective reality, when referring to the way it works: *"Young men have to respect our morals and traditions. They know what is acceptable and what is not. If they deliberately cross the limits, we [senior members] stop them and all the community takes action against them. As a result, this situation becomes a rule"* [R1.2.08]. Another resident said: *"we have principles of social participation here. For example, when you attend a wedding you have to give Nokkout (wedding money gift) to the couple. If you don't, no one will come to your celebrations"* [R3.1.07]. The reaction to such unacceptable manners provides a kind of social sanction against whoever breaks the convention: which resembles a code or set of negative freedom opportunities.[35] Hence, every event or situation is subject to an individual's practical evaluation of the likelihood of the success of his action.[36] This cognitive process brings in different facets of practising home: convention, ethical precepts, wisdom, and regular structures according to which he was educated.

10.3   Coffeehouse at the harah's external boundary: signifier of the boundaries that dictate spatial practices

In that sense, practising home is practising intimacy and authority, familiarity and coherence under strict control, in which seniors direct juniors, and old members are in a higher position than newcomers. Neither the authority nor the convention is limited to the spatial territory of the harah or the internal interaction within the community. Every member is bound by that system even when he is outside his harah, or dealing with people from another locality. When a member is in a social situation that involves others, he is representing the community and its morals. The protection of home partners (other residents) extends outside the boundary gates of the harah. "*If a man of our people [the community] found one of our harah's ladies in trouble anywhere in the city, he should act immediately in her favour, protecting, defending or supporting her, otherwise he will be disgraced*" [R1.1.07; R2.1.07]. The practice of home, then, accompanies members wherever they go.

****

SPATIAL PRACTICES IN THE HAWARI OF OLD CAIRO

The organisation of home could be traced through the investigation of daily activities and spaces within which those activities take place. Home cannot be understood except in terms of journeys and daily trips to and from home as well as being a point of reference for daily activities.[37] Use of space, indoor and outdoor, reveals the way home is perceived and practised. Practising home in a Cairene harah, therefore, is perceived to be an interactive combination of three principal elements; the human factor (action and behaviour), spatial order, and temporal arrangement, which come together in daily activities. Field work and narratives of residents' daily life over an interval of three years revealed five principal activities: sleeping, eating, socialising (indoor/outdoor), entertainment, and work. While socialising and entertainment are fluid practices which could be merged, the other three remain consistent in their settings (time and space). With the exception of work, that may take place outside the harah's frontiers, all the above activities take place predominantly in what I call the *territory of home*. Following the natural diversity of local social groups, the pattern of activities changes from one group to another and from one house to another depending on the social hierarchy and spatial organisation within which activities are arranged.

In nineteenth-century bayt al-Suhaimy, for example, spatial organisation was stretched to provide alternative venues for every activity. The same activity could take place in two parallel venues at the same time. The four mandharahs of *bayt al-Suhaimy* allowed several socialising and entertainment activities to take place simultaneously, as was the case with several harem qa'as. Similarly, both Mustafa Ja'afar and al-Kharazi houses provided parallel venues for social activities. The harem qa'a, on the other hand, housed temporary venues in the same space. It was strictly used for sleeping at night, while it hosted several activities in the morning (guests, entertainment, weaving and trade). Houses of lower order were organised, like the harem's qa'a, on a temporal basis: with the difference that the harem received male guests during the day, a practice that was not accepted for high-profile houses. In comparison to the traditional organisation, the contemporary compact apartment houses could not afford such luxurious use of space. In such houses, each space has to synchronise several activities, utilising a programmed succession and temporary possession strategy. A living room in a one bedroom house, for example, accommodates children studying in the afternoon, sleeping at night, eating during meal times, and becomes a family entertainment venue in the evening. Accordingly, using space organisation as reference to determine social spheres of the home might be feasible in the case of high profile houses of the nineteenth century. Such definition of low profile or contemporary houses is blurred and it is not possible to determine any social sphere. Tracing social spheres in Cairene homes based on the spatial configuration of houses, therefore, became insufficient to explain how the organisation works.

On the other hand, some activities, such as sleeping, have to take place in a specific and restricted setting in terms of timing and control. Others are more flexible and could take place in any space and at any time. Socialising and entertaining could take place at home with family, in the harah with neighbours, and in the city with friends and sometimes with strangers. The diverse nature of activities and venues reveals another problem: that they are not spatially defined or temporarily traced. Therefore, the suggested spectrum of social spheres includes activity patterns for both the spatial and temporal settings. Social sphere, hence, becomes an imaginable territory situated spatially and organised temporarily for every type of activity. The sacred sphere is mentally constructed to secure the venue of sleeping from potential invasion of strangers. In the Cairene's mind, the bedroom (historically and contemporarily) is sacred, locked, and not accessible to guests; it is the safe of the house, where money and jewellery are stored and locked.

The detailed investigation of each activity pattern enables us to recognise the change in domestic or public activities within the social spheres and the significance of their spaces' form and function. Consideration of three principal activities that remained active during the period of study, their practice and rituals, could inform this study as to how the relevant spaces developed. The association of social activities and spatial form, therefore (as a result), is seen to provide an indication of how spaces are determined by cultural values, and daily habits. Sleeping, eating, and socialising (with family or friends) are the everyday activities I found to be of prime importance to the people of al-Darb al-Asfar. Work for them is a regular activity that takes place either formally outside the harah or informally inside the harah, where it becomes a part of home activities. Entertaining is a non-regular activity, but it is a significant occasion during which the socio-spatial and temporal order are reorganised and the space is redefined.

Sleeping represents a particular activity that is enclosed in its own secure and ultimately private social sphere. The intimacy of sleeping, in Old Cairo, has been associated with a secure and sacred space, the harem, or more recently the bedroom. However, the spatial order of activities of this sphere has changed over the past 200 years. The harem was a combined set of spaces which corresponded to diverse activities, including female reception, normal housewifely duties, entertaining by clowns, Quran recitals, weaving cloth, children's area, meeting friends and guests and in certain cases business trading.[38] At least this was the case in high profile houses. Such quarters were relatively large and spaces were highly decorated and largely inaccessible.[39] The objective of the architect, then, was to create a secure environment that retained all the potential and qualities of the front door mandharah but within a secured boundary. In order to secure a safe environment for every wife to maintain her social activities and environment without interrupting others, each harem apartment had to be at the farthest and most secure point. In the house, protection of the harem was emphasised in the lengthy journey through relatively dark and tight corridors, crossing a series of preparatory spaces before reaching the destination: the sacred shrine of woman. This ritual of a journey was essential to crossing the threshold of a sacred life and world.

For people of the lower orders, activities were taking place in a simpler form of spaces that consisted of a small multipurpose utility room (part time harem/mandharah). The harem was the only closed space at home in which the husband could receive a special guest. Harem quarters of 1800 were not, however, prison cells, isolated from the outside world. All harems had a large mashrabiyya overlooking the courtyard or outdoor streets, providing a public domain and communication links with visual and vocal access to other houses. In short, women's privacy, not isolation, was the motive behind the complex organisation of space and activities of the 1800s.

Along with the development of the houses during the twentieth century, the harem quarters were turned into whole family apartments. The previously exclusively female space became the only family enclosure to be governed by negotiations and games of control. The harem reduced in size to be a small bedroom space and mostly is jointly occupied by appliances (like fridges), or a sofa set. In the case of multi-room apartments, parents have their own exclusive and sacred bedroom exclusively for sleeping and intimacy. Other bedrooms lack such privacy and are used jointly by other members of the family. In many cases, however, the contemporary apartment has only one bedroom to accommodate the whole family (parents and kids): in such a situation, sleeping becomes a collective practice. Intimate practices, such as sexual intercourse, have to take place in a planned temporal situation, when other members are out.

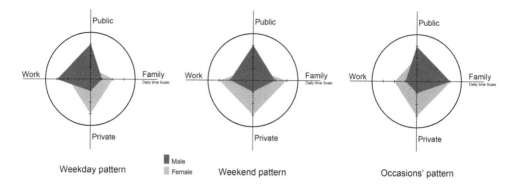

Weekday pattern   ■ Male   Weekend pattern   Occasions' pattern
          ▢ Female

10.4   Daily routine in al-Darb al-Asfar:
daily practice as determined through the three elements of the social spheres:
human (work-family), spatial order (private-public) and temporal setting (daily time scale)

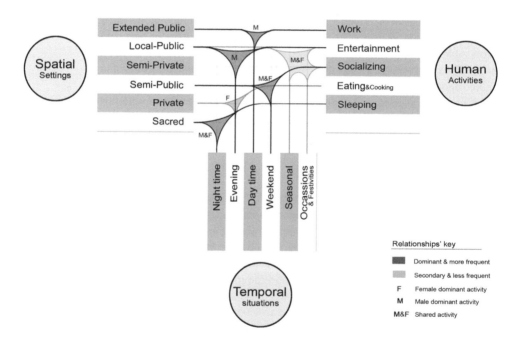

10.5   Dominant patterns of daily activities in al-Darb al-Asfar's social spheres

Note: The diagram traces social spheres through investigation of the association between daily activities and their socio-temporal situations.
These relationships, however, are not exclusive, but the most frequent within the studied context and mainly within the male domain.

Following the reform movement, women's privacy had been interpreted in different terms, allowing them an extended involvement in the public sphere: at work, in the community and at home. Women expanded their active domains into the street while their exclusive sphere was reduced spatially to a single room. After they had lost their position as sacred characters, isolating or securing their spaces was no longer justified or socially viable. The sacred social sphere, the venue of sleeping, shrank from a set of gradual spaces into a small bedroom space, or even a temporal situation in commonly used bedrooms and in certain cases taking place in shared spaces that were not bedrooms at all. Therefore, the spatial transformation of this activity gave rise to huge limitation of the physical setting and increasing complexity of social interaction. A combination of spaces that were highly enhanced and secure was transformed into limited rectangular non-exclusive spaces that are sometimes shared by others. Investigated examples showed the frequent appearance of a new model of extended family setting where the father shares the home with his married son/daughter.[40] This situation required dual sleeping venues of prime importance in the same house and provision of two sacred spaces in the same private sphere: another consistency of sleeping practices and organisation of home despite the huge difference in size from the houses of 1800.

Eating habits and meal settings provide the best evidence regarding the regular rituals of the family; they manifest family order, domination or subordination, authority and control. When associated with the use of space and time, they become relevant to the understanding of the structuring of social space due to the links between the home and locality.[41] In a community on the scale of al-Darb al-Asfar, it is natural to find a range of habits and settings that comply with the background or social level of each family. Collective cross-family meals were found to be a daily ritual of people living in the eastern part, while residents of the western part eat separately. Moreover, settings change from one temporal situation to another and daily meals are different from weekend or festival meals. Normal daily meal times, despite following a routine, are considered flexible and irregular when compared with the strict timing of breakfast meals of the holy month of Ramadan, during which eating rituals conform to a religiously ordered system.[42]

Over the course of the study period, we can reflect on the changes in meal settings and habits witnessed within a changing spatial organisation. The strict order of medieval gender-segregated tradition controlled family customs and order during gatherings from 1800 until 1990. In such situations, males of different families ate more frequently together than with female members of their own family (again: in high profile families). Every meal was taken in two separate gender-based settings, males first and females to follow. Positions around the table were arranged in order of gender seniority and authority, with the younger members surrendering to their elders.[43] Female meals, which followed those of the males, were ordered in a similar manner, with daughters being superior to daughters-in-law. Among the daily regular meals, two meals were taken at home: breakfast after dawn, and dinner, the principal family meal for all social levels, by sunset. Both meals were followed by coffee and took place in either the harem apartment or in the shared family seating area of every sub-family quarter. While breakfast and dinner were marked as family occasions, lunch was usually characterised by gender separation and cross-community unity. Lunch, unlike breakfast and dinner, had different venues, according to the situation: workers and merchants used to eat it at work, while community leaders gathered frequently at one or other of their mandharahs in a collective setting.

The contemporary family, however, has less diverse meal settings and habits. The regular communal meal is dinner, while both breakfast and lunch are taken at workplaces: although those who work in the harah or are retired eat in the local street space. I came to realise that breakfast is becoming another cross-community and gender-based meal. Dinner is being taken in the living room or dining room (in a few cases), while both males and females join the table in an unordered arrangement. The social downgrading of the community allowed the flexible practices of lower order families to dominate the social scene. Superiority and control continue to be in the hands of the male master. However, females are practising power in different fields of action.

Socialisation, as a representation of collective cross-family interaction, occupies the most time in daily activities. In al-Darb al-Asfar, due to the tight landscape and builtscape, it is hard to find an individual alone, either in his apartment or in the street. Social life is the predominant domain of activities. Socialisation, therefore, takes place at two different levels of interaction. The first is a gender friendly venue in the life shared with other family members (males and females); the other is in male-dominated street space along with neighbours, friends or workmates. In the absence of workshops and commercial activities in al-Darb al-Asfar early in the nineteenth century, social interaction conformed to the first two categories: families, neighbours or friends. At the family level, socialising took place in the evening in harem apartments or in the Meqa'ad during the day along with children and other family members.[44] Neighbours and friends were actually socialising in the indoor public spaces: al-mandharah and Takhtabush or even the Meqa'ad, of large houses. Harem quarters served the same purpose for female neighbours, relatives or friends. The structure of community was always present at every gathering. It was represented by the status every individual occupied in such collective meetings. High profile members would lead the gathering and sit in the inner central point. Honouring a guest meant to sit him in the best position, namely closest to the owner. Such representation had been associated with spatial features such as different platforms separating lower from higher order people.[45] That separation did not exist, or at least, was not manifested spatially, in the houses of lower classes.

Today, on the contrary, family members (males and females) hardly gather at home due to the tightness of available space. Socialising (gatherings, meeting friends, and chatting) usually takes place on a gender basis. Females gather inside apartments, or talk through windows, or at the doorsteps of their houses, while males (men or boys) spend most of their time (day and night) in their by-street lanes. Boys and teenagers gather in groups and either play games in the street or sit at the harah's entrances. Men usually gather at a corner, junction or at the front of someone's shop. For the passer-by, the main darb is divided into social groups: each of which sits collectively in a particular spot. Groups maintain reasonable distance between their gathering points. The interesting phenomenon is that every gathering comprises adjacent neighbours (resident, workers, or shop owners). Exceptions, however, exist: especially when families isolate themselves from the "unsatisfactory cultural level" of the common members of the harah. The strange thing is that even though the community was downgraded, social differences appeared to be important to some residents who believe in their superiority to others.

Both work and entertaining represent special patterns of activities, in which individuals experience frequent absences from the ordered organisation of home, spatially and temporally. For men and some modern women, work takes place mostly outside the boundaries of home, while the home remains the workplace for the majority of women. By work here, I mean the routine activities every individual performs to keep the household life running. For the people of al-Darb al-Asfar, a man's job is to earn money while a woman's job is to manage household needs and economy in addition to child rearing. Men are working between two spheres; the local harah space (mainly shopkeepers) or in the city. The first spends most of his working time, provisionally, at home, keeping immediate contact with his family and friends. Therefore, he does not express formality in his dress style and behavioural manners. They resemble the home workers of modern society, but with their work remaining outside the house. The second crosses the harah's thresholds everyday towards a different world, in which personality, character, dress are adjusted. This adjustment requires a journey between the informality and formality expressed through the crossing of the harah's entrance. Such situations did not change significantly between 1800 and 2000. While in 1800, many religious scholars had to wear their formal costume in their trips to work, for local workers and servants there was no difference between home and work clothes. In 2000, professional workers return home in formal costume, but appear in traditional galabiyyas in the evenings and at weekends.

Women's experience of work has become entirely different over the two eras. Women, who were not allowed to work (included most classes, except the poor) in 1800, had gained considerable employment power by 2000. Some women work outside the harah in official jobs (part of their intervention in the public sphere; thanks to the reform movement). They have the same experiences as men, with their clothing and habits entirely transformed for such excursions out of the home. A housewife, who wears galabiyyah and loose scarf covering her head in the evening, is turned into a working lady dressing formally with a smart, fitted scarf in the morning. Informality turned into formality and the loose becoming tight. Domestic duties, on the other hand, have undergone a number of changes. Communal cooking in the extended families of the nineteenth century had disappeared, and their houses either had very small kitchens implanted in other rooms or had kitchens situated outside the apartment. In both situations, the social gathering around cooking has been lost and replaced by reciprocal visits in the morning when the men are absent.

Entertaining involves people participating in a collective and informal practice. The informality of the practice is represented in the temporal departure from the organised daily routine. Entertaining your family means breaking the ordered pattern of daily life: which essentially requires breaking the cage of home. If entertainment is to take place in the harah, then the regularity of both spatial and temporal situations has to be broken. During festivities, the public sphere is reproduced as a factory of fun, exceptional and extraordinary. The street is reorganised to host the venue and people have to reschedule activities in accordance with it. Entertainers arrive, loud songs are sung, the whole street becomes noisy and the dark night turned morning by lamplight. This was manifested in the celebrations of 1800 of weddings, births and mawlids. In those situations the central activity was the ritual in which the journey of the bride, the groom or the newborn child throughout the community with musical companions represented the journey from one status to another. At this moment, all ladies, women and children, emerged from their houses (through mashrabiyya, terraces or windows) to merge with such public entertainment. The only change in the contemporary harah is that of women's appearance in the streets for every practice, which was prohibited in the medieval city. This, I believe, is due to the lack of sufficient interior space, leaving no alternative for ladies but to gather in the local public sphere.

10.6   Video stills of collective community participation during the transport of wedding furniture to the bride's house al-Gammaliyyah

During occasional celebrations, entertainment is mixed with duties and obligation in the public spheres, (the central lane, in particular). While the preparation of furniture for a local bride (described in the previous chapter) constitutes certain ceremonial rituals over the course of three days, the spatial configuration of these rituals provides interesting venues in which both private and public spheres melt together to serve such temporal needs. During the first day, pillows and mattresses are to be weaved at the bride's family home. Due to lack of space in the house, a corner space of the harah's main darb is cleaned for furniture weaving work. Here the local street [lane] works as a representative home, while all the neighbours are expected to attend and participate. The social obligation is combined with joyful and entertaining activities. The following day the furniture is carried by the youth of the harah to the bride's new house. On the second and third days, strings of light bulbs illuminate the bride's house, with an array of chairs lined in the front of the door. The lights guide any stranger/visitor to the host of the occasion from the harah's entrance. The night of the wedding has an overnight entertainment venue (for females in the bride's house, and males in the groom's house). To receive these celebratory activities, both private and public spheres are spatially ordered and adjusted in preparation for the rituals.

NEGOTIATING PRIVACY AT HOME

Complex and sophisticated representation of public life within the domestic sphere was not affordable in the post-1950s houses. Traditional guest-designated spaces such as al-mandharah and al-takhtabush were not affordable in the contemporary houses. Those houses follow a similar spatial system to houses of the lower order in 1800, where temporal adjustments of a single harem qa'a's light furniture were an effective practice to welcome a guest. The formality of the house's image in those reception spaces, accurately configured to represent the master's wealth and standing, was therefore replaced with flexible arrangement of space to meet immediate needs. Extensive service spaces such as kitchen facilities and storage were substantially reduced in size. Sleeping, eating and socialising come to the forefront as essential and basic minimum needs for families' everyday lives. Hence, public activities were sent to the alleyway space, leaving this domain of the private sphere to be displayed only through a set of symbolic elements: a corner sofa, a valuable piece of furniture or portrait. In most cases, such part/part-time spaces in the house are scarcely used by male guests, whose visits are kept minimal. Those part-spaces, part-time venues representative abstractions suggest that the semi-public sphere remains an essential part of the house regardless of its size and quality.

10.7   Theatrical and spatial adjustment of public space for celebration

The publicness of the private sphere becomes a symbolic and social need rather than luxury. As architects, we learn that the negotiation of an internal public sphere takes two forms: spatial, by giving specific space when affordable or temporal, by switching the sacred or shared private space into a part-time public venue.

In contrast to those representation spaces, the spatial territory of semi-private spaces in the public sphere has been problematic to define. The harah's lanes retain many features of privacy through its control of entrances, cohesion of social interaction and shared activities and meals in outdoor space. Considering the harah's central lane as one collective private sphere is rather a romantic judgement on collective interaction which ignores the micro-scale conflicts and social divisions within an inclusive community/group. Collective privacy is composed of small gathering spots, which do not become by-products, cohesive or integrated. Rather, there could be conflict and continuous rebellion against authority. The community is divided into small groups of friends or neighbours (3–5) who gather in particular spots where they assume control in the alley. Shop owners, for example, believe in their right to the public spaces exactly similar to local residents, thereby keeping the game of power politics active.

Throughout the duration of this research, the same people gathered in the same spots, mostly connected to their properties. Workshop owners gather at the junction of lanes within sight of their workshops.

Semi-private spaces are not spatially defined as much as they are socially. Areas adjacent to house entrances become part of the territory, whose interruption would attract fierce defence. An invisible curve is drawn around a shop frontage to form this type of vicinity. The regularity of certain activities such as sitting on stone seats (historically), eating, chatting, gatherings, drinking coffee, and receiving guests at these spots are spatial gestures to affirm identity and define the invisible territory of the semi-private spheres. Passers-by and neighbours would need to recognise this right by offering a vocal salute. Ignorance of this custom implies disrespect to the owner and the household.[46] However, the spatial alteration of the alley's enclosure and accessibility to allow through traffic did not change such cognitive belief in private territory; rather, its practice had a setback due to the following reasons: first, buildings' ground level facades became dominated by shops and workshop doors, at which workers and owners spend much of their daytime with friends or clients. Second, shops' adjacent pavements are used to display and store the workshops' metal products.

Territories are usually defined by the way boundaries are drawn and established. Within large houses, harem quarters were always distanced from each other, to avoid conflicts on authority and territory. With each harem, one lady set the rules, system of relationships, and communication. Once, they moved outside them, ladies were bound by the social conventions and systems of communication set by their master and the society, similar to being oudoors. Due to the continuous interaction between women on daily basis, architects tended to keep them apart, some of them having separate stairs, which turned the harem apartment into an entirely independent unit. Such spatial separation was physically empowered by indirect lengthy trips. The interwoven organisation of Bayt al-Suhaimy, for example, provided a sophisticated process of joining separate quarters together into one unit, in which spatial logic came evidently to follow the interwoven social pattern of the family. If the house was too small to facilitate horizontal separation, as for the bayt Mustafa Ja'afar case, the harem was lifted three levels high. Boundaries, accordingly, did not mean just doors or gates; they were the journeys and experience one had to go through to move from one domain of such prominence to another. These could include a series of circulation links, a private staircase, or even a labyrinth of inclined corridors with many doors. Spatial boundaries are not formed to define the physical territory so much as to disorient strangers and increase journey time to and from each sacred sphere. This psychological preparation emphasises the feeling of meeting a precious person in a precious space.

Such ability to combine ideology, culture, social convention in providing a satisfactory socio-spatial organisation was a distinct ability of local architect-builder experience that is largely missing in contemporary practice. In light of limitations of space, contemporary houses responded to the basic physical needs of the families, with basic territories. The majority of apartment houses represent a quite open harem quarter in modern terms. Each apartment, regardless of its size, is a combination of a few spaces, including small bath and kitchen, connected directly by a central space. Rooms are usually left open and only closed at night. The stairs lead to a series of private apartments arranged vertically in a typical manner. We can, therefore, draw a kind of spatial similarity between the sacred social sphere of 1800 and the apartment unit of 2009.

Social borders between social spheres where individuals' status has to change were represented differently in the medieval and modern models for obvious reasons. In 1800, passing through the threshold of the city or the harah was a lengthy journey with no sign of what was happening at the end of each turn or threshold. The barricaded gates and their spatial features (tight and dark) emphasised this mystery. Similarly, the access to the inner space of the house or the harem quarters entailed a lengthy journey through undetermined irregular non-linear corridors with several intervals and control points. Time and space were interplayed with darkness to emphasise such a crucial and tortuous journey. Guests were left with no sign of the end of the journey and in fear of what might suddenly emerge ahead. Hence, visitors had always to be guided to their destination, otherwise they would get lost. Medieval architects were sensitive to the impact of these psychological experiences on the guests, knowing that extended teams of servants were expected to monitor guests' journeys into the house.

Even within the same space, transitional journeys were at play from a certain position to another. Al-mandharah, for example, was divided into three distinct social levels, the main Iwan, where the host and high profile guests sat; a small iwan, where lower profile guests sat, and the durqa'a, the highly decorated central space which served as an entrance lobby, to which servants movement was restricted. Moving from the durqa'a to the small *iwan* was a big social transition in the servant's standing and the case was similar for guests moving from the smaller to the bigger iwan. Such differences were emphasised by certain physical and spatial arrangements. Iwans, for example, were raised by 20–30 centimetres from the durqa'a's plateau. The durqa'a's ceiling, on the other hand, passed upwards through the building to the roof to provide a high level skylight. Both iwans had double height decorated ceilings with recessed seating areas.

Expectedly, new apartments lack such arrangement of extensive or sometimes even any internal corridors: the essential transitional spaces, with much focus on efficiency and maximum use of a very limited space. That is why internal rooms do not resemble separate spheres; rather, their immediate connectivity with adjacent rooms merges them spatially to perform one function at a time. No chance for reading if the TV is on, and if someone is speaking over the phone, he is heard throughout the house. The stairwell becomes the only transition between the house and the local alleyways. Internal transition was eliminated on account of the inside-out transition being reinforced. Such spatial transformation suggests the opposite correlation between complexity of organisation and simplicity of transition.

But, borders of the social sphere mean much more than a transitional space or edge. They work as the thresholds of two worlds, which include change of rituals, behavioural adjustments, rights and obligations. Architectural representations of borders, however, witnessed a significant setback during the twentieth century, with modern architects/ builders lacking basic knowledge of local socio-cultural customs, as well as intelligible organisation of local activities.

Now, I can summarise a number of essential borders in Old Cairo's hierarchical structure into three distinctive and powerful thresholds. The first is the *Community Border* (Level A) between the city and the inclusive home, the harah. The second is the *Family Border* (Level B) between the harah and the enclosed private space of the house. The third is the *Gender Border* (Level C) inside the house between the inner private (the sacred) and the inner public part of the house. The rituals and practice of crossing each border have experienced diverse changes over time. The change of social spheres was essentially associated with a shift of borders and the process of crossing thresholds. Every time a sphere shrinks, another grows and borders move accordingly.

10.8   Development of social spheres in Cairene harahs over the past 200 years

Notes: Top: relations and interconnections of social spheres in 1800AD. Bottom: relations and interconnections of social spheres in 2008AD.

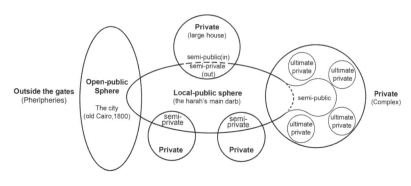

A. Organization of social spheres
and its boundaries (Dead-end harah of 1800)

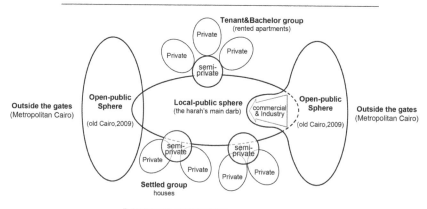

B. Organization of social spheres
and its boundaries (Open-ends harah of 2009)

The family border between the harah's main alleyway and the private sphere of the house has moved from the ground level entrance door as established historically to the apartment door, a double-leaf door that separates the open-to-people stairwell and the apartment-house. The ritual journey is lost. Once the door is open, the family border vanishes as the private and the public merge. The stairwell, which was part of individual houses in 1800 and the early twentieth century, became a public venue open to strangers. Gender borders, which used to separate the sacred private (harem/bedroom) and the inner public (male domains/al-mandharah/reception) have almost disappeared with the elimination of the public spaces from twenty-first-century houses. The concluding summary of the dynamics of social spheres shows the fragmentation of the complex social spheres into small unitary entities of private spheres which connect to the local public domains through means of vertical circulation. Internal borders moved outside the house, with the public sphere extended inside the building as far as the apartment's door. On the scale of the community, the open access of through traffic resulted in the alley's private spaces being exposed and practice of privacy in the alleyspace withdrawing to a more spatially controlled space wherever possible.

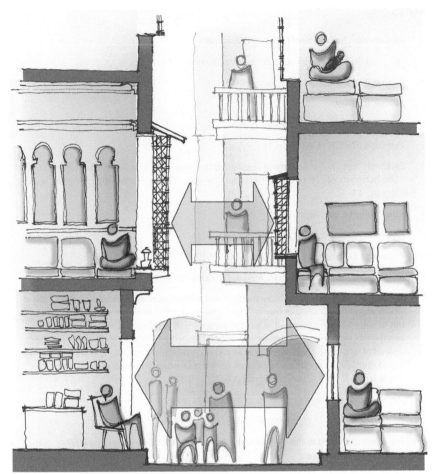

10.9 Typical section of a Cairene harah: a multi-layered arrangement of spheres horizontally and vertically. Ground level: male public sphere; upper levels: female public sphere

What is surprising is the local residents' consciousness of this change. Even though the change was gradual, the residents of the harah marked the 1950s–60s as the turning point of this change. They extensively used the term "was" to describe what happened before this time and "is" and "now" for anything since then. For the local people, their consciousness of change was manifested at this particular point in history, despite the fact that social structure and the introduction of industrial workshops appeared a few decades earlier. It is obvious that their realisation of today's identity comes from the historical reference of the harah. Their understanding of today's borders is reinforced by their memory of their different past. This notion was emphasised by Stephen Kern when he referred to William James's notion of the persistence of the past as *"a function of the fluid nature of human consciousness and belief that the past remained in a dynamic relation with the present"*.[47]

Local residents' explanation of today's values and interaction was accompanied by historical comparison emphasising the loss and the change. There is, for them, a constant interconnection of all past experiences with the present, regardless of how far back they may have occurred. The justification always come from the past, the experiences, the ideals and value system, which no one has the capacity to breach.

NOTES

1   De Certeau, *The Practice of Everyday Life*, p. 198.

2   Uraz and Gulmez, *Impact of Small Households on Housing Design*, p. 72.

3   Stieber, *Architecture between Disciplines*, p. 176.

4   Stieber, ibid.

5   Paddison, *Communities in the City*, p. 195.

6   Paddison, *The City as People*, p. 141.

7   Paddison, ibid. The general model explained by Paddison is identical to the situation of Old Cairo during most of the twentieth century. See also, Beinin, *Formation of Egyptian Working Class*, pp. 14–23.

8   Abu-Loghud, *Cairo*; Raymond, A., *Cairo*.

9   *Population Census for the Year 1937*, p. 2.

10   *Population Census of 1917*, pp. 673–4.

11   Myntti, *Paris along the Nile*, p. 14; Aygen, *The Centennial of Post-Ottoman Identity*, p. 310.

12   Timothy Mitchell summarised the medical and political reasons behind the state's policy which was in favour of open towns: "Open, well-lit streets were a benefit not only to health but to commerce, for they embodied the principles of visibility and inspection whose commercial usefulness was demonstrated at world exhibitions. The dark interior of the city, cleared of its human agglomerations, would become easier to police, and artificial lighting would enable the new shops and places of entertainment to do business into the night". See Mitchell, T. 1988. *Colonizing Egypt*, p. 67.

13   Mitchell, ibid., pp. 96–7.

14   Beinin, *Formation of Egyptian Working Class*, pp. 14–23.

15   The informant refers to the civil laws which fixed the rent value during the lifetime of the tenant of the apartment.

16   Sedky, *The Politics of Area Conservation in Cairo*.

17   See for example temporary building code for Historic Cairo, June 2009; *Cairo Governor's Decision 457* (1999), *Ministry of Culture Decision 250* (1990). Refer to the recommendations of UNDP sponsored projects of a historical nature, such as Gamlia Development Project (1994), and Rehabilitation of Historic Cairo project (1997).

18   *Population Census of 1917*, pp. 673–4.

19   El-Siuofi, *A Fatimid Harah*, pp. 26–7.

20   GOPP, *Al-Darb al-Asfar Rehabilitation Project*, p. 53.

21   Population numbers were stable during the last quarter of the century. This meant the number of new comers equalled the number of those who were leaving.

22   Fargues, *Family and Household in Mid-nineteenth-century Egypt*.

23   Keddie, *Women in the Middle East*, p. 90.

24   Christopher Alexander's pattern of language was designed in accordance with this particular perspective. He gave every resident different groups or combinations of tools, with descriptions, allowing the user to build his own home, according to his/her personal understanding/practice of that idea. See Alexander, *A Pattern Language*.

25   Bourdieu, *The Logic of Practice*, p. 55, as quoted in Heba El-Kholy, *Defiance and Compliance*, p. 201.

26   Bourdieu, ibid., p. 55.

27   Interview with Dina Shehayyeb about al-Darb al-Ahmar's rehabilitation project. August 2009.

28   Bourdieu, *Outline of a Theory of Practice*.

29   Abu-Loghud, Janet, *Cairo*, p. 65; Al-Messiri-Nadim, *The Concept of Harah*, p. 314.

30   El-Kholy, *Defiance and Compliance*, p. 32.

31   Behrens-Abouseif, *Islamic Architecture in Cairo*, p. 40. She referred to earlier travellers who estimated the number of cooked food stalls in Cairo as 12,000–24,000 cooks and 48,000 bakers.

32   Dina Shehayyeb is associate Professor at the Housing and Building National Research Centre in Cairo and one of the leading Egyptian specialists in hawari environments and their socio-cultural context. She did extensive social and architectural studies for the ground-breaking al-Darb al-Ahmar Rehabilitation Project, in Old Cairo.

33   Interview with Dina Shehayyeb, Cairo, August 2009.

34   Bourdieu, ibid., p. 77.

35   Bourdieu, *Outline of a Theory of Practice*, pp. 76–7.

36   Bourdieu, ibid., p. 77.

37   Porteous, 1976. *Home the Territorial Core*, p. 387.

38   Keddie, 1990. *Women in Middle Eastern History*.

39   Lane-Poole, 1845. *The Englishwoman in Egypt*.

40   Four investigated apartments showed this arrangement. However, they are equally distributed throughout the community, with no particular concentration.

41   Sibley and Lowe, *Domestic Space, Modes of Control and Problem Behaviour*, p. 189.

42   Here we can draw a comparison with the British culture's Christmas dinner or Sunday dinner: regular rituals that require the presence of all family members, with certain prerequisite dishes.

43   Mahfouz, *Palace Walk*.

44   Lane, *An Account of the Customs and Manners of the Modern Egyptians*.

45   Lane-Poole, *Englishwoman in Egypt*.

46   Lane, *An Account*, p. 198.

47   Kern, *The Culture of Time and Space*, pp. 43–4.

PART III
Modernity and the
Architecture of Home in Cairo

11

Why Architects Fail in Cairo?

Limited involvement of the architect in the hawari is due to three reasons: the high restrictions of building codes for such a low rise area, limited affordability for the owner (people of lower order), and the presence of an existing low cost reference for local architecture before his eyes. This is no innovation. Opportunity for architects' creativity is very limited. Architects are invited only when the situation is complicated such as a high-rise 12 storey building. [I2.1.09]

Homes, similarly to other types of buildings, were exposed to the influence of globalization and its universal culture which disconnects people from their context. You become culturally neutral, irrelevant to your place. Residents become just users [consumers of space], whose needs are shaped by global culture. This disconnects the home from the context. [I2.1.09]

Building codes and regulations do not reflect real life situations. The latest code requires a ground level garage for a house that lies in a two or three meter wide lane. The regulations are focusing on the style rather than the actual conditions of everyday. [I9.1.09]

THE CAIRENE HOME: A CONTESTED TERRITORY

Walking through a Cariene harah is a movement between three worlds, the social, the physical and the political. People are active players of the social world, whose movement, interaction and rituals define reproduce the notion of home through every day. Physically, Old Cairo hawari are a combination between two entities, valuable monuments that are models of medieval architecture, and a majority of less valuable structures that represent contemporary forms of compact living with loads of problems on their adequacy and sufficiency.[1] Politically, they provide long-standing evidence of the confusion of planning policies and the invariable top-down planning management and action plans. Ignorance and rigidity have been words commonly used to describe the state's policy on development of the hawari communities since Muhammad Ali's time.[2] The intersection between the three worlds could compose a valid and credible picture of the idea of home. The domains representing those worlds, namely, the practice (social), the buildings (physical) and the system (political), are normally in state of conflict on the way forward for these unique entities. But, this was always the case since 1800.

Local Communities in Cairo had been a contested territory in terms of gender relations, practicing privacy, and the more recent human/product dominance. This struggle was primarily in order to gain territorial significance, space and authority in shared spaces. However, the contest between local residents from one side and architects and planning authorities from the other side is arguably of similar significance. At the time when residents were the only active players in this social drama of everyday life, decisions and projects of spatial organisation were made by remote officials and architects. To draw a complete picture of the harah's home, I need to consider the perspectives of those active players. Hence, this chapter tries to explore local people's perception of the idea of home, while it investigates what it means to the architect in professional terms and whether the institutions' policies promote it or not.

Architects view the hawari, as we can see from the first and second quotations above, in a purely economical and physical sense, which is, however, in stark contrast to the perspectives of the official institutions which see them as a touristic and stylistic domain that needs enhancement and cleansing. The locals, on their part, work out their home through adjustment and reorganisation of the available space to meet the needs. So, while architects and institutions are participating in producing homes, residents actually reproduce it in their own terms.[3] We are therefore, faced with a dual reading, understanding and production of the home in Old Cairo. That missing link, I argue in this chapter, can be bridged by putting the perspectives of the three parties together.

LOCAL PERCEPTION OF THE CONSTITUTION OF HOME

If the home were to be constructed around a particular social requirement, it would be security. Living at home is basically a practice of continuity and regularity of someone's life that requires stability, trust and safety. Home, hence, is the anchoring point of, in Anthony Giddens' terms, ontological security; a sense of order and continuity through the individual's everyday experiences.[4] Security, in this sense, has always been a prerequisite and fundamental living need for a successful residential context in Cairo. Two factors normally compromise security and make a particular place appealing to villains: darkness of the alley and the absence of shops and active life.[5] Active life, shops, and continuously lit alleyways work as powerful assets in providing a safe and secure home for its residents. Security is, hence, a by-product of the socio-spatial order that is historically constructed and contemporarily preserved. Heavy presence of young people at the harah's entrances represents an existing system of visual surveillance around the clock. Despite the open ends of the central lane, the public domain is local private property claimed by the community, in the sense that every stranger has to submit to the implemented code of behaviour; No staring at windows or women, no standing at junctions or at a building's entrance unless you are accompanied by a local member. Local leaders prohibit the presence of nefarious people, and the behaviour and the flow of bachelor residents is continuously monitored [R1.2.08].

During the past five years, the course of this study, neither major crimes against residents nor other incidents have been reported. In such a predominantly secure and safe environment, incidents and crimes are significant occasions that are engraved in the memory and would be part of the community's history. The formally-named Zuqaq al-Darb al-Asfar is commonly known as *Haret al-Qateel* (the harah of the murdered). This name was derived from an incident which took place decades ago, in which the body of a milkman was found in one of the houses' ground level hawsh. The association between an incident and a place is a common phenomenon in Cairo, by which ordinary people designate places according to events. Such remarkable response to the incident conforms to the locals' recognition of the scarcity of crimes in their neighbourhoods.[6]

Perception of security is a fundamental and valuable asset to the people of the Cairene harah. It is achieved through a comprehensive strategy of defending space, collective as well as individual spaces. Space is, as per Thomas Dutton, *"one of the society's most fundamental mechanisms through which power is produced and organized to animate cultural traits and practices, social and work relations, and general everyday experience"*.[7] Defending space is essentially a resistance to external challenges: whether physical, social, or even cultural. Extensive presence of local males, mainly teenagers and youths, through the night until early morning, is replaced by that of local workers and employees who fill the harah's alleys in the morning. Such human-surveillance system is a strong sign of active measures for security monitoring.[8] Collective and continuous presence in public space creates active venues of cultural practice and exchange among people from different customs and backgrounds. No time gaps, no moments of silence, during which external cultures and forces could intervene or penetrate. In defence of this space, *El-Harafish,* groups of powerful young men, were formed in the nineteenth and early twentieth centuries.[9] Today, the community youth and teenagers could assemble in few minutes to stop any offences being committed by outsiders. During my field work, I saw that it takes seconds for the whole community to assemble in response to a screaming woman, or to put an end to a serious fight.

Collective protection of the shared local public domain in *al-Darb al-Asfar* is superior to individuals' protection: which becomes a by-product of the former. While the collective presence appears at the entrances and connecting junctions,[10] it is at a minimum at building doors and houses entrances. The toughness and powerful protection at the boundaries with the external world, is replaced with softness at inner boundaries, where individual safety is guaranteed by the local social code. An elderly single lady described the harah as *aman* (safe) and the people as *ahli* (my family), *"who would never hurt me"* [R9.1.09]. While women leave their kids playing in the alleys without direct supervision during the day, shopkeepers and workshop workers leave their shops open, without fear of theft or intrusion, during prayer times. Those practices imply confidence in the existing measures. Marking the context as *aman* (safe) is, therefore, a cognitive construction of reality.

The harah's space of resistance and protection works as an active domain of solidarity and family relations in the practical sense. In the absence of real relatives, family relations are customarily constructed within the locality; family is the practice of being relatives, rather than being actually relatives [R3.2.09]. In essence, a home in the harah emerged as a domain of community-wide social units linked by economic and social ties. In the words of Hajj Arabi, a senior resident and local leader, there is a confirmed cognition of this phenomenon in the most simple and practical sense:

"Our neighbour the late Ahmad Borai'e was living alone for many years. He was not married and his sister did not visit him for thirty five years. He was seriously ill for a long time, during which he was looked after by his neighbour's family (the landlord). We didn't see his sister until after his death, when she came to claim the apartment". However, the presence of ad hoc relatives does not mean a community of isolated members who persuade a sort of family relationships. Rather, brothers, sisters and cousins most likely to live in the neighbouring alleys of the old city.[11] Such relationships are emphasised by Hajj Nabeeh who introduced his brother, a fellow resident in al-Gammaliyyah, to me, saying, *"all of us are relatives: my brother lives here, my son lives with me in the apartment, and Hajj Arabi is my relative (step family) too"* [R5.1.07]. This mode of family relationships reproduced new generations of local lineage. For them, leaving the harah is leaving the family, and therefore, departure from home [R9.1.07].

This association between the home and the family could be seen in the settings of collective meals to which every resident is always invited, and poor workers/ residents do not need to pay. For example, a Ramadhan breakfast in al-Darb al-Asfar, which I personally shared, was sponsored by one landlord. The dining table and chairs were prepared by another, while tea was offered by a third from the neighbouring coffee house. Spatially, this meal was the central event in the main alley for an hour, accompanied by a certain set of rituals; praying al-Maghreb in al-Aqmar mosque, marching to the dining table, drinking tea and sharing memories/stories. Invitees included poor workers, sons and neighbours. Similar practices are spread around several hawari at different times with large hawari's offering sponsored charity fasting meals for the poor, called *Mawa'id Rahman*.

For the locals, this is primarily a family practice. Being part of the family puts every individual under an obligation to support his/her fellow members. This solidarity allows the intervention in each other's personal matters: to a certain extent, of course. They intervene to resolve problems organise and participate in personal events and occasions and to stand by each other in good as well as hard times [R2.2.08, R6.1.07]: *"The harah is my family. I don't see my relatives in years, while I see my neighbours every day. On my death, my neighbours, not relatives, who would support my family"* [R1.3.09].

The structure of Old Cairo as a maze of dense population and local markets within walking distance proximities made it viable market for manual industries, crafts, and food making. Many of vendors reside in local courtyards and abandoned houses, turning them into informal industrial estates. They benefited from the accessibility and proximity of their locations at the heart of the busy old city. Economic viability of the place, for those vendors, is prominent and relevant to the idea of home. Vendors were known by name and their timetables were in accordance with the temporal order of daily activities. Vendors of ready-cooked *foul medames* (Egyptian beans) start the day as early as 6.00am, moving between alleys delivering the breakfast meal to local families [R5.1.08, R15.1.08]. Similarly, other vegetable and sometimes fish vendors circulate between 9.00–10.00am, supplying housewives with their lunch components. At night, milkmen follow the same trajectories. Such domestic needs and economy have acted as veins of life for long time, especially in the pre-refrigerator era. Women did not need to move out of the harah for normal daily supplies. The development of several parts of the old city led the vendors out to remote sites, marking a relative disappearance of such economic pattern, and following the recent restriction on cars and vendors' traffic. With their departure, a significant economical group has disappeared. In the light of that change, we can understand how certain community associates homes with businesses. Stability of the business becomes a result of the stability of the residency and vice versa.

The proximity of several monuments in the old city shapes, to a large extent, the identity of the home as part of the historic Islamic city with that history shaping local value system, morals, tradition and habits. Daily habits and rituals, such as eating, gatherings, public participation and celebrations, are deeply rooted in the history of the place and affirmed by its buildings. Residents do things as they saw them done in their childhood. Daily practice, hence, is a process of reproduction, not production. This process of historical reproduction involves buildings as well as open spaces. Every building resembles a memory in itself and significant moment in history. It marks a particular personality, character or even story.[12] Memorable buildings are occupied with living people and history.

But home is about needs as well. They are bound by the reproduction of cultural values within the proximity of the old city. Some inhabitants don't leave the old city for months and sometimes years: *"All that I want is here, my work, family, food and entertainment are here. Why should I go outside [she means outside the gates of Old Cairo]?"* [R23.1.09]. Proximity here is signified by the homogeneous and integrated patterns of familiar lifestyle, resources and needs. If we add the short lengths of every straight segment of the spinal alley, we can imagine the intimacy between people and spaces. The pedestrian is always in control of space (scale) and in proximity to a visual destination (texture). He or she becomes an expert in the small world of the harah in which he lives and becomes familiar with its architectural details and decorations.

Contrary to common perception, monuments and historical buildings do not hold significant importance of any kind to people's everyday life Old Cairo. Most answers to my question *"do you define your harah in reference to a particular monument or significant building?"* are "No". Being a member of a specific harah and community is a sufficient identity, that somehow undermined by the flow of tourists traffic. Historical structures seem to be marginal part of their life that belong to the Ministry of Culture, on which they have no control. On the contrary, tradition, social norms (*Osool*) and habits are frequent terms that inscribe everyday conversations. Emphasis is on people not buildings. In addition, the proximity of the old city with its wall enclaves meant exclusive access to local markets, foods, services and trades of all kinds. The proximity to historical monuments and artefacts could undermine local social integration, have negative connotation to the idea of community, and hence is ignored. A resident and shop owner protested against the harsh security measures taken for formal visits to these monuments. *"The minister frequently invites friends and guests to Bayt al-Suhaimy's shows and cultural nights, and every time he comes, the security guards close our shops for the whole day. The building's opening day was a nightmare, the First Lady was coming. They closed our houses and shops for three days and asked us to stay somewhere else. They put snipers on rooftops"*. He further expressed his anger at such conditions: *"While they [the guests] enjoy our area, we suffer from the havoc caused by their presence"* [R5.1.07].

ARCHITECTS AND HOMES

> *People dismissed my question as irrelevant when I asked about having an architect who might help in designing the new additions. "What for?" was the answer. "The contractor and the builder (usta) know what should be done".*[13]
>
> Farha Ghannam

The role of architects in the contemporary harah is arguably, problematic and somehow controversial. Even though professional input is desired, the cost and financial consequences of employing an architect is beyond capacity. Official, on their side, cannot stand architects' protests against the state's visions, policies and restrictive codes. Hence, everyone resigns to the easy solution of searching for a local and familiar professional expert, the *utsa*, as we saw in their replies to Ghannam. The officials tend to restrict architects' options with excessive constraints and prescriptive images that confine any construction to defined forms of openings, materials, finishes, colours, and heights.[14] The emphasis is on the facade, regardless of what happens behind it.[15] But, in fact, Cairene architects hold diverse views of the home as a medium of architectural production, with many focus on physical and artistic attributes of buildings, giving less weight on the process of being at home. According to Khaled Adham, *"Architecture for these professionals is fundamentally an artistic, visual, and formal construction".*[16] The consideration of social issues was not on the list of priorities for those who design for affluent clients.[17] However, architects with academic and previous experience in those areas showed better understanding of such sensitive process and a challenge of working under artistic constraints and strict regulations, what would limit potential creative addition.[18]

However, I recognise three different issues that emerged to be essential for architects to consider while working in this part of the city: Architecture as a *practice*, the *heritage* context, responding to people's *needs*.

The practice of architecture in Egypt is determined by its professional education and training system. Architects are trained according to the principles of international style that have little sympathy with notion of heritage, authenticity and historical integration.[19] Architects dismiss the home as a passionate imagination of utopian world constructed on the ideals of tradition and cohesion. In contrast, contemporary practice deals with people's needs under economic pressure and lack of affordability in an attempt to offer the basic capacity of a shelter: *a space to accommodate their needs at the minimum possible cost.*[20] Most available designs reduce the home into a physical box with three dimensional space and services. Having examined this argument, it is worth noting that an architect working in the hawari of Old Cairo is faced with three basic difficulties: first, house owners do not understand the job of the architect in the first place. Second is the long-standing tradition of relying on contractors'/builders' experience of building along with lack of the wherewithal to pay a professional architect.[21] Third, house owner regards architectural consideration of the environment, ventilation and minimum space requirements as luxury that trigger unnecessary cost.[22]

Abdelhalim Ibrahim,[23] the prominent Egyptian architect, stands out as an architect aware of these problems and who envisaged his unique method of the building ceremony, in which the act of building becomes a mean of interactive communications with local residents and users.[24] He has transformed the practice of architecture towards the shared-production of buildings. Abdelhalim's practice is exemplified in the Children's Cultural Park, El-Houd El-Marsoud, in Old Cairo, where he involved the children and local community in a quest to develop an authentic spatial experiences infused by the local character and identity. The involvement of the local aimed at gaining information during design stage concerning the ways in which the building would be used, perceived and reacted to, so he could improve this design in the way he saw best.[25] This approach looks at community/user participation as an informative tool that does not make the design but adds to it. Similarly, Needs Assessment Research (NAR) is a user-driven design process that is based on social research and analysis.[26] This approach constitutes the core of the al-Darb al-Ahmar project in Old Cairo, which is run by the Aga Khan Trust of Culture in collaboration with the Ministry of Social Affairs. This project focuses on developing existing homes for the poor communities and depends on interdisciplinary practice of architectural configuration of actual social needs.[27] Salah Zaki's projects in Old Cairo, which will be discussed in the next chapter, are also other examples of success that relied on historical investigation and social analysis of residents' needs. After all, Architects, who intend to work in Old Cairo, need advanced training that involves in-depth analysis of socio-cultural situations.

Practicing architecture in this Old Cairo becomes a *de-facto* statement and response to the historical Islamic style, full of arches and domes, the dominant character of the medieval city.[28] Formal institutions recognise that working heritage-bound sites requires essential confinement to the surrounding Islamic context.[29] Using domes and arches in this context, some architects argue, is enough to become enlisted as a nominee for the prestigious Aga Khan Award for Islamic Architecture (AKAIA).[30] For some architects this was a superficial replication of Islamic architecture and reduced its complexity into a set of physical characteristics, architectonics, or formal aesthetics developed under different conditions.[31] For others, such intense world heritage site imposes many constraints related to building form, height, and even materials. But the formality of these constraints legally deprive architect from undergoing their own processes of creative exploration and radical designs.[32] For example, one architect protested these restrictions questioning: "*What are they looking at, a reproduction of Islamic architecture in the twenty first century*?!!"

While the architecture of home in Old Cairo, remained open to all practitioners, there was a set of procedures to get the planning permissions. A group of specialists and academics with experience in dealing with the heritage sites was set to review submitted projects and either approve, disapprove or reject design proposals. This body is called National Organization for Urban Harmony (NOUH).[33] Ironically, this was the most awkward and superficial process of planning applications, one could imagine. The proposed design drawings in the old city take two directions; floor plans go to local district engineers (mainly not architects) for checking of compliance with building regulations in term of floor plans. Elevations, on the other hand, are sent to for a parallel but separate approval from the National Organisation for Urban Harmony NOUH), mentioned earlier.

On the other hand, architects, perhaps preoccupied with the view that their mission as of elitist nature, questioned the need for the presence of professional architectural practice in informal local communities, in the first place.[34] They asked the simple question; is there a real need for architects, if local builders are better informed and expert than any distant architect? In a runaway approach, they contend that professional architects, bound by regulations and design standards in areas that retain continuity of building traditions, is a problem-making not a problem-solving strategy. Architects limit the margins of any freedom local residents currently enjoy: through the formal processes of design, strict adherence to the regulations and prevention of illegal building activities.[35] On the other side of the argument, well experienced architects are urgently needed to stop the deterioration of existing homes and built fabric.

11.1 Arches and domes of contemporary architecture in Old Cairo

According to this view, architecture should diversify its ways/norms of practice from being a professionally oriented activity into a more interactive process of decision making. Building regulations and law are always set to seek creative solutions within a framework. Or perhaps, regulatory bodies should refer to actual successful practice for regulating building activities; *law follows practice*.

Architects, hence, need to engage with the users and the community members in order to develop architectural solutions based on the effective interpretation of actual problems and practices. Those solutions are not bound by image, significance or regulations but should be successful, maintainable and associated with people's local identity and needs.[36] Two approaches are involved in the creation of homes in the contemporary hawari of Cairo. The first relies on the general studies made by the United Nations Development Programmes (UNDP) and governmental authorities and aims at empowering local communities through providing them with missing social, medical, education and sometimes professional training services. Those architects' work is based on a prescriptive client's (NGO) brief. The needs here are communal and collective, directed towards the inclusive home, the community in general. The product, accordingly, alludes to the historical style in photographic imagery transcribed through the arches and the inner courtyard. Such an approach considers the local needs in a visual sense, translated by the architect's philosophy into recomposed elements of traditional architecture. The goal here is to satisfy the needs of the public sphere.

Everyday homes are, however, created by architects who conduct comprehensive analysis of their clients' living patterns and social norms through, socio-economic surveys, interviews, as well as daily observations of developments in daily practice.[37] In such practices, the needs are complex and involve economic, social, health and safety as well as architectural issues. In this type of practice, the architectural brief is created to reflect daily life situations and practice (i.e. raising livestock, collective cooking, proper sewage system, restoration of deteriorating structures, etc.), while the project plan targets long-term objectives. Those architects, in particular, believe in the homes in the hawari as more sustainable communities when compared with modern urban compounds. According to one architect [I8.1.09], they attend to the prerequisites of sustainability, socially, economically and environmentally; e.g., social solidarity promotes safety, local support and security. The locals' workplaces are easy to Livestock and locally produced food provide for domestic everyday needs while old buildings constructed with thick stone walls/bricks allow for cooler temperatures during hot summer months. As such, homes of the hawari, have the potential to be environmentally, economically and socially sustainable in the long term.

INSTITUTIONAL APPROACH TOWARDS CAIRENE HOMES

In the context of this book, I looked at Old Cairo as a large area of condensed urban fabric that is composed of a series of socially and physically attached homes, each of which comprises a harah. The state looks at it as a historic site of traditional Islamic architecture and medieval urbanism, a promising site for tourism and economic viability, undermining to a large extent, living homes and associated social organisations. From this perspective, the state dealt with the hawari as a problematic deteriorating area that requires urgent physical restoration of style and plans for socio-economic upgrading, and in some cases social engineering. Based on the findings of several studies made by the UNDP programme of Historic Cairo, a series of legislation has been developed and issued in the past five years.[38] From a brief survey of the recently issued laws and decisions and interviews, I can shed some light on the two directions of current thought on the future of those homes of the hawari: preserving the image of the past (the frozen home) as locked in medieval Islamic architecture forms, and social engineering with replacing the current lower middle class inhabitants, gradually with people of higher social order.

Developing a dense context of so many homes involves diverse efforts that are proved difficult for the government to control as one package. In response, governments took a pragmatic approach that focused on the physical development of its historic assets and adjacent building facades, not floor plans or services; exterior not interiors. In 2002, Abdelbaki Ibrahim, a profound architect and academic, summarised the development of Egyptian policy towards Old Cairo as: *"reduce population density in the city centre, control its peripheries, modernization and preservation of heritage, identity, urban fabric and conserve the diversity and the personality of different districts".*[39] Acknowledging the broadness of these terms, Ibrahim insisted that this policy should be joined by clear strategy and regulations that accept renovation and freedom of creativity, not rigid legislative rules. In a practical implementation of this policy, presidential decree no. 37 was issued in May 2000 to form the National Organization for Urban Harmony (NOUH) as a regulatory and advisory authority for the Minister of Culture with a mission to *"apply the values of beauty to the exterior image of buildings, urban and monumental spaces, the bases of visual texture of cities and villages and all the civilized areas of the country including the new urban societies".*[40]

This mission, however, was in contrast with Ibrahim's comments against the superiority of the image and visual textures which have become a prisoner of medieval design ideals. Although architects acknowledged the urgent need for action to stop deterioration in Old Cairo, they have criticised the work of the NOUH for its superficiality: as an artistically-driven mission whose ultimate concern is the image not the content, regarding the exterior of building in particular.[41] Those claims are legitimate for the documented problems of the hawari are essentially socio-economic rather than artistic or architectural, which are consequences.[42] For the people of the hawari, such aims are useless for a few reasons. First, communities of homes are living practices, not physical products. Economic hardship and limited resources are the main obstacles to keeping these houses in good working order. Second, the mission divides the same building into two separate entities, the exterior (to be approved by the NOUH) and the interior plans (to be submitted to district engineers for building permission). NOUH's work, accordingly, is to craft a stylistic two dimensional image that could be irrelevant to the building plans, a radical approach rejected by many architects.[43] Third: committees of the NOUH are predominantly made up of architects and urban designers, who ignore the complex social and economic consequences of their decisions. Moreover, many arches and domes used in design proposals appeared in stark contrast with Islamic style principles and proportions.

These comments sum up the concerns over abstraction of the built fabric into a set of panoramic images composed of attached facades and openings. *"Local residents would never spare a pound to spend on developing the facades if they cannot afford proper interior spaces for their family, therefore, the state has to spend huge sums of money for the sake of a beautiful, fake image"*, one architect said.[44] When I took the issue up with some local residents in al-Darb al-Asfar, the response was harsh and full of anger, confirming the architect's statement.

* * * *

Exploring local perspectives (residents, architects, officials and intellectuals) of homes in Old Cairo revealed different attributes each gave to its architecture. In social terms, the harah was a big inclusive home, in which resident looked for social ties, mutual support and solidarity at difficult times, which is associated with the space that is well secured and contained. Architects, on the other hand, looked at the harah either as a home for historical buildings, in which contemporary practice has to come to terms with credible response to the medieval character and valuable buildings. Practice, is bound by certain roles and chiefly dictated by conservation policies that aims at the image images, where power politics and economic viability are at the centre. Officials, on their part, impose their vision of the hawari as a context that represents historical homogeneity and displays images of the past through legislation and regulation that allows them the upper hand in architectural production. Against this backdrop, the architecture of home emerged as a battleground between the authorities, interested architects who took the challenge to and developed new methods through interaction with local residents.

However this conflict of visions, strategies and objectives is peculiar to Old Cairo, one could trace several similarities with other cities. Much work has been done to address problems of similar settings such as informal quarters and marginalised communities in urban and rural contexts. Authorities and state institutions policies seem to follow the predominant top-down structures with provisions aims to reshape the landscape of built environments and spatial experiences. However, similar to Cairo's case, these policies proved to be the least successful approach to improve the living experiences at these homes. For obvious cultural and political reasons, residents do not trust these institutions nor communicate their needs with them. While the state wants to clear Old Cairo's homes for better investments and economic opportunities, local residents frequently violate building regulations and built beyond their anticipated permits, which is basically the reason it is informal. Governments, bound by their legal systems and procedures, need to adopt new structures to deal with the unique situations such as Old Cairo. While NOUH was an attempted answer, it was not enough to challenge deeply rooted problems, inherited for centuries. Many of the buildings renovated or supported by formal renovation programmes are superficially decorated without seriously solving these problems. There is an obvious need to be critical about those governmental policies that deemed fragmented despite the availability of many development programmes and documents by international organisation.[45]

It was quiet interesting to learn the architects dismissing their role in the building of homes in Old Cairo, without enquiring critically about the shortage on their side. If architecture is about, according to Kim Dovey, about the imagination and invention of spatial futures,[46] so how could architects rule such contexts and challenges out of their agenda? If knowledge is a quiet limited on this front, then we need to look at other deeper structures that could help us understand this underlying system of living and the way they translate into architecture. The work of Teddy Cruz on the American-Mexican border areas is quite informative in that sense.[47] When conventional forms of architecture and building design failed to address local needs, then architects fail to do their job. They would need to develop new forms of understanding, comprehending and contributing creatively to this condition and challenge. Cruz's Casa Familiar project in the district of San Ysidro with the development of 12 homes, a church converted into a community centre and a garden is perhaps a good example of these forms of new structures. Innovative in scope, forms and method, Cruz's Manifesto is about forcing policies and practices to become progressive, flexible and inclusive of unconventional forms of living. Architecture's main job, for him, "*is not only to reveal ignored socio-political territorial histories and inequalities within this polarised world, but also to generate new forms of sociability and activism*".[48] His case is that architects need to be radical and think about housing differently, "*not as units thrown on the landscape, but strategically and politically about density and social structure*".[49] In this sense, Cruz develops contemporary method of Hassan Fathy's vision of helping those informal structures to find a coherent way to meet their needs. The architecture of home emerges to establish this discipline through critical enquiry in such condition as Old Cairo.

Obviously, there has been a lack of concern on the side of architects over how deeply history and memory are rooted in everyday practice, or in the spatial order of the hawari homes. The analysis of everyday practice, visions and traditions revealed the power that historical continuity and memory exert over present situations, social relationships and spatial practices. Among Cairene architects few were interested to dig deeper into the value system, and social networks and structures to developed meaningful architecture of home. Those are perhaps equally critical to Teddy Cruz and the pioneer Hassan Fathy, about the role of architecture and its methods of innovations. Of these, the work of Salah Zaki and al-Darb al-Ahmar rehabilitation project led by the Aga Khan Trust of Culture hold substantial credibility. The critical process of architecting home, hence, will be discussed in the next two chapters.

NOTES

1   This view represents the dominant perspective of most interviewed architects and officials.

2   Those terms are frequently used in principal textbooks about Cairo such as Janet Abu Logud's *Cairo: 1001 Years of City Victorious* and Andre Raymond's *Cairo*, and Susan Staffa's *Conquest and Fusion*.

3   This recalls Michel De Certeau's discussion of a concept city as one that is planned and designed by institutions and corporations (defined as strategy), while walkers in the city carve their own production and consumption of the same space on their own (defined as tactics). See, similar reference in Chapter 6. De Certeau, *The Practice of Everyday Life*, pp. xix, 91–5.

4   Dovey, *Framing Places*, p. 139; for ontological security see Giddens, *Modernity and Self-identity*, pp. 35–40.

5   Adham, *The Building Border*, p. 224.

6   Residents interviews: R1.1.07, R3.1.07, R21.1.08.

7   Dutton, *Cities Cultures and Resistance*, p. 5.

8   Residents interviews: R1.1.07, R3.1.07, R6.1.08.

9   Andre Raymond, Harafish.

10 Those internal junctions, according to De Certeau, are spatial borders that require protection and monitoring as people cross them frequently. See De Certeau, *Practice of Everyday Live*, vol. 2, p. xx.

11 While the study is not a quantitative survey, I can confirm the presence of related people in at least four different buildings. Many others are living in simple forms of extended family settings.

12 Interviews with residents. Codes: R1.1.07, R6.2.07.

13 Farha Ghannam describing the practices of the people of the hawari of Bulaq after their relocation to al-Zawiay al-Hamra in Cairo. Her description spotlights the perception of the architects' role in such local cultures as a luxury and not essential. See Ghannam, *Remaking the Modern: Space, Relocation and the Politics of Identity in a Global Cairo*, p. 172.

14 Refer to Egyptian building code, special section for Old Cairo, issued in 2009 (Appendix C).

15 Building façades, in particular, have to be reviewed and approved by National Organization for Urban Harmony (NOUH).

16 Adham, *The Building Border*, ibid., p. 234.

17 Salama, *Architecture Reintroduced*, p. 28.

18 Interview: I2.1.09.

19 Hassan Fathy in his criticism of architectural education in Egypt.

20 Interview coded I2.1.09, I3.1.09.

21   Hassan Fathy argues that people have been building houses for themselves for thousands of years, and it is only recently they have begun to consult architects about the design of their houses. See Fathy, *Architecture for the Poor*, p. 115.

22   Interview: I2.1.09, I3.1.09, I6.1.09.

23   A prominent Egyptian architect, who has received several international awards.

24   Abdelhalim Ibrahim, *The Building Ceremony.*

25   Nabil, *Reconciliations and Continued Polarities*, p. 77, as quoted in Adham, *The Building Border*, ibid., p. 223.

26   Shehayeb and Abdel-Hafiz, *Tradition, Change, and Participatory Design*, p. 1.

27   Interview: I3.1.09, I6.1.09.

28   Interview: I1.1.09.

29   There is always a reference to Islamic architectural style in all building regulations concerned with building in Old Cairo. However, the translation of this clause is interpreted by most professionals as arches and domes [I8.1.09, I2.1.09].

30   Adham, *The Building Border*, ibid., p. 234.

31   Interview: I2.1.09.

32   Interviews: I8.1.09, I4.1.09, I1.1.09.

33   Interview: I2.1.09.

34   Interview: I10.1.09, I2.1.09.

35   Interview with architect: I8.1.09.

36   Interview with architects: I6.1.09, I1.1.09.

37   Interviews: I8.1.09, I4.1.09.

38   Sedky, *The Politics of Area Conservation in Cairo.*

39   Ibrahim, *Urban Harmony: A Dream Yet to Come True*, p. 3.

40   Mission statement of the National Organization for Urban Harmony (NOUH), as mentioned on the authority's official website. Translated from Arabic. See http://www.urbanharmony. org/en/en_target.htm (accessed 1st November 2009).

41   Interviews: I1.1.09, I2.1.09, I6.1.09.

42   UNDP and SCA, *Rehabilitation of Historic Cairo: Final Report.*

43   I witnessed, myself, some examples of this kind, where façade drawings were sent separately to the NOUH specialist architectural consultants to approve. "*It usually takes three to four changes to develop an acceptable facade*" [I8.1.09].

44   Interview: I1.1.09.

45   Sutton, K. and Fahmy, W. 2002. The rehabilitation of Old Cairo, *Habitat International* (26), pp. 73–93.

46   Abdelmonem, 2012. *Peripheries Dialogue*, p. 20.

47   Murray Fraser interview in: Abdelmonem, 2012. *Peripheries Dialogue*, p. 20.

48   Teddy Cruz in an interview with Beatrice Galilee, of *ICONEYE Magazine*, November 2008. See Galilee, B. 2008. "Estudio Teddy Cruz", *Iconeye*, Issue (056).

49   Cruz, ibid.

12

Architecture of Home: From Theory to Practice

THE DILEMMA OF THE PROFESSION AND THE PRACTICE OF HOME[1]

Having confronted perspectives of active players and institutions involved in the production of homes in Cairo, I shall take on the practical process behind this production interrogating architects' contribution to the making of homes in an age predominated by professional practice, modern technology and thought. Homes lie at that intersected territory between the local people interests and the government's policy, between deeply-rooted building tradition and professional intervention, and finally, between individuals' needs and those of the community and the traditional fabric. This position continues to cause confusion of priorities and objectives that leave homes off the development agenda and therefore lead to the deterioration of built fabric. However, they are the only priority for those who occupy for decades. After all, it is home, whose residents "are ready to invest their own resources to improve their living conditions".[2]

While the rehabilitation of old buildings emerged to dominate the scene in Old Cairo, it provided limited response to the socio-spatial practice of making new homes. Many questions remained unanswered; how professionals can architect homes in these localities; what kind of knowledge and skills do they need; and what professional adjustment is required? In short, how can architects work with the idea of home that is peculiar to Old Cairo? Here, I try to contribute to answering these questions by linking the gained knowledge of everyday practice in the home and its patterns of activities to the desired socio-spatial practice of architecture. What Mary Douglas identified as the systematic organisation of home is treated in an enhanced architectural practice that aligns to the social structure, networks of the inhabitants rather than physical requirements.

Architects dealing with homes in Old Cairo face a critical dilemma about how to approach this context and such unique fabric. Here there are two propositions; Old Cairo as a place for innovation, creativity and imagination, confined with the ideals of the practice, or Old Cairo as a historical site with a valuable image of the past that should not be interrupted. In the first, architects would be eager to challenge and contest existing forms and fabric with radical solutions that dwell on new functions, forms or use. The latter is more conservative approach that required deep attention to what is behind the image to remain part of the present and avoid running the risk of imitating the past. Of course, architecting homes in Old Cairo, a prominent world heritage site, would require a careful balance between both propositions, being imaginative without losing the connection with the past. After all man related to his locality through dwelling according to Martin Heidegger; "*Man's relation to locales, and through locales to spaces, inheres in his dwelling. The relationship between man and space is none other than dwelling, thought essentially*".[3]

But, let's interrogate these positions rationally. In its October 2009 issue, the *Journal of Architectural Education* featured an article by the New York architects Reese Campbell and Demetrios Comodromos that proposed a radical vision for the future of the hawari (alleyway communities) of Old Cairo: a "speculative skyscraper that verticalizes the complex interrelationship of informal social networks and urban/civic form".[4] Campbell and Comodromos's proposal, which I referred to in the introduction, was based on Stefano Bianca's theory that traditional Islamic urban form took shape around prototypical patterns of behaviour.[5] Based on this work, Campbell and Comodromos may have been aware of the complex association between spatial organisation and social and behavioural patterns in Old Cairo. However, their design reduced the community to a spatial and morphological abstraction, in a layered stratum of services and land uses that ascended in social significance to the mosque at the top.[6] They then claimed the design would "simulate the complex social interactions and norms present within the medieval fabric of Islamic Cairo", utilising accepted characteristics that had been "reinforced in the subconscious of the population over centuries".[7]

Campbell and Comodromos's claims notwithstanding, their plan for a vertical arrangement of services – mapping street patterns and irregular plot shapes onto the external form of a skyscraper – involved questionable decisions regarding the socio-spatial complexity of Old Cairo's everyday life.[8] In this sense, their proposal recalled Zygmunt Bauman's notion of liquid modernity, in which a distant authority (the architect, in this case) decides the destiny of people living thousands of miles away – sometimes without ever having visited them.[9] Perhaps, work like this may contribute to design theory, but it is equally representative of a contemporary practice in which the architect is increasingly isolated from the dynamics and peculiarities of context.[10] Such intellectual interventions also must be scrutinised with regard to practicality. In this case, one might ask: Is it even relevant to life in Old Cairo, with its inherited practices; or does it belong more to the architects than to the setting?

The thesis by the German philosopher Martin Heidegger on the act of dwelling appears precisely relevant in this case. For Heidegger, the relationship between man and space was about the act of dwelling. He saw this as a process of making spaces that reflected man's understanding of his position in the world – a comprehensive act, loaded with inherent relationships between man, locale, and produced space.[11] Architects, in Heidegger's view, were principally concerned with mathematical space (physical settings), while the act of building (making space) was similarly of little interest.[12] It was the understanding of locale that was key to man's process of dwelling. It would thus be naive to suggest that everyday life of Old Cairo as analysed earlier could be abstracted into a simple form or arrangement of services based on a visual or stylistic taxonomy. Indeed, sociologists and anthropologists have undertaken intensive investigations in Old Cairo and on its periphery to understand the working of its communities, to which this book attempts to contribute.[13] From the local resident perspectives, as we saw, the harah is a notional collective home that is a strong hold territory, fiercely defended by every resident. However, from the opposite point of view, however, the harah can be understood as a product of architectural decision-making – a social phenomenon built out of physical forms. The hawari of Old Cairo are thus like any other urban structure, built house by house, building by building. This sense of physical place is what allows the history of the community to be linked to its buildings, and every inhabitant's memory to be indelibly inscribed in space. This association is particularly powerful given the strange ability of spatial memory to conjure up a dense web of images, particularly in association with areas adjacent to one's house.[14] Architects tend to see their role as being to devise innovative forms, and they often play down the importance of context and its everyday power. At least this is what one might take away from the proposal by Campbell and Comodromos, mentioned above: even when the harah is the context, architects tend to advocate highly artistic and intellectual products.

SUSTAINABLE LIVING, CONTINUITY AND THE NOTION OF CHANGE IN OLD CAIRO

Before taking on such task of making a building in the hawari, one must understand the basis on which such sustained communities stand. At the heart of sustainable living patterns is the fact that change is inevitable in human life, both culturally and socially. Here I define sustainable living as the ability of a community to manage its resources and available spatial settings to elaborate new systems and organisations that respond to changing needs and the challenge of time. During the process of change, complex constructs (such as home) may be decomposed to its preliminary elements, then reconfigured and reorganised in new forms suitable to emerging needs and demands. This process may be slow, unnoticeable, and in constant flux.

In the context of Old Cairo, hawari communities developed, over centuries, the sustainable notion of a collective home, in which boundaries between individual houses were seen as less significant than the collective territory. While the boundaries of this shared home were historically barricaded and closed with gates, in the contemporary context borders are more likely to be determined by patterns of activity and by points at which behaviour and reactions change. This implies a territory that is built mentally and practiced socially in the minds of its holders.[15] Relaxed communications between men and women, accepted modes of dress, and mutual support during hard times are, for example, all basic principles of this agreement that were described by interviewees during fieldwork. According to the residents, noncompliance with this code invited a tough response, and could result in collective exclusion of an individual and his/her family.[16]

Practices developed over time, hence, become regularities, which may define transitions from one social sphere to another. This condition may be apparent in the way women and men dress, how freedom of movement and interaction is circumscribed, and how they present themselves in the public sphere. According to one expert in the social structures and networks in the old city, Dina Shehayyeb:

> Within Old Cairo, the boundary of home is determined by the way women, in
> particular, are dressed. They move freely with their home-style informal clothes,
> within the area they consider a home. Crossing this envisaged boundary requires the
> dress code and behavior to change. These boundaries, however, are not physically
> or spatially distinguished. Rather, they are marked cognitively based on a particular
> social reference, such as a coffee house, where strangers monitor every passer-by,
> especially women.[17]

12.1   Inherited
practice: eating
in the alleys
and streets of
Old Cairo, early
twentieth century

Courtesy of Rare Books and
Special Collections Library
at the American University
in Cairo.

On the other hand, these practices and regularities maintain an unbreakable link with the past. It could even be argued that they have worked against change because they represent a system resistant to compromise. However, the sense of a home territory in Old Cairo could likewise be seen as determined by Pierre Bourdieu's system of objective potentialities. Thus, knowledge of the absolute possibility of people's reaction to an action might control a person's momentary decision-making.[18] According to Bourdieu, such a "socially constituted system" of "cognitive realities and structures" controls what people do in successive situations during everyday life.[19] One older resident expressed the power of such a local system of objective reality this way: "Young men have to respect our morals and traditions. They know what is acceptable and what is not".

Activity patterns are also affected by the potentials inherent in different spatial layouts. Thus, in comparison with the relatively large traditional courtyard-centered house, the contemporary compact apartment does not afford the luxury of a large multipurpose space. In one such apartment, a resident mentioned that each space therefore had to accommodate several activities according to a strategy of programmed succession and temporary possession. This, however, accorded to inherited customs and living styles. Historically, bedrooms in Old Cairo might have been used in several ways: at night solely for sleeping, but during the day to host other activities such as guests, entertainment, weaving and trade.[20] Especially in the houses of the lower social orders, women's areas might thus be used to receive male guests during the day, a practice not generally acceptable in more high-profile houses.[21] Evelyn Early has described such a pattern of active social spheres in one family, where the wife assumes control over the house space, as her castle, where she "spends free time with her women neighbours, and feels content, not neglected", while the husband "comes home only to eat and change his clothes".[22] In this extended version of home, men typically meet, socialise, and sometimes eat with their male neighbors within the alley.[23]

12.2   Flow of products in the alley space

NEGOTIATING CONTROL IN THE COMMUNAL SPACE

The social theorist Max Weber has asserted that it is only in praxis (acts, courses of action, and interaction) that it is possible to trace the essence of a community, group or society.[24] Praxis thus involves in the very activities of everyday life that local actors see as holding no significance of any sort. Michel De Certeau has also written of the association between spatial practices and the quality of space.[25] By investigating simple activities and the way space is organised to accommodate them, it is possible to trace the way the public space is utilised to suit basic social needs and the essence of community.

Eating meals, drinking coffee, and smoking *sheisha* (a waterpipe) are typical activities performed on a daily basis in the alleyways of Old Cairo. Most of those interviewed in the alleyways (men) said they took their meals (mainly breakfast and sometimes dinner) in front of their shops, workshops or houses. A movable dining/drinking table, previously stored away, would be set up to allow this to take place without interrupting public movements. Outdoor space could thus be adapted to provide a sufficient alternative to missing indoor social spaces, which might formerly have been used to host similar activities. Interestingly, the extended sense of home allowed the intrinsic qualities of a private atmosphere to be maintained in open outdoor space. On several occasions, passers-by volunteered to participate in my discussions and interviews in the alleys.

If they were also residents of the harah, they believed in their right of intervention once a conversation was taking place there. In general, we observed a complex pattern of space use in alleys throughout the day. In the morning, the alley would be overwhelmed by industrial activities and workers. Their impact was evident through the noise of machinery, gatherings in front of shops, and the flow of products (Figure 12.2). However, the dominance of work activities receded in the evenings and on holidays, when local residents took control of the space. Thus, even though work hours might extend into the evening at most shops, the claim of industrial activities on the space was no longer exclusive. Instead, the alley became a venue for interpersonal communication and negotiation.

During events such as weddings and funerals special arrangements were made for the entire community to be mobilised, and for the public sphere to be transformed to serve the needs of a particular family. Rituals on these occasions required a physical capacity beyond the capacity of individual apartments, requiring that private space be extended into the alley and into neighbours' houses. These uses of outdoor space were particularly associated with the character of many harah as lower-income communities. Among the city's upper-middle-class population, such events might take place in specially designated but costly indoor spaces such as hotels, community centres, or social clubs.[26]

Such merging of spaces both reinforces the notion of a collective home and supports social integrity and cohesion. However, in architectural terms, it challenges the assumed conventional spatial order of contemporary houses, derived principally from the expectation that each residential unit will be independent and self-sufficient. As such, the architectural image and physical characteristics of Old Cairo have little to reveal to architects or outsiders about the actual practices of daily life. For its inhabitants, these are structured around individual, mutual and collective social interaction, restrained habits and behaviours, and historically rooted traditions and moral values. In comparison, the spatial layouts and house forms are marginal to the constitution of a sense of home. To a large extent, the system of living in Old Cairo today is able to adapt to spatial limitations through a system of synchronised activities, part-time spaces, and merged venues (Figure 12.3).

Architecture, unlike other design activities, is a situated process determined by a specific site and a certain socio-cultural context. As such, it cannot be isolated or limited in terms of influence, as discussed earlier. Moreover, architecture is also not like social sciences, which limit their scope of inquiry to constructing subjectivity; rather, according to Susan Bickford, the creation of the built environment involves the generation and entrenchment of a form of intersubjectivity.[27] To practice architecture is thus to elaborate an environment that governs such social interaction and communication. Linda Hutcheon has argued that by "its very nature as the shaper of public space, the act of designing and building is an unavoidable social act".[28] Hutcheon has further argued that architecture reinstates a dialogue with the social and ideological context in which it is produced and lived. Successful architecture, accordingly, requires that the practitioner understand the nature of the environment and give proper consideration to the everyday lives and social norms of potential occupants.

With few exceptions, most architects during the twentieth century did not take these issues into consideration in the design of buildings/houses in Old Cairo.

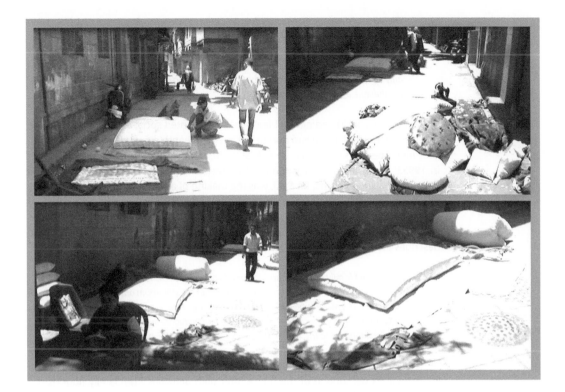

Of course, not all of them took such highly intellectual or theoretical positions as Campbell and Comodromos. But their work still demonstrated a separation between architecture as a profession and people's practice of home. According to the chief housing engineer in the planning department of Hai-Wasat (the district in Old Cairo), regulations enacted over the past two decades would require new buildings to contain a garage at ground level, despite the fact that the area's alleyways could scarcely accommodate the passage of even a small car. One response to such problems is for people to use unauthentic drawings to gain needed permits, and then build something entirely different. New regulations imposed by the National Organization for Urban Harmony (NOUH), formed in 2000, have further restricted the possibility of innovation and creativity. Concerned primarily with image, they have introduced strong restrictions on building façades, in an attempt to mandate typical openings, styles and materials.[29]

12.3   The use of harah's space for the preparation of mattresses for a wedding, an event shared by all

PROFESSIONAL ADJUSTMENTS:
A CUSTOMISED SOCIO-SPATIAL PRACTICE FOR OLD CAIRO

The emergence of socio-spatial practices in Old Cairo owes much to the prominent architect Salah Zaki Said, whose work during the 1990s reflected, for the first time, consideration for a sense of home that combined physical characteristics with lived experience. Said tried to integrate these concepts in practice by establishing a socially responsive architecture, whose main purpose was to ensure outcomes that reflected the pattern of people's lives in local traditional contexts – or, as he called it, "the lived space".

> *The study of domestic architecture is actually the only way to relate to everyday life*
> *of the people. Naturally we can tell about the customs and habits of the people easily*
> *by studying the nature and organization of living spaces in domestic architecture …*
> *We need to give stress to and find out about the roots of Egyptian architecture, not*
> *only by studying large monuments but also by studying people's habitat and domestic*
> *architecture in general.*[30]

Said's work considered old buildings that represented social and cultural values to the living patterns of its hawari during the late nineteenth century and early twentieth centuries. The house of Sokkar in Bab al-Wazir area was rehabilitated by Said's team in 1995.[31] Later, the project extended to develop another four houses in the same area. The making of home, for Said, was the making of a useful lived space, facilitating people's activities in secured and safe environments, which at the same time conserve the cultural value of the building. The home, as such, is composed of social elements (residents) and economical value (cost) as principal elements of its survival.[32] Said describes the house of Sokkar in terms of its activities pattern as:

> *… the main family floor is in the first floor. Ground floor is usually for shops on the*
> *street; the mandhara of guest reception room and the storeroom opening onto the*
> *court area. The Mezzanine is for kitchen and storage facilities as well as living of the*
> *servants. The main floor on the first floor level is entered through the stair way either to*
> *the guest room or to the sala, or the main living space in the middle of the house.*[33]

Although differing in his description from architectural historians, Said's words recall Edward Lane's description of the house after daily activities. The "family floor" is the main level in which most of routine daily activities and rituals take place. Said displayed his awareness of such detailed narratives of living patterns, what informed the comprehensive analysis of daily needs and their spatial systems. The team's response to this information was to develop a spatial layout that responded to these patterns. Said's major contribution, then, was to resolve structural problems, rather than impose a predetermined design style. His loyalty to the practice of architecture as lived experience led him to work at the fundamental and organisational levels to restore the building as a useful component of the shared home.

Said's work was pioneering in the emergence of an architecture of home in Old Cairo wherein the agenda stemmed from local everyday routine. This practice of making homes, as a consequence, supported Said's approach to the preservation of cultural history in the form of valuable buildings as lived history. Said understood the intrinsic nature of traditional communities in Old Cairo and the association between private and public spheres.

THE STRUCTURE AND SPATIAL ORDER OF AL-DARB AL-AHMAR HOMES

Departing from the comprehensive approach toward design services favoured by many architects, al-Darb al-Ahmar rehabilitation project led by the Aga Khan Trust for Culture (AKTC) aimed to empower local inhabitants by supporting them financially and technically to restore the structural safety of houses and reorganise their living spaces to better suit their needs. This would be achieved only through a continuous open-ended process of enhancing people's lives and activities within both the private and the public spheres:

[The strategy] consists of improving the area's physical assets through greater public and private investment and raising family incomes through small business loans and employment generation programmes. No all-encompassing projects and no far-fetched social engineering agendas are required; rather, what is needed is an incremental improvement of what is already in place, and a strengthening of the available social capital and positive economic trends.[34]

The trust manifested its intervention as reviving the urban history through the local process of making homes, by developing, rehabilitating old buildings or building entirely new houses. The implicit target is to put this quarter on the map of Cairene homes' production rather than consuming the past.[35] It is interdisciplinary in a way that is inclusive of social, cultural and economic aspects through adopting long-term strategies that capitalise on the existing inhabitants and their conditions/qualifications.[36] By 2008, al-Darb al-Ahmar development project became one of the exemplar projects of its type, with more than 100 buildings being developed, restored or newly built with the assistance of an extended team of specialists. This long-term project worked on three planned stages: defining the needs of current inhabitants and built fabric, maintaining professional adjustment by including multi-disciplinary teamwork, and finally improving the safety and the quality of developed buildings by incorporating creativity and contextual integration.

Occupied principally by population on low incomes, the homes in Darb Shoughlan alleyway form a similar social unit to al-Darb al-Asfar, forming an inclusive form of home that merges the spectrum of social spheres. The residents are predominantly workers living on often unstable incomes from skilled work like carpentry[37] and work within the vicinity of al-Darb al-Ahmar. The instability and insufficiency of family income impacted on the maintenance and the quality of living spaces. Housewives had to maintain local domestic resources/supplies such as raising livestock, using shared services such as water supply and drainage points in the local street. The 2003 surveys show the local inhabitants to be the poorest in Egypt with average annual income level less than 200USD, which means less than a dollar per day.[38] However, the inescapable deterioration of the buildings was exacerbated by the imposition of unrealistic rent controls, counter-productive planning constraints, limited resources and access to credit.[39]

The organisation of homes in this context is composed of blurred organisation in which public and private spheres are merged in a disordered mix of activities due to the lack of space and affordability. Many families occupy one or two multi-purpose rooms as their living spaces while sharing the services (kitchen and bathrooms) with other families. Social spheres are, therefore, confusing and disorganised: as is exemplified in the absence of boundaries, borders or thresholds among different spheres of activities. Residents in the Zuqaq al-Ezzy building complained about the lack of privacy in the most intimate spaces; their cooked meals (representation of affordability) are exposed to their neighbours, and privacy of women (represented in the frequent visits to toilets and kitchen) is violated.[40] While solidarity and participation are essential assets, for them, absence of privacy, when needed, is a problem. Such blurred organisation/hierarchy of spheres and lack or privacy caused havoc among resident families (with high potential for conflicts and frustration) and is an implicit reason behind ignorance, deterioration and lack of concern about their built environment. The public sphere, on the contrary, was an essential participatory venue for daily practices, for females spend significant time in the alley space washing dishes, cooking, and chatting. Women are used to moving between their local social spheres (public and private) without change or enhancement of their clothes/costumes.

The case of several houses in Darb Shoughlan provides a good illustration of the project's goals. Bayt no. 5 in Haret El-Ezzy, for example, was replanned, with some spaces omitted or merged, and others added.[41] Realising the strong association between the houses and residents, community and inhabitants, the project used traditional social ties as a resource for positive change. This strategy was intended to embody the essence of place, identity and culture, based on everyday patterns and interaction. According to the project documents: "Preserved and respected for their intrinsic qualities, the monuments, old buildings and traditional open spaces must be integrated into the everyday life of the residents and reconnected to the complex, multidimensional social and cultural character of the area".[42]

Architects, joined by an extended team of social workers and economic specialists, using Needs Assessment Strategy (NAS),[43] interviewed the inhabitants and monitored the daily activities of the community with the aim of determining the socio-spatial practice and social progress.[44] The interviews revealed the residents' appreciation of the context and community participation at the public sphere level, while they were concerned about the lack of stable, secure jobs, sufficient privacy in their habitual units, both public and private. They stressed the need for controlled boundaries between their private units of habitation to avoid undesired exposure (particularly for women) to other people.[45] Architects responded to such appreciation of the context by scrapping the proposals for radical clearance and sanitising of the district.[46] For them, importing inhabitants does not match with the idea and practice of home, which relies on the loyalty and intimacy between people and people and people and buildings. Moreover, they have translated the concerns about blurred organisation and lack of privacy into a set of requirements to reorganise homes into ordered and organised activity-based spaces.

Following the investigation, a continuous high demand for homes became apparent among existing local residents, who, for various reasons, are unlikely to leave the harah.[47] As a result, the socio-economic study and interviews with local residents concluded that unless there is rectification of the blurred organisation, the current patterns of deterioration will persist:

> If the present pattern of abandonment and disinvestment persists, it can only pave the way for further deterioration and the eventual demise of irreplaceable social, economic, and cultural assets. It will also deprive the district of the critical mass of inhabitants needed to sustain the area's social and economic life.[48]

As part of the author's fieldwork, everyday situation in one of the project sites in al-Darb al-Ahmar, could be portrayed as follows:

> A family living in a single room, their immediate neighbors live in two bedrooms and both share the same kitchen which lies outside their apartment. The only clean water source is in the alleyway among many livestock. One bathroom in bad condition exists at every level and is shared by a few families. The sanitary services have collapsed, and waste water is running into the street. The building is deteriorating due to the flow of waste-water that attacks ground-level load-bearing walls. Residents have no other place to move and don't have enough cash for the repairs.[49]

In such situations architects must liberate themselves from the constraints of design standards at the same time they must reject the clearance option and work toward the production of feasible homes.[50] The spatial order in such multifamily living units is frequently confused, and the idea of home is blurred. For example, intersecting patterns of movement may hinder the privacy of supposedly private paths (bedroom-bath, living-kitchen). In one house targeted by the AKTC work the logical reordering of such a confused system involved planning each unit to comprise living/sleeping space(s) augmented by two basic service spaces: a kitchen and a bathroom.

Reflecting typical arrangements in Old Cairo's communities and accepting the absence of a complete suite of social spaces in each house, the architects engaged in the AKTC effort, however, could also capitalise on the inherited social organisation to make use of alley space as a haven for public and social activities. Such professional adjustment represented a creative strategy to deal with an unconventional and loaded situation. But it required comprehensive knowledge about everyday patterns of activities, family structures, and shared as well as individual needs in Old Cairo. Based on gathered information, every level in the house was ultimately re-planned to suit residents' needs and to provide each family with their required level of privacy while capitalising on the harah for semiprivate activities.

Architects had to work out autonomous organisation, in which the house unit avoids intersections with others, and allow the private sphere to be contained spatially. For example, every unit should include a separate kitchen and bathroom. The rule is that each unit should be composed of living/sleeping space(s) and two basic service spaces, a kitchen and a bathroom (Figure 12.4). While the former space(s) should enjoy proper natural lighting and ventilation, latter services should be sufficiently ventilated through a light well. In the absence of spaces for social and public activities, architects capitalise on the use of alley spaces as a haven for public and social activities, a traditional and cultural practice deeply rooted in everyday life. In reordering the home, the house had been analysed in terms of its principal elements, activity spaces, based on actual everyday practice. It defined every family's paths and pattern of activities. Intersected paths and issues of privacy are identified, and problem areas and potential solutions are explored through discussions with the residents. Following this communication exercise, principal organisational adjustments are agreed and architects are given the go-ahead with their drawings and technical developments. The modification, retrofitting and restoration budget is set and the residents and funding agencies agree on their shares, and the method of payment. During this process, we expect certain proposals to be favoured and others dropped, based on affordability for the residents.[51]

This socio-spatial architectural practice required a comprehensive knowledge about dominant architectural style, family structures and shared as well as individual needs. A record of individual as well as collective activities and living patterns was made for every building, following interviews with their residents. Departing from this gathered information, every level was re-planned to suit residents' needs and to provide every family with their required level of privacy while capitalising on the harah for semi-private activities. As a result, every level has had a distinct layout according to the needs, while structural walls are kept to support the building's structure. Such practice conforms to the process of design of Hassan Fathy and Peter Hübner in terms of the close investigation of socio-cultural structures and needs of the potential residents, followed by professional adjustment of their work. However, it differs in that it is a process of production in an existing, long-established and grounded context, not entirely new construction as in those two cases.

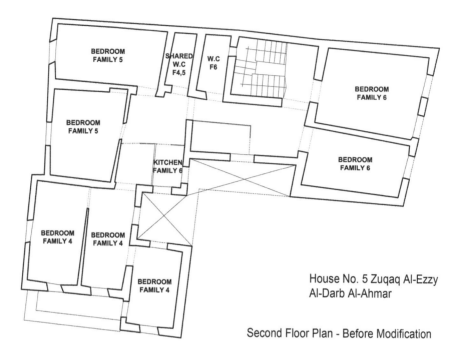

BEDROOM
FAMILY 5

SHARED
W.C
F4,5

W.C
F6

BEDROOM
FAMILY 6

BEDROOM
FAMILY 5

KITCHEN
FAMILY 6

BEDROOM
FAMILY 6

BEDROOM
FAMILY 4

BEDROOM
FAMILY 4

BEDROOM
FAMILY 4

House No. 5 Zuqaq Al-Ezzy
Al-Darb Al-Ahmar

Second Floor Plan - Before Modification

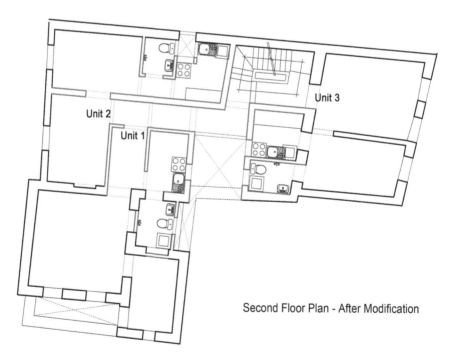

Unit 2

Unit 3

Unit 1

Second Floor Plan - After Modification

12.4   Professional adjustment:
planning second-floor organisation to provide autonomous housing unit with private services

Courtesy of Aga Khan Trust for Culture (2009).

Critics of such practices could raise issues of stylistic quality, architectural image, and the collective character of home. In this regard, the buildings, although not noteworthy as individual works of architectural imagery, are of historical significance for two reasons: their simple façade treatments are representative of an important type of Cairene architecture; and they establish a coordinated order as we note from their modular patterns, entrances with stone arches, and ground-floor sandstone walls.[52] But what mattered to the residents was the flow of indoor-outdoor activities they allowed. Even personal safety and the hygienic quality of the environment were secondary concerns to social cohesion and the presence of a supportive and secure community. Such priorities were ultimately a professional obstacle for the architects, who saw little creative benefit to rehabilitating and reproducing homes without also being able to have a stylistic impact.[53] Critical analysis of the final product, hence, remains problematic for our inability to agree on appropriate criteria to evaluate them.

Any evaluation, hence, should be based on how effectively family needs and activities were maintained between the indoor and outdoor spaces. This only can be traced in the long term and through periodic observation and recording of people's behaviour and interaction in the produced spaces. For example, in Christopher Alexander's Mexicali, this interaction was regarded as highly active and suitable for the residents immediately after completion, while the project was deserted within a short period afterwards. Successful spatial organisation can be detected when social practice is sustainable and active on the long term, as only this makes the community and the inclusive home a viable living environment. Moreover, we learned that people in Old Cairo usually alter the uses and organisation of designed spaces to facilitate and enhance their daily lives and activities under changing conditions.

Creativity is envisaged by the spatial engineering of the previously mixed and blurred individual spheres, which users stress as a priority. It appears at the level of strategic decision in, for example, returning facades to their original styles and features with neither additions nor new creations; also as managing the proper venues for private and public spheres for families lacking clear spatial organisation for their activities. Architecturally, successful design, in terms of the facades or forms, is translated into contextual enhancement and improvement in the quality of the collective scenery, rather than changing it (Figure 12.5). The quality of the new-old home, hence, could be seen through the integrity of those homes in existing living patterns and daily activities without challenging particular social or cultural values.

Modifications of space organisation are usual practices that allow continuity and adapted homes to be in action for a long time. To evaluate this practice in Old Cairo, two useful tools are helpful, but not definite, as short-term indicators; the first relies on the residents' immediate perceptions, activities and interactions within the new homes. This is traced through the spatial activities taking place (indoor/outdoor) and kind of internal modification of space-activity combination. The second requires professional assessment of the performance of that home, through independent specialists' inspection according to predefined criteria. As we are yet to have either tool implemented in this project, neither tool can detect the quality of the production in pure architectural terms. The combination of both interviews and observation over a certain period of time could, in that sense, be a useful tool to review architectural production as a socio-spatial practice.

12.5
Redevelopment of façade at houses nos 15–17 Atfet Hozayn, al-Darb al-Ahmar

Note: Left: before upgrading, right, after completion.

Courtesy of Aga Khan Trust for Culture (2009).

＊＊＊＊

During the Egyptian Revolution in January and February 2011 police were largely absent from the streets of Cairo. However, during this time communities across Egypt mobilised to form *lijan sha'biyyah* (public patrols), whose job was to guard residential areas against attack by criminals and gangs. These public patrols were a creative and immediate response to a sudden collapse of the national system of security. It was surprising how quickly the patrols were formed and how efficiently they managed to maintain security across a nation of more than 85 million people. The practice of collective defence of a shared home was clearly still present in the collective memory of Egyptians, and in reviving it, they were merely recalling a deeply rooted tradition at a time of need. In the absence of former determinants of social hierarchy, everyone had a role to play in ensuring local security, with businessmen, doctors and intellectuals attending to their duties and shifts.

Through the constant practice of home, we produce and consume the spaces in which we live. This happens through the frequent rearrangement, merger, and division of available space. By looking closely at the pattern of daily activities and the way furniture is synchronised, architects may discover the practice of contemporary home. In Old Cairo this revealed the notion of part-time arrangement, an efficient system of space management that is at work on a daily basis. The research showed the system of part-time usage to be especially practical when spaces are not sufficient to accommodate all activities at one time. In these examples, space and time essentially became associated within the organisation of the social sphere.

An understanding of part-time spaces could be beneficial to the design of new high-density residential environments. Acknowledging the flexibility of the social sphere could liberate architects from restricting spatial requirements and enable them to design shared social venues, multipurpose spaces accommodating the temporal synchronisation of daily activities. This could be an especially important strategy in designing residential communities for working families or households where work is accomplished in the home.

The hawari of Old Cairo provide a comprehensive and historical construct of the idea of home, represented and manifested in the dynamics of everyday life and its socio-spatial associations. To remain positive agents of change, architects need to learn the history and processes by which Cairene homes have evolved in response to everyday needs. Present professional knowledge is lacking in terms of making lived spaces that are peculiar to a traditional context such as the old city. A new architecture of home in Old Cairo, thus, needs to embrace a collaborative socio-spatial practice, in which architects learn the dynamics of local contexts and help provide effective responses to daily needs. In this sense, creativity and innovation in architecture might be more strategic and more responsive.

In the wake of the Egyptian revolution of 25th January 2011 and subsequent decline of the power of the police state, there were sad developments in large parts of the old city. Several brick apartment-towers quickly aroused to damage the traditional fabric and skyline in several protected and preserved quarters, diminishing the achievements of UNESCO projects of rehabilitation and those of the Agh Khan Trust for culture. the rise of these apartment buildings by opportunist developers and landowners would add pressures to the practice of architecture in this part of the city. It further highlights the ills of such top-down management system, the sudden absence of which would culminate in devastating circumstances. It also makes a stong case for urgent action to facilitate the engagement of architects in local contextual issues and acknowledging local needs and demand. Only through such engagement, there will be awareness of the risks associated with these uncontrolled developments. Similarly, programmes of capacity building and support of local interest groups seem to be the best option to instigate local order and management as was the case for centuries.

NOTES

1   Earlier version of this chapter was published as an article in the *Journal of Traditional Dwellings and Settlements Review* as: "The practice of home in Old Cairo: Towards socio-spatial models of sustainable living", 23(2), 2012, pp. 35–50.

2   Francesco Siravo reporting the outcomes of the survey conducted by AKTC at the beginning of the al-Darb al-Ahmar project. See Siravo, *Urban Rehabilitation and Community Development in al-Darb al-Ahmar*, p. 178.

3   M. Heidegger, "Building Dwelling Thinking" (1951), in D.F. Krell (ed.), *Martin Heidegger: Basic Writings* (London: Routledge, 1993).

4   R. Campbell and D. Comodromos, "Urban Morphology + the Social Vernacular: A Speculative Skyscraper for Islamic Medieval Cairo", *Journal of Architectural Education*, 63(1) (October 2009), pp. 6–13.

5   S. Bianca, *Urban Form in the Arab World* (Zurich: VDF, 2000); and Campbell and Comodromos, "Urban Morphology + The Social Vernacular", p. 7. Egyptian architects and historians have criticised Bianca's theory for its abstraction of medieval Islamic urbanism into a few morphological typologies.

6   These included a mechanical substation, burial chambers, and hospitals below ground; and a *madrassa* (school), green bridge, housing, and mosques above.

7   Campbell and Comodromos, "Urban Morphology + The Social Vernacular", pp. 9–10.

8   An approach based on the postmodern abstraction of the past was highly noticeable in the article, with the authors using the term "vernacular" several times when speaking of their design rationale.

9   Z. Bauman, *Liquid Modernity* (Cambridge: Polity Press, 2000), p. 11. Bauman's concept of distant authority refers to "the end of the era of mutual engagement: between the supervisors and the supervised, capital and labour, leaders and their followers, armies at war".

10 Here I do not suggest that architects should not target change; rather, I incline towards Schneider and Till's view of the architect as an agent of change in the framework of architecture as "spatial agency". Change should not be driven by the architect's personal bias as much as by concern for the real needs of the users. For more, see T. Schneider and J. Till, "Beyond Discourse: Notes on Spatial Agency", *Footprint*, 4 (Spring 2009), pp. 97–111.

11 Heidegger, "Building Dwelling Thinking".

12 Ibid.

13 See, for example, F. Ghannam, *Remaking the Modern: Space, Relocation, and the Politics of Identity in a Global Cairo* (Berkeley: The University of California Press, 2002); E. Early, *Baladi Women of Cairo* (London: Lynne Rienner, 1993); and H. Hoodfar, "Women in Cairo's Invisible Economy", in R. Lobban and E. Fernea (eds), *Middle Eastern Women and the Invisible Economy* (Gainesville: University Press of Florida, 1998).

14 M. Yaari, *Rethinking the French City: Architecture, Dwelling, and Display after 1968* (Amsterdam: Rodopi, 2008), p. 306.

15 Delaney, *Territory*, p. 71.

16 Interview by author [R1.1.07, R4.1.08].

17 Interview with Dina Shehayyeb, Cairo, August 2009. Shehayyeb is an Associate Professor at the Housing and Building National Research Centre in Cairo and a leading Egyptian specialist in hawari environments and their socio-cultural context.

18 H.A. El-Kholy, *Defiance and Compliance: Negotiating Gender in Low Income Cairo* (Oxford: Berghahn publishing, 2002).

19 P. Bourdieu, *Outline of a Theory of Practice*, translated by R. Nice (Cambridge: Cambridge University Press, 1977), p. 77.

20 N. Keddie, *Women in the Middle East: Past and Present* (Princeton, NJ: Princeton University Press, 2006).

21 N. Hanna, *Cairene Homes of the Seventeenth and Eighteenth Century: A Social and Architectural Study*, translated by Halim Tosoun (Cairo: al-Araby Publishing, 1993).

22 E. Early, *Baladi Women of Cairo*, p. 68.

23 The fieldwork in Old Cairo between 2006 and 2009 confirmed this pattern of activities in the alley space.

24   J. Eldridge, *Max Weber: The Interpretation of Social Reality* (London: Michael Joseph Publishers, 1971), p. 26.

25   M. De Certeau, *The Practice of Everyday Life*, Vol. 1 (London: University of California Press, 1984).

26   Y. Elsheshtawy, "Urban Transformation: Social Control at al-Rifa'I Mosque and Sultan Hasan Square", in D. Singerman and P. Amar (eds), *Cosmopolitan Cairo: Politics, Culture and Urban Space in the New Globalized Middle East* (Cairo: The American University Press, 2006), p. 303.

27   S. Bickford, "Constructing Inequality: City Spaces and the Architecture of Citizenship", *Political Theory*, 28(3) (2000), p. 356.

28   L. Hutcheon, "The Politics of Postmodernism: Parody and History", *Cultural Critique*, 5, "Modernity and Modernism, Postmodernity and Postmodernism" (Minneapolis: University of Minnesota Press, 1986), p. 181.

29   Prime Minister Decision no. 2,003 for the year 2007 was the first to consider particular requirements for Old Cairo, even though it remained very general in its terms and conditions and lacked technical precision.

30   S.Z. Said, "Cairo: Rehabilitation with People Participation", *Alam Al Bena'a*, 216 (1999), pp. 6–9.

31   Said, ibid., p. 8.

32   Said, ibid., p. 8.

33   This organisation is different from El-Dar, which is to a large extent similar to the investigated houses of al-Suhaimy and Mustafa Ja'afar. See, Said, ibid., p. 7.

34   AKTC, *Cairo: Urban Regeneration in the Darb al-Ahmar District*, p. 2.

35   Sutton and Fahmi, *The Rehabilitation of Old Cairo*, p. 76.

36   AKTC, *Cairo*, ibid., p. 3.

37   Siravo, *Urban Rehabilitation*, ibid., p. 177.

38   AKTC (2005), *Cairo*, ibid., p. 6; People spend 50 per cent of their income on food items. The minimum rent, in this case was essential to family survival. Neglecting maintenance and guaranteed deterioration is a consequence of those lower rent strategies.

39   Siravo, ibid.

40   Interview by author [l3.1.09].

41   Interview by author [l1.1.09].

42   Aga Khan Trust for Culture: Historic Cities Support Programme, *Cairo: Urban Regeneration in the Darb al-Ahmar District: A Framework for Investment* (Rome: Artemide Edizioni, 2005).

43   Shehayyeb, *Tradition, Change, and Participatory Design*, p. 1.

44   The diversity of local activities, live interaction within public spaces as well as the sense of solidarity and support were seen altogether as viable assets that have supported the community for a long time.

45   Interview [l6.1.09].

46   Siravo, *Urban Rehabilitation*, ibid., p. 177; the project team was opposed to the plans calling for radical clearance and sanitising of the district, which, according to them, pose "*another threat to the survival of the historic urban fabric*".

47   Local support, family solidarity and shared resources are of principal importance to local families who refused to abandon their harah. Affordable alternatives are on the deserted periphery of Cairo and do not provide social support or solidarity in hard times: which influenced residents' desire to remain in the area at all costs. Anyone who is forced to leave the harah to get affordable housing, usually awaits the first opportunity to move back to the area [I8.1.09; I6.1.09; I3.1.09].

48   Siravo, *Urban Rehabilitation*, ibid., p. 183.

49   Author's first-hand description of Zuqaq El-Ezzy, in al-Darb al-Ahmar.

50   F. Siravo, "Urban Rehabilitation and Community Development in al-Darb al-Ahmar", in S. Bianca and P. Jodidio (eds), *Cairo: Revitalising a Historic Metropolis* (Turin: Umberto Allemandi and C. for Aga Khan Trust for Culture, 2004), p. 182.

51   A similar process of budget justification used to take place in the traditional process of building houses in Old Cairo.

52   S. El-Rashidi, "The History and Fate of al-Darb al-Ahmar", in Bianca and Jodidio (eds), *Cairo*, p. 61.

53   Interview by author [I2.1.09].

13

Architecture and the Construction of Memory[1]

> The reality and reliability of the human world rests primarily on the fact that we are surrounded by things more permanent than the activity by which they were produced.
> Hanna Arendt, *The Human Condition*[2]

In the opening scene of *The Palace Walk*,[3] Naguib Mahfouz, the Nobel Prize-winning novelist, portrays a night view of Old Cairo through the eyes of a wife awaiting her husband's return at midnight in the early years of the twentieth century.[4] She looks at the alley from a small opening in "*a cage-like wooden latticework*" balcony overlooking an ancient building housing a cistern and a school. Looking left, the alley becomes narrow and twisting with busy coffee houses at street level, while to the right it is engulfed in darkness. Ending the scene, Mahfouz hints to the association between ancient buildings and the memory of the wife: "*There was nothing to attract the eye except the minarets of the ancient seminaries of Qala'un and Barquq, which loomed up like ghostly giants enjoying a night out by the light of the gleaming stars. It was a view that had grown on her over a quarter of a century*".[5]

As a novelist who lived in Old Cairo during his childhood, Mahfouz was conscious of the influence the presence of historical buildings has on shaping the memory of the inhabitants: like the daily silent dialogue between the wife and the historical minarets. The presence of such historical imprints is, however, inherently loaded with more than the mere picturesque value of the image;[6] while the centuries-old minarets represented traces of a society that had vanished, its heritage and traditions remained part of everyday activities and relationships. The woman was, for Mahfouz, the perfect character to demonstrate his criticism of certain inherited social constraints, such as gender inequality, the tyranny of man, and female imprisonment. The combination between the images of ancient buildings and the recurring night view of the imprisoned wife, hence, was intentional in the sense that everyday narratives in the old city are informed by the continuous presence of the past and the evidence on that past was its architecture. Architecture, in this context, acts effectively as a reminder of not only the traditions and culture of the past, but as an agent for everyday narratives of men and women.

But, in reading Mahfouz's text, we have to be careful of such negative interpretations of the past. The presence of historical structures could have negative connotations for the imprisoned wives, but it could also be a powerful element of collective memory that reasserts identity, social status and resilience against the challenge of time.[7] The very existence of these buildings is seen as a sign of the immortal nature of the place that is as important as the state of the individuals. Monumental space, according to the French philosopher Henri Lefebvre, offers individuals an image of their membership of society, an image of social visage that collectively is more powerful than any personal representation.[8] Mahfouz's interpretation, hence, touches upon the deep values architecture holds beyond its image and experience. In fact, it recalls the above assertion of Hannah Arendt that while we rely on material objects that we produce ourselves, they last much longer than we do and are therefore able to inform following generations.

No matter the evidence of its durability, some critics remained unconvinced of the significance of architecture as a conveyor of cultural values and meaning down the generations. In the Art of Forgetting, Adrian Forty rejects the claim that memories formed in the mind can be transferred to solid material objects.[9] For Forty, objects themselves cannot mysteriously communicate knowledge or experiences in a way that will influence the collective memory. Nevertheless, it is also argued that the presence of buildings of the past reinforces national identity, reasserts specific value systems, and that is precisely why architecture becomes a significant target at times of ideological change or times of war.[10] In this sense, architecture and space work as meaningful cultural products of a group of people at a defined place and time in response to certain social needs and systems. There is no doubt that space can be a signifier. It is what it signifies that is subject to interpretation, distortion, disjunction or substitution.[11] Here arises a question about the extent to which historical imprints, as seen in inherited spaces and buildings, influence the social practice of a particular community.

Implicitly, Mahfouz's Trilogy (Palace Walk 1956; Al Sukkariya 1957; Palace of Desire 1957) considered such multilayered relationships as inherent to the development of the Cairene society. He went on in the Trilogy to display contrasting images of the traditional harah (literally, alleyway) community, with its conservative culture and social constraints, on one hand, and the more liberal and vibrant domains of the modern quarters of Cairo on the other.[12] So, how relevant is the presence of old structures to everyday social practice in Old Cairo? In other words, how does architecture work as an agent of cross-time communication within one specific place? This article, in responding to this question, explores this relationship between architecture, memory and everyday social practices through determining the way architecture moderates community experiences and communicates narratives among generations in haret al-Darb al-Asfar in Old Cairo.

ARCHITECTURE AND THE CONSTRUCTION OF MEMORY: TIME, SPACE AND PRACTICE

Does, as Aristotle once claimed, memory relate to the past?[13] Or is it about the narratives of human experience in the present? It could be argued that memory intrinsically contains connotations of the present and its contemporary relevance is greater than its relation to history.[14] There is a consensus of research that declares that memory is a contemporary cognition embodied in everyday activity, rituals and even social interaction.[15] Memory exists in the present to inform our habits, institutions and way of life, not just as some remote archaeological presence: *"Memory, we could say, is the fabric of a community's way of life"*.[16] Architecture, in that sense, becomes the most durable part of this fabric, through objects of remembrance, but is paradoxical and contested when it comes to collective memory:[17] Does it belong to the past, or is it part of the present? On one hand, buildings are representative of their age (material, construction systems and style) and also provide the physical evidence on certain social systems and rituals (room layout, accessibility, spatial configurations). The harem quarters, house layouts, and the wooden lattice windows are, in this sense, coherent evidence of the way women were protected as well as controlled in medieval houses in Cairo and other places in the Middle East (Figure 13.1).[18]

On the other hand, buildings are seen as signs of forgetting old times, of a society disappearing and leaving behind only its physical manifestations: buildings and spaces. But, is the memory embedded in the space or in the mind of its inhabitants? For the French sociologist Maurice Halbwachs, renowned for his concept of collective memory, the inhabitant's memory becomes indelibly inscribed in space, given *"the strange potential of spatial memory (as opposed to temporal memory) to conjure up dense web of images. The process of localization of memories takes place … because I have often crossed and in every direction the area adjacent to my house"*.[19] For a long time we believed that our memories could be transferred to solid material objects, which would come to symbolise these memories and, by virtue of their durability, preserve them in perpetuity.[20] However, such mysterious embodiment of memory within material objects and spaces has been questioned by many scholars and researchers.[21] Spaces cannot magically embody memories by virtue of their existence, just as the city cannot disclose history without supporting narratives.[22] But, in fact, in terms of societies, material objects retain limited if any significance in perpetuating collective memory. Rather, societies retain their memory through continuous and sustainable performance of acts, rituals and normative social behaviour.[23]

But, the question, perhaps, needs to be rewritten: is memory about history or about everyday practice of living? Is it about the desire for remembering or about the fact of forgetting? In his introduction of *The Art of Forgetting*, Adrian Forty questioned the value of buildings as objects of remembrance and as agents for the collective memory of society.[24] Artefacts become a memory when they cease to exist in people's minds; once they are forgotten they then need to be materialised into objects, artefacts, or even buildings. The best case by which to illustrate Forty's argument is, probably, Joelle Bahloul's investigation of the spatial system of Algerian houses as instilled in the minds of migrant Algerian Muslim/Jewish families in France.[25] Uprooted and dispersed abroad, former neighbours constantly re-created their previous way of living and social structures through the architecture of their houses in terms of internal boundaries and shared spaces.

In-depth investigation of the way collective memory is constructed suggests that it could be seen as a mindset of layered or overlapping events that correspond to a specific place, time and people. It is, according to Paul Connerton, *A Social Performance*.[26] Architecture shares with memory both time and space that, as basic philosophical categories, are particularly appropriate to a reading of cultural history, because they are comprehensive, universal and essential.[27] Knowledge of the past shapes the guidelines by which present activities and living conditions are measured and appropriated. However, while this knowledge can be communicated through books, documentaries, and articles, only the space-story combination (event/incident) is capable of expressing it at the popular, everyday and intergenerational levels. While every age and generation has a distinctive sense of the past, people, especially in traditional quarters, view their history (through the proximity of its everyday existence) as a guardian of stability in the face of rapid technological, cultural and social change.[28]

For the contemporary inhabitants of Old Cairo's enduring communities, adhering to inherited traditions and social norms is a strategy of conscious resistance to the overwhelming power of modernity that threatens to challenge local authority, social structures, or value systems. Ancestral houses provide the physical evidence that materialises old experiences and situates events and incidents in their specific space and place. Bayt al-Qadi (the house of the state judge at #11 Haret al-Darb al-Asfar) is, for example, the primary reference for high profile figures of al-Darb al-Asfar community during the mid-twentieth century; while the monumental Bayt al-Suhaimy (seventeenth–eighteenth centuries) is the undisputed example of the wealth, lifestyle and social constraints the community once had. To the same effect, places connected to incidents or a distinct group of residents are registered as situated memory that is used to reinforce shared history and value systems. The lane where a notorious murder was recorded in Haret al-Darb al-Asfar during the twentieth century was renamed by the people: *The Lane of the Murdered* (*Haret al-Qateel*) and Bayt al-Kharazi was remembered for once being occupied by rural migrants with a specific but vulnerable lifestyle.[29] Commemorating incidents by renaming places is a unique way to bring the past into everyday practice of the present and a clear statement of the changing meaning of the place. It is, according to Hilde Heynen, when a tragedy occurs that places seem to gain new, uncanny connotations as sites of unprecedented events.[30] Buildings and places of the past are just spatial and physical containers that situate verbal memories that are meaningful for the group and to some extent are constitutive of the group's identity. In these events consciousness of the past is perpetuated, albeit unintentionally and non-physically, through local strategies of active remembrance that create distinctive monuments.

HOME AND COMMUNITY IN THE MEMORY OF OLD CAIRO

> *Remembrance of concrete experience is structured in terms of two main fusing dimensions: domestic space and family time. Events are not remembered simply as they were experienced by the family and the domestic community. Memory draws the boundaries of the family and domesticity by shaping within them local, regional, and international events. The domestic and family world makes up the woof of remembrance, of memory. The house is inhabited by memory. Remembrance is moulded into the material and physical structures of the domestic space.*
>
> *Bahloul ibid.: 28–9*[31]

If the remembrance of concrete experience is an act to resist forgetting, then the desire to remember is invoked by precious values/elements that would be lost if forgotten and which provide substance to the existence of communities. According to the interviewees in Old Cairo, despite the poor services, quality of living, and financial deprivation, Old Cairo's hawari were incontrovertibly their homes, which they could never abandon. The merging of notions of home and community in relation to the harah occurred quite frequently in conversations with the locals, but, at the same time, this rationalises the desire to remember the past. According to Bahloul, memory can shape the boundaries, interactions, morals within private and local events, and it is more discrete in the domestic environments. Bridging the gap between the home and community, therefore, could be significant to the understanding of *what to remember* and how architecture *moderates* this process of remembrance. So, how does a harah become home?

The Cairene harah establishes its justification as home by responding to its occupants' fundamental needs in everyday reality; a home-like "collectivity" that is more than "the sum of its individuals".[32] While the medieval harah was a mixed community of multiple social classes, contemporary hawari, in contrast, are dominated, as we discussed earlier, by lower class population, and, increasingly, by commercial and industrial activity in their narrow alleys as discussed above. The interchangeable roles of privacy and frequent movement between the private and public in the harah invoke a strategy of blended boundaries in which private and public memory and history are anchored together in one inclusive domain of home. Samia Mehrez noted, in *The Literary Atlas of Cairo*, that the representations of private spaces in Cairo are usually anchored in particular public spaces – a neighbourhood or an area – whose history is not only written into the private spaces themselves but is also responsible for shaping and defining their very existence as private.[33]

THE SELECTIVE CONSTRUCTION OF MEMORY IN OLD CAIRO

At the community level, collective memory encompasses a distinct meaning system as it fosters and reproduces a unified cultural identity estranged from its territorial basis. *"Remembering the house in which an uprooted culture originated and developed involves reversing history and sinking symbolic roots into a vanished human and geographical world"*.[34] When people find little significance in their physical surroundings, they recall the images of old homes and their layouts: the spatial images that are so important in shaping the collective memory of a group. However, the influence of memory cannot be traced directly or rationally in the physical form of space. Rather, it could be interpreted through spatial order, organisation and sometimes in artefacts and decoration. There are pieces of furniture that recalled styles and fashions from the first half of the twentieth century in houses occupied by families with extended histories in the locality, as highlighted earlier. Those inherited elements, like the poetic images, reverberate the past through a progressive form of development that is very distinct from the nostalgic representation of old times.[35] The past is experienced in everyday spatial practice of the present, and to some extent guides what should be practised in the future. The now and then are experienced simultaneously, they feed into each other, there is no before and after. Linear time appears to be frozen, and replaced by the recurring order of the everyday.[36]

The medieval urban maze, of extremely short, broken, zigzag streets with innumerable dead ends,[37] is becoming busy and full of commercial traffic and relatively tall apartment buildings. The harah today is a site of paradox, a contest between the medievality of its domestic environment and the modernity of its busy commercial and industrial life: between the past and everyday practice of remembrance, supported by the presence of old buildings, on the one hand; and the forces of the present mobility and uninterrupted movement of workers on the other. Interestingly, this paradox has resulted in socio-spatial segregation between those who live near main traffic routes and those populating inner parts of the alley. A descendant of an extended family reinforced this segregation in stating: *"This is our building that we inherited from our grandparents, I have lived here since the 1930s ... we are different from them (formerly rural families who reside in inner areas) we don't party with them but we attend some of their formal occasions*, because we have to do that" [R23.1.09].

Although central to its present character and layered experiences, the collective memory and history of the old city as we know it today is incomplete. A significant part of that history has been lost due to the selective destruction of Old Cairo's built heritage: especially destruction of buildings from the Shiite era (tenth–twelfth centuries AD) under the Sunni Ayyubid rule during the twelfth century. The Shiites' iconic structures and mosques were entirely destroyed, with the principal aim of eliminating any traces of the traditions, rituals and heritage of the Fatimids. In preventing the adaptation of Fatimid structures to new functions, the Ayyubids showed understanding of the meaning of the very existence of buildings. They aimed to carve a customised memory by erasing any traces of the Fatimids' existence, and to achieve this goal only the total destruction of their built heritage would suffice. After all, architecture, according to Robert Bevan, is a *"persuasive tool"* that by its *"selective retention and destruction can reconfigure its historical record and the facade of fixed meanings brought to architecture can be shifted"*.[38]

This negative approach towards the past was not exclusively a medieval attitude. Muhammad Ali's project for Modern Egypt emerged in the early nineteenth century, with the explicit intent of breaking the continuity with the past. Ali forced people to abandon the mashrabiyya (lattice wooden window) and promoted European-styled Rumi Windows.[39] Painting buildings in white and breaking down the hawari gates were further steps in a comprehensive and consistent process of eliminating the traces of the past in the present. This was coupled with the imposition of new typologies of house design that were inspired by European models. Such negative attitudes still existed in the late twentieth century, although they were manifested implicitly through deliberate neglect of the maintenance or conversion of the valuable buildings of Old Cairo, which resulted in widespread deterioration and collective destruction.[40]

In general, such attitudes towards the past were not uncommon. They emanated from elitist views of modernity which opposed the *negative* traditional values of the past as represented in the hawari of Old Cairo. This was a consequence of a longstanding dichotomy between tradition and modernity that was accepted, according to Hania Sholkamy, by the majority of Egyptian social scientists, aligning themselves with examples from American sociology, or perhaps, inheriting similar ideas from French-educated Egyptian elites of the late nineteenth century in whose view the creation of a modern city required traditional quarters to be abolished or ignored.[41] Both the Egyptian government and planning institutions in Cairo perceived the hawari of Old Cairo as a problematic context, at odds with the desired modern image of the city.

However, recently this understanding has changed and the traditional quarters have assumed centre stage in the construction of modern Cairo as an authentic city with a long historical continuity expressed through its built heritage. But this raises the critical issue of whether historical continuity and memory are contradictory notions. Historical continuity reflects continuous existence, with no separation from the past, while memory implies remembrance of what is otherwise forgotten; distanced from current everyday practice, in short, something that does not exist anymore.[42] In this sense, Forty's argument about forgetting is relevant. The continuing existence of medieval houses in a contemporary harah is, arguably, similar to the presence of the traditional quarter within the modern city in being informative in guiding new forms of development and in some respects reasserting the authenticity of the community and its social systems.

We can, accordingly, refer to three principal paths through which the contemporary hawari have been guided by the continuing existence of their past; and where architecture has been instrumental in moderating the experiences of that past. These are: spatial transformation, the continuity of social practice, the value of the historical fabric. Each of these paths links one aspect of contemporary social practice to its past.

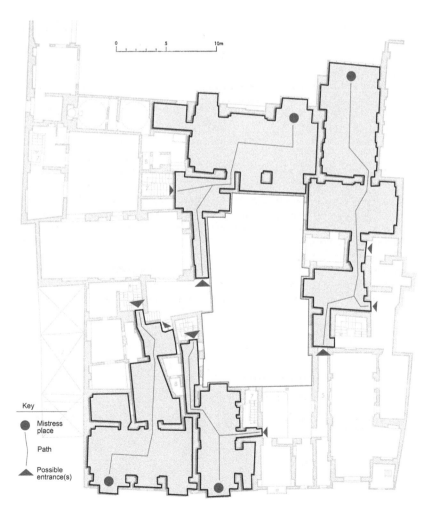

13.1 The labyrinth of spatial experiences in historical courtyard houses to get to the harem inner spaces

0   5   10m

Key

● Mistress place

⌐ Path

▲ Possible entrance(s)

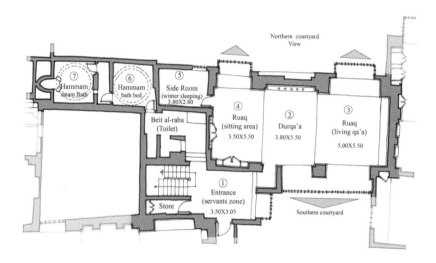

13.2   Detailed spatial organisation and furniture layout of one of the harem quarters

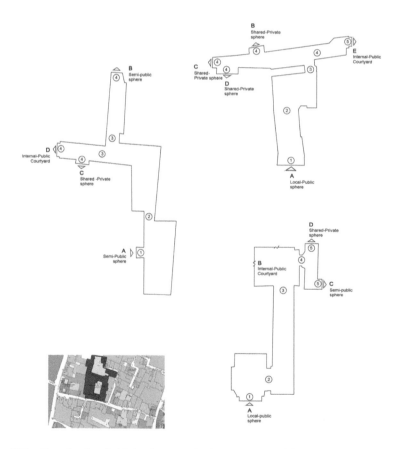

13.3   Complex thresholds between semi-private spaces and the most sacred space,
the harem, as mapped in bayt al-Suhaimy

SPATIAL TRANSFORMATION

Historical continuity is naturally embedded in the way some housing typologies vanish and others emerge and the presence of these typologies together in one site becomes instrumental in tracing gradual evolution within the same value system and cultural constraints. The layout of medieval houses in Cairo, however diverse, was developed as a complex form of spatial relationships that signified the thresholds between public spaces, which were the domain of men, and the women's sacred and private spaces.[43] This can be seen in Bayt al-Suhaimy's sophisticated hierarchy of spaces that structures a labyrinth of spatial experiences that are mysterious to strangers (Figures 13.1 and 13.2). The ground floor mandharah had relatively simple access from the courtyard when compared with the harem, which could only be reached by means of a long journey, involving negotiation of a staircase, lengthy dark corridors, and a series of spaces, which sometimes extended as far as the third floor (Figure 13.4). Such well-planned spatial complexity was a reflection of the Ottomans' conservative culture (sixteenth–eighteenth centuries), shaped primarily by gender segregation.[44]

At the turn of the twentieth century, this complexity was relaxed and in one case the harem became a simple room, situated more accessibly on the first level (Bayt al-Kharazi 1882), and in another, the courtyard-centred form was abandoned entirely to make way for more compact houses.[45] In the latter, the courtyard has been reduced to a hawsh; a small, covered multi-purpose space attached to the house entrance.[46] Such changes in form and layout – from inward looking courtyard houses to multi-storey, outward looking apartment buildings – followed a mainstream ideological reform, in which women, unlike the sacred character in Mahfouz's writings, were granted a dignified position in the public scene.[47] Their former, closeted existence was no longer practical, and the rationale behind the introverted spatial organisation collapsed, giving way to the extrovert, more vertically-stacked apartments. The changing social structure, caused by the departure of high class residents and the economic difficulties, arguably made such transformation inevitable. But it was the forces of modernity that drove the rich out of traditional quarters, leaving financial and social gaps that were to be filled by low class rural migrants.

The dynamic mobility and spatial transformation of the harah's houses are not simply a gift of modernity. A pattern of similar processes of dividing and/or merging properties, often following inheritance and divorce payments, existed throughout their medieval history.[48] Medieval socio-economic conditions allowed rich people to buy and merge smaller properties, or exchange them in order to gain a bigger property.[49] At the same time, division of large properties, due to inheritance, to create small compact houses was also a well-established practice.[50] The investigation of archival records of houses in Old Cairo showed that most of the contemporary small plots were a by-product of inheritance, divorce, or subsequent division of ownership and reconstruction.[51] Figure 13.4 illustrates different scenarios of the spatial transformation of a typical courtyard house as traced from common practice.[52] Building no. 8 in al-Darb al-Asfar is quite an accurate example of such a form of transformation.

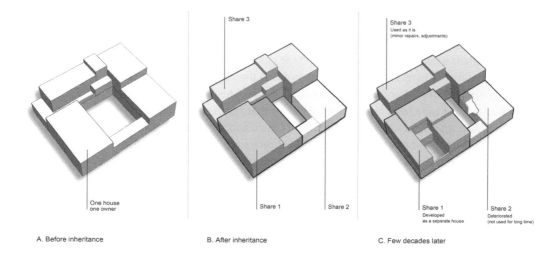

Share 3

Share 3
Used as it is
(minor repairs, adjustments)

One house
one owner

Share 1   Share 2

Share 1   Share 2
Developed   Deteriorated
as a separate house   (not used for long time)

A. Before inheritance   B. After inheritance   C. Few decades later

13.4   Scenarios
of the spatial
transformation
of a typical
courtyard house

CONTINUITY OF SOCIAL-SPATIAL PRACTICE

In response to spatial transformation, social practice in the Cairene harah has remained largely faithful to the inherited value system of male/female filtered interaction, however, in a much more relaxed and modern form. While men and women meet and talk in the alley more frequently, male guests are not allowed in the house, reserving it for female activities and social gatherings, while men take to the alley for gatherings, chatting and interaction. In this way, local inhabitants have learned to manage their spaces in efficient activity-space-time relationships. While new spaces of the present have to accommodate many social activities, the flexibility of their temporal arrangement was informed by the experiences of the past. Domestic space, according to Bahloul, is designed as a space of social and cultural inscription structured by the collective and symbolic organisation of its residents; *"the social and cultural world is organized in terms of metaphors provided by the house's physical layout"*.[53] In this context, house models and configuration of physical spaces recall certain socio-cultural systems, such as women's position in society, social hierarchy, tolerance about male-female interaction, or, in certain cases, progressive or conservative ideologies.

The socio-spatial association in the practice of architecture stems from its validity as part of urban complexity wherein a single building attends to many different needs and has implications beyond the realm of its users.[54] By virtue of its existence, a new building has already had a bearing on the history of its context, making the present experience different from that of the past, and hence adds to the memory. Demolition of a long standing feature, such as a building, a gate or a dead-end of an alley influences local activities and living patterns in a similar manner. When the western dead-end of haret al-Darb al-Asfar was opened to the central trade route in the early twentieth century, it became just feasible for already congested commercial activities to intrude into this residential territory. Industry took over the alley space and there was a resulting polarisation of traditional public activities between day (industrial) and night (residential), east (exclusive) and west (accessible), in a very particular temporally and spatially synchronised system.

Whilst the engineered open space in the front of the monuments to the west end of the alley and the informality and historical narratives of the eastern side manifest the differences between the traditional and the modern, in a way they inform each other. The former worked as a space fronting iconic and historical structures that were no longer used (memory). It had little relevance to the locals and more relevance to strangers. The stark contrast, from most exclusive (dead-end) to the most accessible, had in effect stripped the space of its historical value and quality that had for so long enriched its use as a semi-private social space.[55] As a result, it has been largely deserted by the residents and hardly any events take place there nowadays. It is open to the outside world and very difficult to control or protect; it is not a home. The latter, on the other hand, retained its relative qualities of control and security through its narrow configurations, but more industrial activities were incorporated to comply with the modern need to generate income.

Younger generations, on the other hand, influenced by the power of modernity, tend to break with the past and to challenge the authority of the elderly leaders. To control such a tendency, social action, reaction, behaviour, and management of conflicts are measured by senior leaders against the norms as set by previous experience. The balance between individual freedom and community duties and constraints can be seen clearly in the actions of those who listen to loud music and dress like westerners but without compromising their duties towards others: "*our men have to protect our women, even if they see them far away from the harah. Our code enforces this as a principal duty … The younger generation like the customs of foreigners (westerners), listening to loud music, wearing strange shirts, and having strange haircuts, but they know that certain duties and responsibilities cannot be avoided*" [R1.2.08]. Collective participation in ceremonies such as weddings (including furniture preparation and transportation) and funerals is another form of obligation [R4.1.07]. For example, during my field work in 2009, I saw that it did not take more than a few moments for all the community's men and youths to mobilise in response to the screams of a woman.

To some extent, some traditional practices have been reproduced in modern forms. The mashrabiyya, the wooden lattice window, that worked traditionally as a social barrier to protect women from intrusive stares (sometimes known as architecture's veil[56]) have evolved into new but similar systems of screening in the contemporary apartments. Its popular replacement, the Rumi-styled window, (louvered leaves), was introduced in the nineteenth century and recently has been adapted to suit contemporary dimensions, along with loosely fitted exterior fabric curtains on balconies. When the pre-modern water stores (such as Sabeel Qitas, located at the harah entrance) ceased to operate, shop owners began to provide rows of water bottle as a modern alternative (Figure 13.5). The rules of social behaviour and visual contact in the shared alley, on the other hand, remained largely the same: staring at windows is forbidden and strangers' movements are monitored throughout the alley. The inherited concern for security is satisfied by the continuous guarding of junctions with main streets, while homes maintain security through the active presence of the local residents and workers at gathering spots (Figure 13.6).[57]

13.5 Modern reproduction of the *Sabeel* (water store): rack of water bottles with the Old Sabeel appearing in the background

THE VALUE OF THE HISTORICAL FABRIC

The inherent qualities of the historical fabric of Old Cairo were continuously subjected to different interpretations and paradoxical relationships. This was apparent in Timothy Mitchell's discussion of the Egyptian display at the Paris World Exhibition in the 1880s.[58] At this exhibition Europeans were able to see Cairo only through an accurate scale model of a winding street of Cairo, made up of houses with overhanging upper levels and a mosque. The model was a panorama of images that were not only carefully designed by the French to mimic buildings and streets of the old city, but also they were full of donkeys, dancers and dervishes imported from Cairo to capture the essence of the real environment.[59] However, the display caused a lot of controversy and showed a huge lack of understanding of the Egyptian way of life. The interior of the mosque, a highly spiritual and religious institution, for example, was set up as a coffee shop with dancing Egyptian girls and whirling young men and dervishes.[60] For the Egyptian visitors, it was mischievous, degrading and disrespectful to the local values, culture, and meanings inherent in buildings. The exhibition, Mitchell states, "*seemed all to belong to the organizing of an exhibit, to a particularly European concern with rendering things up to be viewed*".[61]

The paradoxical interpretation of values showed that while outsiders value the image of the past as seen in buildings, the inhabitants attribute value to their meaning as useful venues for their social, ritual or religious practices. About 120 years later, Old Cairo today still cannot escape the outsider's overview of its fabric as a display of interesting images of the past rather than inherent contents and meanings. Architects and policy makers still perceive the hawari as a visual display, whose valuable asset, worthy of conservation and reproduction, is the image; people-free buildings. For them, its value lies in the display of the rich past as a separate entity, and its significance to contemporary everyday life is disregarded.

0 5 10   20m

● Gathering spots   (R) Resident   (W) workers   (C) Children   (S) Strangers

Narrow streets, attached houses, and dead-end lanes are characteristics of the medieval fabric that facilitate verbal communication, social integrity, security and collective defence through the production of semi-private spaces.[62] They hide, in effect, internal differences and tone down socio-economic variations. The difference in size of nineteenth-century houses is inconceivable from the outside, despite the contrast in size and quality of their interiors. Each building appeared as an integral part of an uninterrupted street facade. The typical style of window and door types (mashraibyyah, and then Rumi-style) in houses of all social classes contributed to this integrity (Figure 13.3). Similarly, contemporary buildings' exteriors do not reflect the wide variations in people's wealth and economic situation that are clearly noticeable and explicit in interior spaces (Figure 13.4). In addition, the dense nature of the urban fabric and its narrow street networks promote local proximity that allows for the dynamic flow of human traffic and subsequent attachment to a multiplicity of relatives, friends and family members, proximity of work, markets, and resources (food, tools, and services). When residents are moving house or seeking an upgrade, an adjacent harah within the old city is preferred. A senior leader interviewee relocated with his family from Wekalet Bazarra'h in Darb Qirmis in the 1950s to the adjacent al-Darb al-Asfar, due to the similarities in values, traditions and popular culture.[63] He dismisses any possibility of relocating elsewhere:

13.6 Gathering spots in the local alley as surveyed between 2007–12

> In the harah, we all are one network, neighbours and relatives. We marry each other. The world is one network. The son of this family is married to the daughter of this or that or those … we cannot live outside the harah. Here we are born, live and going to die. The harah is part of us, and we are part of it.[64]

When compared with other parts of Cairo, the persistence of some traditional customs and social-cultural systems cannot be separated from the integrity of the local socio-spatial fabric, as seen in Mahfouz's contrasting images. Its power of survival gains its strength from the relentless repetition of social practices which, through repeated performance, have become fact and a calendar for life.[65] When profound traditions and rituals prevail within a locality, social training becomes easier and everyone learns from the typical occurrences and relatively unchanging course of events. However, such bonding with the fabric proves the existence of implicit, albeit ambiguous and immeasurable, relationships between people and buildings, and between residents and situated culture. As per Joelle Bahloul:[66]

> The things of the past meticulously re-created here [at home] are those experienced in daily life, in the crowded intimacy of the most familiar objects and people, those one sees on rising in the morning and on going to sleep at night. This remembered past is lodged in the monotonous repetition of the necessary acts of concrete experience. The memory that invents it and rewrites it is the product of this relentless repetitiveness.

ARCHITECTURE AND REPRODUCTION OF MEMORY

> The whole factual world of human affairs depends for its reality and its continued existence, first upon the presence of others who have seen and heard and will remember, and second on the transformation of the intangible into the tangibility of things.[67]
>
> Arendt, The Human Condition

On 28th January 2011, the world witnessed the Egyptian youth, in revolt, targeting the headquarters of the established ruling party in Cairo, which had for long exhibited a corrupt regime's abuse of power, while on the same evening, they surrounded the adjacent Egyptian Museum, using their bodies to protect it from destruction and theft. The museum preserves their identity, collective memory and provides income,[68] while the HQ of the ruling party is a reminder of the oppressive past, to be forgotten. However, this was not about the building per se. It was about the activities and practices it accommodated. Both acts were symbolic of buildings as places that mediate experiences that should be either forgotten (by destruction) or remembered (by protection).

Old Cairo, in some sense, shares many qualities with the museum and is known by a number of scholars as the open museum of urban memory of Cairo.[69] The Cairene harah as a home is becoming a contested territory that is being fought over by two opposing forces. The forces of remembrance, associated with senior members and situated in old buildings, and the forces of forgetting, associated with the younger generation and their dynamic lifestyle. The latter, obviously, display limited if any interest in the static state of memory. Forgetting for them equals liberation from the constraints of the past and their traces as seen in old buildings. The negotiation of power between these two forces determines the identity and character of the community as well as the influence historical structures bring to bear on their social practice. Forces of memory and presence of old buildings helped the senior members to bring their knowledge of the past alive in the practical expression of the present and chiefly in the way they manage the space, both in private as well as in collective public order.

This discourse somehow leans on the way memory informs buildings, spaces, practices and appropriates activity-space-time relationships. While contemporary spaces remain the product of the present, social spheres and their temporal flexibility are guided by the memory of past experiences (Figure 13.7). Urban experience and conditions are essentially constructed out of a living repository of human experiences through successive waves of events, practices and buildings, which consequently construct inherent memories.[70] Architecture, in this context and by virtue of its durability and multilayered complexity, tends to be an agent of cross-time communication that materialises the socio-spatial relationships and mediates experiences, values and morals among different generations. Architecture, as process and product, hence, works as agent of continuity, which in conjunction with the narrators, brings the full experience of the past alive in the present and helps guide future generations. And this is precisely what allowed Mahfouz to express his criticism of the cultural and social constraints of the past, using its monumental structures and their persistent survival as a metaphor for traditional values that have remained in control of the minds of Old Cairo's inhabitants.

13.7 Inherited spatial practices: the use of public space during wedding ceremonies and its associated festivities

NOTES

1   Earlier version of this chapter was published as an article in the *Journal of Architecture* as: "Architecture, memory and historical continuity in Old Cairo", 17(2), 2012, pp. 163–89.

2   H. Arendt, *The Human Condition* (Chicago, The University of Chicago Press, 1958), p. 95.

3   *The Palace Walk* was the first volume of Mahfouz's Nobel prize-winning trilogy that depicts the social practices in the hawari of Old Cairo at the turn of the twentieth century. The novel was a debate about the strict traditions of the past as represented in the tyranny of the family's master, exclusive nature of daily activities (male and female domains), and the powerful institution of community.

4   Mahfouz lived in the Palace walk during the first decade of the twentieth century as a child and then moved to the modern quarter of al-Abbasiyyah. His novel depicts many of his experiences while living in this part of Old Cairo.

5   N. Mahfouz, *The Palace Walk*, English trans. by William M. Hutchins and O.E. Kenny (Cairo: The American University in Cairo Press, 2001), p. 2.

6   Maurice Halbwachs developed a credible argument that buildings as objects shape the spatial framework within which events and activities take place, and therefore, are relevant to the collective memory as part of the society. See M. Halbwachs, *The Collective Memory*, English transl. F.J. Ditter, Jr. and V.Y. Ditter (New York: Harper & Row, 1980).

7   M.C. Boyer, *The City of Collective Memory: Its Historical Imagery and Architectural Entertainments* (Massachusetts: The MIT Press, 1994); N. Rabbat, *The Culture of Building and Building Culture*, Arabic Reference (Beirut: Riad El-Rayyes, 2002).

8   Henri Lefebvre, *The Production of Space*, translated by D. Nicholson-Smith (Oxford: Basil Blackwell, 1991), p. 220.

9   A. Forty, Introduction, in A. Forty and S. Kuchler (eds), *The Art of Forgetting* (Oxford: Berg, 2001), p. 2.

10   G.D. Rosenfeld, *Munich and Memory* (London: California University Press, 2000); R. Bevan, *The Destruction of Memory* (London: Reaktion, 2006).

11   Lefebvre, ibid., p. 160.

12   E. Early, *Baladi Women of Cairo: Playing with an Egg and a Stone* (London: Lynne Rienner, 1993), p. 62.

13   He wanted to distinguish it from the [perception] of the present. See: A. Whitehead, *Memory: The Critical Idiom* (Oxford: Routledge, 2009), p. 24.

14   Forty, ibid.

15   *Memory* is a mental system that informs our everyday activities. Basic human activities and manners are learned though memory, just as a child learns to speak through his/her memory. Psychologists acknowledge that once the mental system of memory stops working, the human action becomes confused and problematic. At least this is how we learn to perform as humans as per Hanna Arendt's argument: "*Without remembrance and without reification which remembrance needs for its own fulfilment … the living activities of action, speech, and thought would lose their reality at the end of each process and disappear as though they never had been*" (Arendt, ibid.: 95).

16   W.J. Booth, *Communities of Memory: On Witness, Identity and Justice* (USA: Cornell University Press, 2006), p. xiii.

17   Bevan, ibid.

18   Beshir Kenzari and Yasser ElSheshtawy, "The Ambiguous Veil: On Transparency, the Mashrabiy'ya, and Architecture", *Journal of Architectural Education*, 56(4) (2003), pp. 17–25; Rabbat, ibid.

19   M. Yaari, *Rethinking the French City: Architecture, Dwelling, and Display after 1968* (Amsterdam: Rodopi, 2008), p. 306.

20   Forty, ibid.

21   Rosenfeld, G.D., *Munich and Memory: Architecture, Monuments and the Legacy of the Third Reich* (London: University of California Press, 2000).

22   Humanity had witnessed the ancient temples, artefacts and pyramids of ancient Egypt for thousands of years, but only when we gained access to their narratives in the nineteenth century, did we learn about their value systems and narratives of everyday lives. Only then, did their architectural productions become meaningful and comprehensible.

23   P. Connerton, *How Societies Remember* (Cambridge: Cambridge University Press, 1989), p. 22.

24   Forty, ibid., p. 7.

25   J. Bahloul, *The Architecture of Memory: A Jewish-Muslim Household in Colonial Algeria, 1937–1962* (Cambridge: Cambridge University Press, 1996), p. 28.

26   Connerton, ibid.

27   Stephen Kern, *The Culture of Time and Space: 1880–1918*, 2nd edition (USA: Harvard University Press, 2003), p. 2.

28   Stephen Kern, ibid., p. 36.

29   According to the documents of al-Darb al-Asfar rehabilitation project, the building housed 33 families of migrant villagers who worked in the food industry and used the building for their donkeys, food storage and production. It was consistently described as alien to the history of the community and aroused hostile reactions among many residents.

30   H. Heynen, "Petrifying Memories: Architecture and the Construction of Identity". *Journal of Architecture*, 4(4) (1999), pp. 369–91, p. 348.

31   Bahloul, ibid., pp. 28–9.

32   Jenkins, ibid.

33   Samira Mehrez, *The Literary Atlas of Cairo: One Hundred Years in the Life of the City* (Cairo: The American University in Cairo Press, 2010), p. 211.

34   Ibid.

35   A. Game, "Belonging: Experience in Sacred Time and Space", in J. May and N.J. Thrift (eds), *TimeSpace: Geographies of Temporality* (London: Routledge, 2001), pp. 226–39.

36   Game, ibid.

37   Jomard, as quoted in Janet Abu-Lughod, *Cairo: 1001 Years of City Victorious* (Princeton: Princeton University Press, 1971), p. 65.

38   Bevan, ibid., p. 13.

39   A. Abdel-Gawad, *Enter in Peace: The Doorways of Cairo Homes, 1872–1950* (Cairo: American University in Cairo Press, 2007), pp. 29–30.

40   A. Sedky, "The Politics of Area Conservation in Cairo". *International Journal of Heritage Studies*, 11(2) (2005), pp. 113–30.

41   Hania Sholkamy, "Why is Anthropology so Hard in Egypt?", in S.K. Shami and L. Herrera (eds), *Between Field and Text: Emerging Voices in Egyptian Social Science, Cairo Papers in Social Science*, (22)2 (Cairo: American University in Cairo Press, 1999), p. 141.

42   Heynen, ibid.

43   E.W. Lane, *An Account of the Manners and Customs of the Modern Egyptians*, 5th edition (London: John Murray, 1860).

44   Ismail, *The Popular Movement Dimensions of Contemporary Militant Islamism*, pp. 368–71. Previous times, such as the Mamluk era, were marked by the free movement of women and their participation in public events. Sedky quoted Van Ghistele, in describing women's freedom during the Mamluk era, saying *"One sees women coming and going and paying visits to their folk".* He, furthermore, traced the origin of the isolated women's quarter, the Haremlik, and Mashrabiyya lattice window to the Ottoman culture, imported with their army in the sixteenth century. See Sedky, ibid.

45   Houses # 1,2,3 Zuqaq al-Darba al-Asfar are clear case studies in this regard.

46   El-Siyoufi, ibid.

47   There is a raft of studies on the Egyptian reform movement during the 1890s–1920s. Albert Hourani, for example, studied the ideological reforms in detail in his seminal work, *Arab Thoughts in the Liberal Age 20th Century*. See Hourani, A.A. 1962. *Arabic Thought in the Liberal Age, 1798–1939* (London: Oxford University Press). Muhammad Emara has focused on the changing position of women during this period in his discussions of Qasim Amin's iconic book; *The Liberation of Women* (1899). See M. Emara, *Qasim Amin: Complete Works* (Cairo: Dar al-Shorouq, 2008).

48   L. Fernandes, "Istibdal: The Game of Exchange and its Impact on the Urbanization of Mamluk Cairo", in Behrens-Abouseif, D. (ed.), *The Cairo Heritage* (Cairo: The American University in Cairo Press, 2000), pp. 203–22.

49   S. Denoix, "A Mamluk Institution for Urbanization: The Waqf", in Behrens-Abouseif (ed.), *The Cairo Heritage* (Cairo: The American University in Cairo Press, 2000), p. 197.

50   N. Hanna, *Cairene Homes of the Seventeenth and Eighteenth Century: A Social and Architectural Study*, Arabic translation by Halim Tosoun (Cairo: al-Araby Publishing, 1993).

51   Robert Ilbert in *The House and the Urban Fabric*, as quoted in Fernandes (ibid., p. 203), questions whether one should try to decode the organisation of a city and its structure through the use of text in waqf documents as well as via physical evidence. For reference examples, check archival documents no. 194, record 500 (1281H); no. 54, record 25 (1307H; 1889AD); no. 64, record 24 (1307H ;1889AD). *Sigellat Masr AL-Shari'yyah* (Egypt Legal Records) (Cairo: The National Centre for Archival Documents).

52   In several cases, the multi-room, multi-apartment house was inherited by different branches of heirs or sold to independent individuals who, subsequently, sold their shares separately. Once a married couple divorced, or the husband died, the wife would get the ownership rights of part of the house (usually her harem apartment) as part of the settlement. The new part had become, therefore, an independent unit turning a previously sacred sphere into a new, separate, private sphere with its own subdivisions. The internal courtyard, as a result, is transformed into an outdoor semi-private sphere with entrances to separate houses. Within a few decades, the whole house develops through three scenarios. One heir, for example, was able to tear down his/her part, and rebuilt a new, compact, independent house to accommodate his increasing extended family. Another retained the existing form of his share, making some modifications and additions to accommodate his extended family. A third heir could not afford either solution, nor did he have a family or heirs. He fled the building or left it to go to ruin after his death, without lineage to develop/sell it.

53   Bahloul, *The Architecture of Memory*, ibid., p. 28.

54   Refer to T. Dutton, "Cities, Cultures, and Resistance: Beyond Leon Krier and the Postmodern Condition", *Journal of Architectural Education*, 42(2) (1989), pp. 3–9.

55   This change was introduced by the British authorities in 1908 for the purely functional purpose of allowing the European tourists easy access to the monuments of the harah. There is no evidence or information to suggest any desire to change the local traditions or systems. Therefore, although the consequences of this social change were not planned, they were desirable.

56   B. Kenzari and Y. ElSheshtawy, "The Ambiguous Veil: On Transparency, the Mashrabiy'ya, and Architecture", *Journal of Architectural Education*, 56(4) (2003), pp. 17–25.

57   The simple principle of urban security that Jane Jacobs applauded as effective and sustainable (Jacobs 1961).

58   T. Mitchell, *Colonizing Egypt* (Cambridge: Cambridge University Press, 1988), p. 1.

59   Mitchell, ibid., p. 2.

60   Ibid.

61   Ibid., p. 2.

62   This draws on the studies of Amos Rapoport and on environmental psychology in particular: which relates spatial configurations and tightness of outdoor spaces to social and cultural implications. See A. Rapoport, *The Meaning of the Built Environment*, ibid., p. 178.

63   Interview with resident [R1.1.07].

64   Ibid.

65   For this argument refer to Pierre Bourdieu, *Outline of a Theory of Practice*, English translation by R. Nice (Cambridge: Cambridge University Press, 1977).

66   Bahloul, ibid., pp. 28–9.

67   Arendt, *The Human Condition*, 1958, p. 95.

68   Heynen, ibid.

69   Mitchell, ibid. See also E. Sandweiss, "Framing Urban Memory: The Changing Role of History-museums in the American City", in Eline Bastea (ed.), *Memory and Architecture* (Mexico: University of New Mexico Press, 2004), pp. 25–48, p. 26.

70   Sandweiss, ibid.

14

Closing the Loop:
Gender, Education and Sustainable Homes in Cairo

ON GENDER AND THE REFORM MOVEMENT IN TWENTIETH-CENTURY CAIRO

The *women question* was at the centre of Egyptian reform movement led by intellectual elites who just returned from their European scholarships by the late nineteenth century. Inspired by their education and cultural experience in Europe, Egyptian elites had extensively questioned the marginal and oppressed position of women in the predominantly patriarch culture of households that excluded them from education, work and public life. It was apparent that the elites' call for the liberation of women and their rights to education moved the women from their peripheral position to be at the centre of modernising Egypt. These debates had inevitably extended to question the spatial organisation of houses and the way the harem quarter was isolated. For the elites, this was no longer a suitable spatial order for modern Egypt. The intellectual debate on the position of woman within society had destabilised the traditional form of courtyard house, inviting radical change towards a form of shared tenancy in multi-story apartment buildings. Women's social sphere of activities had significantly changed and the tolerance towards women's exposure in the public scene combined with the shrinkage of domestic areas resulted in more integrated environment between private and public spaces.

The shift was a challenging moment in the architectural history of Egypt, during which several factors including economy, migration and social change were effective and well-studied. However, the position of women at home was largely understudied, despite being a symptom and product of this change due to the centrality woman plays in the family sittings. Furthermore, little scholarly activity had looked at the influence this change had on house planning and design in the old city. Based on historical accounts, records and the survey of surviving houses undertaken in Part II, house forms and spatial organisation during the few decades before and after the turn of the century had largely transformed from isolated harem wings to more co-habitation into extrovert arrangement. Daily patterns of activities between indoor and outdoor spaces, the way women and families communicated and socialised as well as collective behaviour at occasional coincidences, showed a fundamental shift that perhaps declare a changing culture of home that permitted the subsequent change in housing form.

The change was naturally a result of a combination of several factors, including the changing of the social structure and economic affordability as comprehensively discussed in the previous chapters. Yet, those changes would not be possible without changing attitude and culture that allowed these new forms to operate. In fact, community leaders, influenced by the reform movement, repositioned themselves gradually to accept women as active members in public life, and therefore paved the way for more flexible approaches towards house planning and design. In response to the limited spaces for homes, women became frequent players within the public space and became central in organising public activities. In this chapter, I shed light on the gender issues and their impact on the spatial organisation and practice of homes. I look at the process by which the changing position of women, at the turn of the twentieth century, had driven the spatial transformation of house design from the traditional introvert courtyard-centred forms into vertically-organised extrovert arrangement.

FROM CULTURAL REFORM TO THE DESIGN OF HOMES

Women's position in society has been looked at as an indicator according to which the liberal values of modern societies were measured in a conservative region such as Egypt and the Middle East. Cairo, due to its proximity and links with Europe, was the centre of active development towards modernity throughout the nineteenth and twentieth centuries. Whereas Muhammed Ali's reign (1805–48) was seen as the period during which modern Egypt was founded, it is the turn of the twentieth century that seemed the period during which Egyptian society started to come to terms with the values of modernity and liberal society. This was explicit in the persistent call for cultural and ideological reform by the intellectual elites in local newspapers since late 1870s that continued for some 50 years defining an era of enlightenment and reform that challenged the inherited medieval thought.[1] The main driver behind this debate was the western-educated Egyptian elites who returned home from their Educational missions in Europe and were appointed to lead national and cultural institutions.[2] The returning elites believed that without changing the mentality of the people, leading the nation to modernity is almost impossible.[3] The desire was to emulate the European model of knowledge and philosophy within the confines of their national territory and with respect to their local culture and tradition.

As medieval culture and its social conventions were centred on the family and the position of women, the harem system, the reformers considered women as the principal issue for the reform movement. While schools of both sexes (females and males) had been available since the early nineteenth century, the radical sheikhs insisted that women should stay at home at the service of their men, secluded from strangers in harem enclosures.[4] The reformers claimed that in order to attend to the requirements of modernity, Egyptians had to open their minds to the knowledge, scientific and philosophical progress of the west,[5] and the indication of this was to accept women as equal members in a liberal society. They correlated the freedom of women to the social freedom and the progress of the nation.[6] The Apex of conflict with the conservatives came after the publishing of Qasem Amin's *Tahrir al-Mar'aa* (*Liberation of Woman*) in 1899 that was a protest over the radical constraints on women's education and work.[7] The book was introduced by Shaykh Muhammad Abdu (the highest Religious figure, the Egyptian Mufti), what gave it extensive media attention and criticism from conservative and radical leaders.[8]

Interestingly, both the reformers and the radical scholars of the time agreed on the centrality of women to the progress of the nation, but each from entirely different position. The reformers found no contradiction between Islam and western modernity based on women's equal rights to education, work, and participation if public life. On the other side, the conservatives believed that women should remain isolated at home and focus on their domestic duties. The formers' thoughts and principles spread throughout the growing number of periodicals and privately-owned printing presses by the second half of the century.[9] Byron D. Cannon argued that social and cultural change was actually taking place in the previous two decades and the book was just articulation of the current stream of thoughts.[10] He referred to articles appeared in *Al-Lata'if* magazine during the 1880s, handling the same issues, although in less revolutionary ways. The harem system, the principal feature of medieval culture, however, came to an end, as Berth Badran claimed, by 1923 when Huda Sharawi, the well-known Egyptian feminist, drew back her veil in public. Later, many had replicated her approach and high class women eschewed the veil as part of their dressing style.[11]

Regardless of the disagreement on how successful the movement was, its remarkable achievement was undoubtedly seen in the public discourse about women's position in society and the introduction of the *new woman* as an active member in the public domain.[12] A new woman was, unlike the traditional one, educated, active, made good use of time, was not prey to superstition and irrational thinking, while retain her commitments to her family and children.[13] As educated, active and productive, the new women should not be limited to the harem, the isolated quarter in the house. The new woman was the measure to which the society was compared to its European counterpart, as how far it attended to the compulsory requirements of modernity. Furthermore, Egyptian sociologists argued the reform movement has shaped not only the new woman, but also the new man, who likes a well presented and ordered home, has good taste, and admire pleasant structures.[14] The new man was sensitive, emotional and would only choose his partner through direct interaction and emotional companionship, something that was rejected entirely by the conservative culture. In fact, the new man and woman were pictured as European characters in eastern context.

The debate between the old and the new and it subsequent confrontations continued during the first half of the twentieth century in an attempt for each side to change (reformers) or to retain (conservatives) inherited constraints of women's education, movement and active involvement in the public sphere. Regardless of how quick each part of the city accepted the introduced cultural reform, the Cairene society, at the turn of the twentieth century, was ready to implement change to its urban structure that inevitably required new architectural forms that accommodate the needs of the new woman and her subsequent interactions and activities in the public life. Cairenes families were open to move to European quarters and their extrovert apartment buildings that are more dependent in its views, exposure and operation on the interaction with neighbours and outside facilities.

The new perspectives and tolerance towards the movement and interaction of women within the society resulted in simpler and less complex house layout. Extensive measures of isolating the harem were no longer compulsory. The model of independent, self-sufficient courtyard house was no longer a successful or affordable model to follow. This was evident in the increasingly relaxed and simplified organisation of the houses built between 1880s and 1920s and the incremental number of extrovert apartment building construction, either in the old city or the new quarters equally. Few large houses in the hawari of Old Cairo, retained the traditional courtyard houses but with much easier access to rooms around it, including the harem. Transparency became acceptable in the house and corridors have been lit and direct towards the rooms, while windows started to have glass panes, as we see in Bayt Al-Kharazi (1881AD). In short, the idea of home was changing in the mentality of local people. While the home remained the women's castle and a sacred place, they were increasingly welcomed in the public sphere and outdoor environments. Homes became more fluid and flexible in terms of practising everyday life and in terms of integrating indoor and outdoor activities.

The reference to the delicate taste and sensitivity of the new characters required for modern Egypt, would essentially, require a changed home. Hence, the debate between the old and the new and it subsequent confrontations continued during the first half of the twentieth century and naturally extended to the territory of home that required according to the reformers, and equal change in its structure. Qassim Amin, for example, criticised the complexity of gender-segregated houses in Cairo in the Liberation of Woman: *"Look at us, you find our house is divided into two parts, one for men and another for women. When we need to build a house, in effect, we spend what is enough for two houses [two attached houses] …. . This includes furniture for each of the two houses, two teams of servants, one for men and another for women"*.[15] In addition, he applauded the Western middle class house that is *"well organized, more beautiful than its oriental counterpart, even though the European spends much less that the Arab"*.[16]

The attack on the traditional forms of houses was coordinated among other reformers. Almost a decade earlier, Ali Pasha Mubarak, the minister of Public Work, and in justification to the initiation of municipal system that limited design and construction of houses to trained professional, launched a comprehensive criticism for traditional form of living ruling it as inefficient and unhealthy. In a bold statement, Mubarak stated:

> Today people have abandoned old ways of construction in favour of the European style because of its more pleasant appearance, better standards and lower costs. In the new system, rooms are either square or rectangular in shape. In the old system, living rooms together with their dependencies were disordered corridors and courtyards occupying a lot of space … most of the spaces lacked fresh air and sunlight, which are the essential criteria for health. Thus humidity accumulated in these spaces causing disease … facades never followed any geometric order thus looking like those of cemeteries. In the new system facades are ordered and have good familiar look.[17]
>
> Ali Mubarak, Al-Khitat al-Tawfiqiyyah, 1888

Mubarak's criticism, in fact, was designed to make case for western forms of living that follows simpler order and geometric principles. But, to accept new forms, you must dismiss the long rooted and accepted models as problematic and wasteful, while being unhealthy. In fact, one sentence in particular was striking; he dismissed the *"disordered corridors and courtyards occupying a lot of space"*, which he knew as peculiar to local tradition and part of the local social systems in Old Cairo, centred around the movement of women and ensuring adequate level of privacy. Indeed, as official and reformer, he was promoting western lifestyle centred on compact forms of apartment buildings that rely on the independence of the living unit, disconnection from wide street boulevards and geometrically ordered array of windows that expose the interior of the house to the outside world.

Side Elevation

Front Elevation

0 2   5   10

First Floor Plan

Ground Floor Plan

Kitchen · Room · Hall · Room · Living · Room

Kitchen · Shop · Shop · Room · Ent · Room · Shop

MODERN WOMEN AND THE MAKING OF MODERN CAIRO

14.1 Early classical period apartment buildings

Following such series of campaigns and legislative structures in what was called state building, Cairo was set to receive the new models of apartment house buildings as part of its fabric, giving way to the emergence of new typology that combined ground floor shops and stores with higher level floors of housing units, mainly in the new quarters. Even though, one would claim the presence of this typology earlier in the old city, it was a mostly a result of adaptation of the old structures, as discussed in Chapter 8. The Europe-inspired model became more acceptable in the early decades of the twentieth century, when women's movement and accessibility were no longer fundamental issues in house planning or design. Initially started as houses for foreigners, their rich local counterparts soon became resident of these buildings. In fact, this was one of several approaches Egyptians tried to reform their home living to prove their ability to stand modern in the urban scene in Egypt.

At a time when Egypt was still under the British colonial rule in 1910s–20s, the discourse on modernity; "how modern were Egyptians" and "the reform of Egypt's homes", was central to the process through which middle-class Cairenes defined themselves as modern and progressive in dissidence against dismissive tone of the colonial rulers.[18] Lisa Pollard argued that Bourgeoisie male Egyptians of 1919 attached national meanings of solidarity to their domestic habits as part of their national identity; *"Men's marital behavior and domestic habits appeared central to demonstrating Egypt's readiness for self-rule".*[19] In order to become modern, you need to act like one and your attitude to home and family is therefore pivotal. In fact, Pollard listed a host of readings and educational textbooks around 1905–11 that communicated the new home and its manners of modernity to children through simple words and questions. For example, a reader titled "*Reading and Pronunciation*" asked questions such as, "*What do proper homes need?*" and "*How do we build proper houses?*" which were answered that modern lifestyle required, similar to those used by the reform movement, "*order, cleanliness and ventilation*". For them, "*dark or crowded quarters were listed as belonging to another world, a premodern world which had to be done away with such that a new era of modernity could be ushered in*".[20] This could justify particular architectural transformation and predominance of certain forms of the time. In that sense, parallel to the decline of mashrabiyya as old fashion, open terraces overlooking large streets boulevards represented the symptom of modern living in Cairo and other cities such as Alexandria.

The process of structured education on modernity and modern living, indeed reached to the professional training of emerging engineering schools that soon became apparent in the work if their graduates. By the 1930s, design drawings for houses largely followed this trend. The liberation of women from the harem wings, facilities by national modernists, eased the burden of large houses and forced a rational space organisation. In fact, house designs of two prominent Egyptian architects during the 1930s showed how dominant modern open-building forms were. Hassan Fathy's early work and Ramses Wissa Wassif's designs of homes of the time are testimony on that. In (Figures 14.2 and 14.3) Hassan Fathy, known for his admiration of traditional building techniques at later stages, displays a fairly outward-oriented floor plan of apartment-office buildings with large balconies and opening directly open to the rooms' interiors. Access to the house unit and from it is largely exposed to public view. Even though designed outside the old city, it denotes a departure from the mentality of enclosures and manipulation of space advocated by the traditional forms of the hawari. If Fathy's designs were outside the historical core, Ramses Wissa Wassif's Diploma Project of the 1930s, was in the Old Cairo. Wassif's work showed a degree of uncertainty and tension between the value of the contextual fabric and the ideals of modernity of the time with the use of concrete and square spaces arranged around an open courtyard (Figures 14.4 and 14.5)

However, with the colonial rule coming to an end, the socialist agenda took over with more focus on social equity and the ideals of mass production of housing leading the notion of home to be industrious and economic in principle. This resembled a fundamental departure from social domain of interaction to be driven by authoritative state policies management of public properties, with serious violation of local social cohesion and breakdown of the socio-spatial association, witnessed in the old city. During Nasser's projects of social housing, the spatial organisation of home denied any opportunity of privacy and local control on the built environment (Figure 14.6). Moving from state-led social housing projects to private developers of the late 1970s, the modernist typologies of isolated and disconnected apartment buildings dominated the urban scene in Cairo. During that period, the idea of home as social construct represented, to a large extent, a withdrawal of socially cohesive communities and the making of home as way of building society. While the position of women was largely central to the idea of home in Old Cairo, new houses of modernity exposed the vulnerability of the family and their processes of communal and contextual existence, at least in new neighbourhoods that were bombarded with masses of concrete giants that are economic models of consumption.

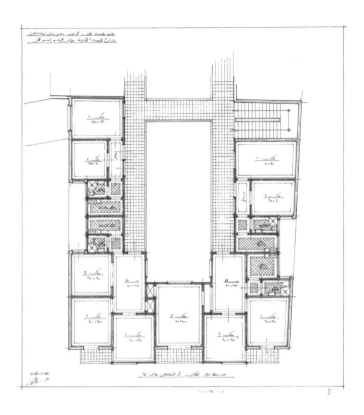

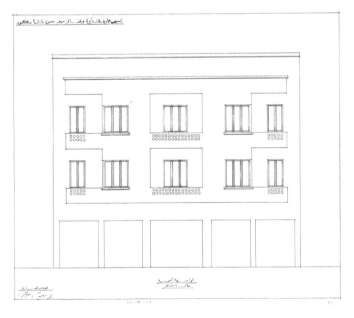

14.2   Hassan Fathy early design with its modern character: apartment-office building, early 1930s
Courtesy of Rare Books and Special Collection Library of the American University in Cairo.

14.3 Hassan Fathy early modern façades: Villa Hevsni, Giza, 1934

Courtesy of Rare Books and Special Collection Library of the American University in Cairo.

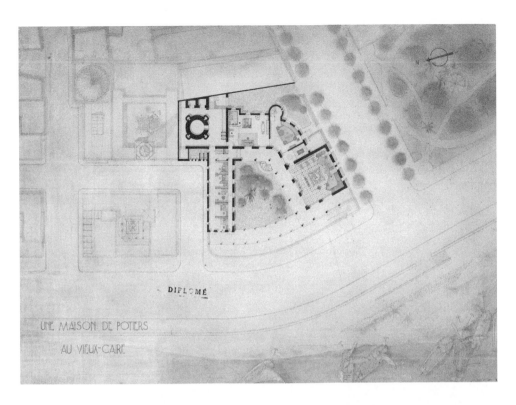

DIPLOMÉ

UNE MAISON DE POTIERS
AU VIEUX-CAIRE

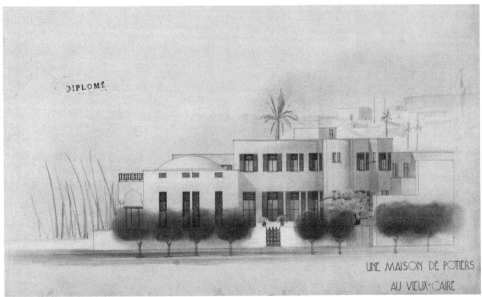

DIPLOMÉ

UNE MAISON DE POTIERS
AU VIEUX-CAIRE

14.4–14.5   Departure from the closed courtyard house in
Ramsis Wissa Wasif's Diploma project in Old Cairo, early 1930s

Notes: Site plan, Ground floor plan and Main elevation.

Courtesy of Rare Books and Special Collection Library of the American University in Cairo.

14.6   Social housing project of Abdel Nasser's era in 1960s

14.7   The jungle of concrete apartment buildings that shape the urban scene of Cairo today

THE MAKING OF HOMES IN EDUCATION

> 'Knowledge disseminated in the design studio is often packaged in the form of
> precedents or generalizations drawn from, at best, a limited number of instances –
> rather than from first principles. This is evident in virtually all texts, theses, treatises
> and papers on architectural education … .[21]

The waterfall house of Frank Lloyd Wright, Venturi house, Philip Johnson's Glass House, as well as Le Corbusier's Villa Savoye are almost mandatory case studies in architectural history course. Collectively they form the fundamentals of twentieth-century architecture and enrich debate on the ideals of modern architecture, each through genuine interpretation of the way of living at home.[22] However, they are iconic in their imagery and picturesque qualities, but telling less, if any, about how people used these buildings and how they responded to the actual life of their occupants. After all they demarcate an architecture driven by art movement and new social ideal that largely departed its role in building communities. The focus has always been on the architect's thoughts and process of thinking and production. Since the start of the twenty-first century, we wake up to find a century-long tradition leaving much of the urban landscape out of its agenda. The work of Peter Hübner and Teddy Cruz, started to bring those areas into the agenda. However, these currents are yet to bring back the studio to its contextual problems and concerns.

The architecture of the home, hence, requires an architect who is able to communicate, learn and develop his/her objectives, strategy and proposals according to knowledge situated into the context. As we have seen earlier, Egyptian architects, graduates of modern architecture-led programmes, are not interested to engage with the context of Old Cairo and its complex tradition of building homes. Despite the emergence of Environmental Behaviour Studies in the 1960s, they remained marginal in their influence on the design curricula in the Arab Countries, including Egypt.[23] The home stems from the same ground as those studies when the social and behavioural aspects of architecture are reconsidered as a necessity and to get the user to inform design development.[24] On the other hand, despite their extended history of practice, participatory design approaches to the production of homes are still limited to certain research or scholarly investigations. With vast areas of Cairo having developed as hawari in Old Cairo or as informal districts scattered around the city, it is surprising how limited their forms and issues are represented in the curricula.

A lack of concern towards the notion of home as centred on people's activities and lives is prevalent in the professional training in Egypt, leaving much of the problems of slum and informal communities in the city to be dealt with by activist and non-governmental organisations. The desire for separation from the traditional master-builder approach has undermined the need for continuous socio-spatial practice and dialogue with end users.[25] This actually reflects on the notion of architecting homes, as discussed earlier, revealing that architectural schools focus on the production of designers not architects. Young Egyptian architects are mostly trained in a similar manner to any European architects. What, thus, does the modern movement have in common with the hawari in Cairo? How distinctive is the training of architects in Cairo? How could we expect young graduates to build such understanding of the Cairene culture of homes?

Throughout the interviews, academics emphasised the need to develop students' basic knowledge and skills during early years of architectural education before starting to design buildings.[26] However, such basic knowledge and skills need to be contextually situated to tackle different socio-political or cultural agendas. With a brief review of completed undergraduate projects at Cairo University, for example, one can easily grasp the contrast between the imagery of virtual productions in the studios and the real life problems of Old Cairo.[27] Design projects are mainly of intellectual or artistic ethos with emphasis on forms and rational design of cultural centres, museums, commercial complexes: which are scarce in the urban context or architectural practice. Old Cairo is rarely considered as subject for graduation projects and is limited to a few projects scattered throughout the programmes. On the other hand, the home is not a subject for architectural investigation of design topics in Cairo. Housing design modules are driven by strategies of mass production of housing.

This could only be read as a separation from reality that could possibly have contributed to the expansion of squatter communities (estimated as 45–55 per cent of the area of Cairo) as an informal face of Cairo. In the shadow of such growth of informal practice, Henry Sannoff argued that the need for an interactive human/environment learning system has never been more urgent, while dismissing the piecemeal solutions to environmental and contextual problems.[28] He advocated community design that includes community architecture and social architecture and depends on elaborating architects' skills in his/her human/environmental approach.[29] Young architects, in order to re-bridge the gap with reality, are in need of special training that allows them to cope with the changing culture and situation of such contexts as Old Cairo and its hawari.

Current teaching approaches do not support the interactive learning initiatives. Every school of architecture has its own prescriptive agenda, which in the case of Cairo does not particularly respond to its local authenticity but search for ideals from abroad, in a quest perceived as a promotion of an elitist profile of the practice. Margaret Wilson's research on architectural design studios reveals a correlation between preferences within each school of architecture and that these preferences are linked to a particular style, which indicates significant problems with the studio concept of design education.[30] Architectural students are *"taught what to like"*,[31] meaning that architects are means of reproduction of teachers visions rather than become genuine in their own rights; a pre-dominant tutor-centred teaching strategy. On the contrary, a creative studio culture aims at supporting students' independence, dynamic and proactive approach to problem solving, as well as being genuine to contribute to contextually situated problems and societal concerns.

This training cannot be carried out through the traditional educational tools centred on teacher-student teaching and knowledge delivery through lecture format: a system that is currently tagged as employing ineffective educational tools. Architectural education, in the light of local challenges and changing social phenomena, is a shifting territory from teaching to self-learning through promoting inquiry-based and interactive participation in learning.[32] The architecture of the home, in this context, requires a radical change in the process of education. Such socio-spatial practice moves the influence from the studio and design tutors to the community, context and residents. Surprisingly, this approach was appreciated by interviewed architects and academic staff alike. A leading academic confirmed the need to acquire social researching and communications skills: *"definitely, architects need those [social] skills in*

order to be able to positively contribute to the local built environment in old Cairo" [I1.1.09].
Another agreed that those skills are missing in the majority of architectural curricula in
Cairene universities. In his view, academic programmes cannot sufficiently fulfil such
special training, which could be developed through practice after graduation as area
of speciality [I2.1.09].

In addition, contemporary political conditions stress, at least formally, the
participation of the local people in taking decisions about their built environments,
including architectural production.[33] Driven by the revolutionary waves of the 25th
January 2011 revolution, there has been a substantial rise in proactive activists
practices and architects who are increasing determined to tackle urban problems of
housing in Cairo. Rejecting the elitists approach, these grassroots movements and
groups moved to practically contribute issues of informality and urban deprivation
through interdisciplinary community projects and developments. How effective these
approaches are is yet to emerge, nonetheless, they denote a shift in the practice and
education in Cairo, driven by new forces of young generations. Even though, the
Cairene experience of this process is still at an early stage, it suggests the emergence
of new situations in the near future where the architect has to engage directly with
wider representatives of the local community, communicate his/her ideas as well as
relate his/her thoughts and concepts to them. Situations like this are more dynamic and
continuously changing according to each project.[34] It, moreover, implies that architects
will need to get to terms with and comprehend the cultural and social aspects of their
society and create strong connections between those aspects and their designs.

ON SUSTAINABILITY:
SUSTAINABLE LIVING AND THE NOTION OF CHANGE IN OLD CAIRO

At the heart of sustainable living is the fact that change is inevitable in a human being's
life; culturally and socially. Here, I look at sustainable living as a process by which a
community manages its resources and available spatial settings to elaborate new
systems and organisations that respond to its changing needs and the challenge of time.
During the process of change, every complex construct, such as home, is decomposed
to its preliminary elements, then reconfigured and reorganised in new forms suitable
to emerging needs and demands. This process is slow, unnoticeable and in continuous
fluctuation. In the context of Old Cairo, hawari communities developed, over centuries,
the notion of collective home, in which boundaries between individual houses are less
significant than the collective territory of the shared home. While the boundaries of this
collective home were historically barricaded and closed with gates, in the contemporary
context, the pattern of activities and the points at which behaviour and reactions
change allow the borders to be determined. It is more or less a territory that is built
mentally and practised socially in the minds of its holders.[35] Relaxed communications
between men and women, dress code and mutual support during hard times are, for
example, the basic principles of this code as addressed by all interviewees. The non-
compliance with local social codes invited could result in collective exclusion of that
member and his/her family.[36]

But, the coherence of the production of homes in Old Cairo is essentially an enduring
process, in which many variables and conditions come together. The changing form,

size, spatial organisation, and construction materials of homes reflect many aspects of adaptation and continuity under various pressures. It was apparent that locality has developed new forms and organisation of space that respond to their socio-economical as well as psychological needs at certain points in time. This can be described, I argue, as a strategy of sustainability, but a sustainability of the process of production and response to the local needs. The hawari are, in this regard, a reliable example of collective homes that keep changing their internal units to keep its validity and respond credibly to its occupants' needs for extended periods of time and under different situations. Replacing large houses with smaller apartment buildings, the emergence of workshops and retail spaces and the change of construction systems from stone blocks and plastered bricks to concrete structures resemble another strategy of adaptation to changes in the needs, economy and technology, respectively.

For me, this inquiry emerged from a personal desire to investigate sustainability in Old Cairo. However, during the course of my work in Old Cairo, I learnt that sustainability is a living pattern and practical strategy that keeps developing on a daily basis and in response to changing circumstances. Recent studies and research on sustainability do not provide an adequate response to one essential question: what we should do if all our knowledge and standards of today change in 30–40 years or more? How valid will their recommendations and findings be or how long will they stand? Researching the harah, the home and enduring communities, however, explores active mechanisms and deeply-rooted strategies of sustainability through everyday practice without a predefined agenda or pre-set requirements. The investigation of the home reflects on how complex and comprehensive human processes are for sustaining local communities and built fabric. Here, I shall shed some light, but briefly, on how the harah responds to the basic requirements of the modern term of sustainability.

While sustainability is perceived as an enduring process involving capability of maintenance, little work was conducted on understanding the process and coherence of maintenance; rather the focus was on the final product in the Cairene harah. In the research domain of sustainability, sustainable architecture is perceived chiefly in environmental and technological terms, with ultimate attention paid to energy efficiency or renewable resources and materials.[37] Studies about socially responsive, socially sustainable architecture are perceived as methods of participatory design rather than a tool to understand how such architecture of the home changes and develops over time. Today, the human aspect of sustainability, relevant to the context of the hawari of Old Cairo, is relatively marginal in studies of sustainable architecture.[38] This, essentially, was reflected in the institutional policies and management of the built environment in Cairo, which for a large part depended on urban renewal strategies. In such strategies, sustainable communities could be achieved by relocating the people of deprived areas into newly built low income housing neighbourhoods on the outskirts of the city; in other words, social engineering.

A new approach to developing Old Cairo is proposed by the UNDP comprehensive study of the current economic and social situation: suggesting change of the land use and occupations of people living in the area in the long term. For example, they suggested the support of tourism-related businesses and increase of office space and outlets. Such studies were prepared in separation from the past and lacking either local knowledge or the historical narratives on how each harah developed and responded to change. Even though the study was meant to be for a sustainable future, its recommendations respond to today's needs and current economic patterns.[39] This approach inclines towards the understanding of sustainability as a scientific process with defined and standard requirements, while ignoring the changing needs of people, and the mechanism of change and development inherent within the communities as well as professional practice itself (what is ideal today, could be a failure in few decades time).[40]

The lack of local and community-based knowledge and experience, which demonstrate physical and social adaptability to dynamic and extreme situations, as we saw in Old Cairo, could hinder the effort to build a sustainable environment at its first test of radical and extreme change. Researchers, in other fields of study, recognised adaptability as the precursor to reliability that is under threat in large-scale developments, which possess rigid physical and social structures that are not easily adapted to a dramatic climatic or social change.[41] Modern large-scale developments have provided no more reliability over space and time and over the long term; they are more vulnerable to extreme events than community practices.

Architecturally, Old Cairo's communities provide an example relevant to this debate.[42] While the community and their residents coped efficiently with the dramatic social and economic change by virtue of adaptability, building plans have been changed to provide space for economically viable shops/stores and managed smaller spaces to make use of existing built fabric to accommodate smaller, low-income families. The courtyard-centred house plan was abandoned when the local culture was found to have changed and when it became unaffordable. On the other hand, the masrabiyyah: the architectural and social element and need, has taken modern form, while the continuity between indoor and outdoor privacy of the home has remained similar. Similarly, the harah's economy was always connected to the local market and its demands, a connection that would have been broken if the UNDP's Historic Cairo Project's recommendations (focusing on large scale development) had been adopted. Large-scale developments, similar to the UNDP Historic Cairo Development Project, faced limited success, when they used top-down policies that in most cases lacked the knowledge of local mechanisms and needs.

Studying historical narratives of enduring communities and their paths from the past to the present, hence, reveals real assets that helped such a distinctive locality to survive. It might be useful to review the notional studies of sustainability as the production of a predefined, secure future, in favour of incorporating more in-depth studies of the existing assets of sustainability in historical and community-based practices. The principal proposal of the rehabilitation of Islamic Cairo project: to connect the hawari's local economy to the global economy and tourism business, is being questioned in terms of its stability in the long term. Redirecting the activities in Old Cairo's residential communities towards tourism-related activities, relates local production to the international markets and business, whose stability is vulnerable against any global problems (especially when compared with the current long-term domestic consumables and utensils industry).[43] On the other hand, such outsider-dependence requires frequent intrusions to what is a home for many people and these might hinder social activities and gatherings.

The local economy of Cairo emerged as a need and was integrated in the local economy of Old Cairo, either in terms of production or consumption. This has its accordance and relevance to modern studies of sustainability, which stress the proximity of production to consumption, workplaces to residential areas.[44] With a brief look, we find that the majority of the harah's population don't use public transportation or motor vehicles as their needs and jobs are situated within the proximity of their homes. This was demonstrated in the analysis of al-Darb al-Ahmar. To this extent and when combined with local ecological solutions, shared ownership, and the use of traditional materials and systems, the hawari, perhaps, resemble a sufficient starting point to achieve genuine sustainable community targets with high potential for being eco-neighbourhoods.[45] In addition, local security is maintained by the local residents and their social interaction without essential presence of enhanced measures.

Certain projects of sustainable development are currently being administered by local NGOs in Old Cairo, the aim of which is to provide social support, empowerment, medical and educational as well as training programmes to local communities, women and families. It was apparent, from the interviews with key players in those organisations, that the effort is taking two principal directions: first, providing some professional training to jobless people, mainly women; second, helping local families and their children to engage in useful social gatherings and activities. Local residents considered these efforts and their involvement of outsiders as unnecessary in the light of their male-orientated economy and their deeply rooted traditions of local social cohesion and engagement. Therefore, they did not show up frequently to the activities organised by those organisations such as the professional training courses and local gatherings, parties and music nights at Bayt al-Suhaimy.[46] Families acknowledged the presence and the good intentions of local organisations like FIDA, while they denied any serious participation in any of its activities.[47]

NOTES

1   Hourani, *Arab Thought in the Liberal Age, 1798–1939* (Cambridge: Cambridge University Press, 1983).

2   Such as Ministry of Education, Ministry of Public Works and others.

3   Refer to Qassim Amin's *Liberation of Women*; Ahmed Lutfi Al-Sayyid Articles in Al-Jarida.

4   Shaarawi, *Harem Years: The Memoirs of an Egyptian Feminist*. Translated and introduced by Margot Badran (New York: Feminist Press at the City University of New York), p. 11.

5   Hourani, ibid., p. 103.

6   This was identical to the view of Gamal El-Din El-Afghani's articles in *Al-lata'if* magazine during the 1880s.

7   Amin, "Tahrir al-Mar'aa" (The Liberation of Women), Arabic Title. In Emara, *Qasim Amin: Complete Works*, p. 322.

8   Muhammad Emara justified his argument, that Amin had little knowledge about strong religious references used in his book. He referred to several statements by their contemporaries that Abdu supported the book and was assumed to have written large parts of it while he was the Mufti, the highest scholarly post in Egypt. See Emara, *Qasim Amin*, pp. 117–27.

9   Hourani, ibid., p. 97.

10 Cannon, *Nineteenth-Century Arabic Writings on Women and Society*, pp. 469–70; women were already admitted to certain fields of education. For example, in 28th September 1886, 13 female students graduated with diplomas from the School of Birth as was announced in the formal newspaper of the time: *al-Waqa'I al-Masriyyah*, Issue 28th September 1886, p. 924.

11   Shaarawi, ibid., p. 7.

12   Amin, *Al-Mar'a al-jadida*, 1900. See Mohammad Imara, *Qasim Amin: Complete Works*.

13   ElSadda 1998.

14   ElSadda 2007: 32.

15   Amin, *The Liberation of Woman 1899*. See Mohammad Imara, *Qasim Amin: Complete Works*. Translation is made by the Author.

16   Ibid.

17   Ali Mubarak (1823–95) was the First Minister of Public Works in Egypt, and was one of the most influential Egyptian reformers in the second half of the nineteenth century. He was educated in France (1844–50) and led the Egyptian Ruler, Khedive Ismail's project to build European Cairo. This statement was written in his book, al-Khitat al-Tawfiqiyyah. The translation is taken from Khaled Asfour, *Identity in the Arab Region: Architects and Projects from Egypt, Iraq, Saudi Arabia, Kuwait and Qatar*, p. 151.

18   Pollard, L. 2000. "The Family Politics of Colonizing and Liberating Egypt, 1882–1919", *Social Politics*, 7(1), pp. 47–79.

19   Ibid.

20   Ibid., p. 58.

21   Akin, *Case-based Instruction Strategies in Architecture*.

22   Neil Levine summarised the characteristic features of his work as: the open plan, dynamic space, fragmented volumes, natural materials, and integral structure, which established the basic principles of what we call modern architecture. See Levine, *The Architecture of Frank Lloyd Wright*.

23   As a reaction to the failure of modernist theories in addressing the housing problems, squatter problems and the deterioration of historic cities, see Salama, *An Exploratory Investigation*, pp. 35–7.

24   Prak, *The Visual Perception of the Built Environment*; Salama, *An Exploratory Investigation*, ibid., p. 25. Prochansky, *Environmental Psychology and the Design Profession*.

25   Ashraf Salam denoted this problem by saying: "*While the contents of environment-behaviour courses seem to address the balance between theories as abstract knowledge and the contextual particularities of local context, it is evident that studio description in all programmes does not indicate whether knowledge delivered in a lecture format is integrated into design assignments in the studio. Thus, it can be argued that knowledge contents are offered in a fragmented fashion*".

26   Mahgoub, *Design Studio Pedagogy: From Core to Capstone*, p. 193.

27   This information is based on analysis of annual magazines of students' work at Cairo University during the years 1991–2000. See for example: *Annual Magazine*, Cairo University, Department of Architecture for the years 1995 and 1997.

28   Sannoff, *Community-based Design Learning*, p. 21.

29   Sannoff, *Community Participation*, ibid., p. ix.

30   Lawson, *How Designers Think*, p. 8.

31   Wilson, *The Socialization of Architectural Preferences*.

32   Abdelmonem, *Shifting Boundaries*.

33   While this stand is formally announced and supported by legislation that allows people to withhold any projects, in reality other laws allow the minister of housing to disregard any complaints or objections for the public benefit. For more details on such superficial policies and laws, see Sedky, *The Politics of Area Conservation in Cairo*.

34   Despite the variety of projects introduced in architectural schools today, real life situations face architects with new challenges and design approaches with which they are not familiar or experienced. For example, cyber homes, science parks, etc. emerged in recent years as new typologies of buildings.

35   D. Delaney, *Territory: A Short Introduction* (Oxford: Blackwell Publishing, 2005), p. 71.

36   Interviews with residents [R1.1.07, R4.1.08].

37   See Williamson, *Understanding Sustainable Architecture*, p. 4; Stang and Hawthorne, *The Green House: New Directions in Sustainable Architecture*.

38   See for example, Steel, *Sustainable Architecture: Principles, Paradigms and Case Studies*; Gauzin-Muller and Favet, *Sustainable Architecture and Urbanism: Concepts, Technologies, Examples*; Pitts, *Planning and Design Strategies for Sustainability and Profit*.

39   They used the typical scientific analysis from other sustainability research, focusing on documentation of current problems and with views on decisive actions to improve the situation.

40   Pérez-Gómez, *Architecture and the Crisis of Modern Sciences*, p. x.

41   Woodland and Hill, *Water Management for Agriculture in Tunisia*, p. 241.

42   Woodland, ibid.

43   Governmental figures show that after any unexpected incidents, such as a terrorist attack or increase in travel costs, numbers of tourists decline sharply within a short time. As a result, tourism-based businesses are considered as less stable than everyday businesses for providing a sustainable source of income.

44   Some governmental reports and studies considered the stability of local economy and the proximity of workplaces to houses and homes (easy access to local facilities and jobs) as essential prerequisites of sustainable communities. For the former, see Great Britain (2005) *Sustainable Communities: A Home for All*, by the office of the Deputy Prime Minister; for the latter, see Barton, *Sustainable Communities*, p. 112; Barton, *Shaping Neighbourhoods*, p. 28.

45   Barton, *Sustainable Communities*, ibid., pp. 128–34.

46   During my several visits to those local centres, I was unable to witness any activities taking place there, which suggests that they are not as frequent as intended.

47   Interviews [R1.2.08, R3.1.07, I20.1.08].

Afterwards:
Towards a Progressive Practice

The popular perception of the hawari communities in Cairo, as influenced by state media, has been problem areas with limited resources or potential for development. Successive Egyptian governments and planning institutions in Cairo approach city planning as a top-down endeavour, in which the communities and structures are difficult to control or develop without long-term investments. This dismissive view has dogged the old city for a long time with devastating impacts on the quality of living that have been exemplified in underdevelopment, lack of investment as well as absence of distinct architectural practice or quality production.[1] This book critically reviewed this perspective by shedding light on the comprehensive processes of everyday human production and consumption of space. It challenged the view that the value of the spatial organisation of the hawari lies in their remnants of medieval heritage, or traces of history. Rather, it took the organisation and everyday life of the hawari seriously and dealt rationally with their mechanisms and response to changing needs. In other words, this book was an attempt to learn the dynamics of architecture of home, from bottom-up, through understanding ideals, values systems, social structures and socio-cultural developments that resulted in changing forms of living over time. It tried to bring architecture back to everyday pattern of living as its core drivers.[2]

My aim was to understand and clarify the processes of the architecture of home and its flexible organisation of spaces in Old Cairo through interrogating the way changing everyday life influenced spatial organisation and transformation in the homes of the Old City, loaded with centuries of continuous development and habitation. These environments suffered the preconception that its architecture was a mere physical manifestation of Islamic laws that had ceased to be relevant in contemporary discourses of urban and planning theory and practice. This project maintains the belief that even though Cairo has moved on from medieval conventions and traditions of pre-eighteenth century towards modern administrative structures of the twenty-first century, local peculiarities such as value system, social networks, and spatial practices remained in connotation with the memories of the past as well as being influential in reproducing socio-spatial systems that are largely satisfactory to its residents.[3] Hence, the architecture of home its spatial order today is as relevant to activity patterns and practices of everyday life in Old Cairo, as it was for centuries.

Throughout the chapters of this book I traced the everyday practices of home and its architecture, underlining inquiry into the meaning and requisites of home socially, culturally and cognitively. We learned that the idea and concept of home, in spatial and social terms, encompassed a system of interconnected spaces, in which outdoor open spaces complement the indoor-closed domestic spaces in fulfilling each residents/inhabitants needs. Open spaces accommodated private activities when closed spaces could not. Public meetings take place in inside houses, whenever and wherever available. In certain localities, the local street stands up to fulfil activities and needs particular to the domestic sphere. The architecture of home, therefore, is essentially a progressive discipline of architecture that deals with such fluid situations and events in socio-spatial manipulation of the physical space to reach a contingent system in which space-activity-time are coherently synchronised.

I utilised the context of the hawari of Cairo to interrogate and test such comprehensive and historical construct of the idea of home that provides evident dynamics of everyday life and its socio-spatial associations that endured centuries of fundamental social, economic and political change. To be able to contribute and integrate positively, architects need to learn the history and processes by which homes evolved as an architectural response to everyday needs. The experience and professional knowledge of the architect do not guarantee the successful production within such contexts as the Cairene harah. The architecture of home in Old Cairo, in this sense, embrace a collaborative socio-spatial practice and knowledge, in which architects learn local dynamics and work with residents to provide effective responses to their daily needs.

Looking at architecture as a notion of everyday socio-spatial practice, the practice of home was deconstructed as a series of social spheres to render visible the socio-spatial associations within everyday life. Daily rituals, temporal moments and occasional events exhibited explicit impact on spheres of activities that transcended the physical boundaries of indoor-private and outdoor-public spaces at any given period. While the sizes of houses, their capacity and spatial complexity change over time, the practice of home continue to have involved consistent and coherent use of a distinct system of interconnected spaces. Having investigated and analysed the development of Old Cairo's social and spatial structure over two centuries, the architecture of home succeeded to liberate the spatial practice of everyday from the limitations of the physical characteristics of houses: allowing the local population to redefine their socio-spatial organisation and relationships on a daily basis. The harah was shown to be a meaningful and defensible territory, as home, including an interconnected system of spaces whose purposes are redefined according to daily and occasional needs.

Studying the development of Old Cairo hawari throughout the city's urban history helped to liberate the social and spatial history from the mythical and spiritual influence of religious laws that were central to several scholarly efforts such as Jamil Akbar, Besim Hakim, and Stefano Bianca's investigations of medieval Arab cities. It was crucial to develop understanding of the domestic environments in Old Cairo as a set of spaces and practices of everyday life, alienating passionate and idealist notions of the past that is always pure, authentic, and religious and resemble God's well, that infiltrate popular Arab culture and literature for decades. This was informative in extending our ability to realise the multidimensional meaning and influence of spatial order and architectural form of such organic system of the Cairene homes and hawari. Amalgamation of these theses with the justification of spatial rationale provided interesting tools that helped explain archaeological patterns of space in different quarters of Cairo and reveal inherent meanings and systems embedded in their form and components. It became apparent that the socio-cultural context exerts significant influence on daily rituals and, accordingly, appears to inform of typological arrangement of the spatial models of homes and their architectural image.

In this theoretical and epistemological discourse, the journey through the threshold appeared to retain all the transitional qualities (in time and space) that reconstitute people's identity and state when crossing from one social domain to another. Internal organisation appeared to reflect set of daily rituals manifested spatially in the binary opposition of front-back, private-public, male-female and their components and the socio-spatial barriers between them. This approach was verified locally when practical investigation of medieval houses' mashrabiyya revealed its essential social and cultural functions: while providing aesthetic representation of architectural quality and character.

I wish at this point to synthesise this book's argument and contribution into five principal themes

READING THE ARCHITECTURAL HISTORY OF MEDIEVAL CAIRO

Investigations of Islamic architecture and buildings in Old Cairo owed much to the scholarship of K.A.C. Creswell and his volumes on Muslim architecture in Egypt. However, Creswell's work, a typical orientalist approach, has set a tradition for surveying structures and relates them to historical, mythical, and spiritual encounters of their relevant periods. Such approach tends to jump to conclusions and discounts evidence of ordinary people's activities, process of production and everyday lives, as subjects of overarching power. Similarly, but in a much more complex form, surveys of house typologies in al-Fustat by Aly Bahgat and Gabreal Baer provided valuable information about the spatial organisation and street patterns. Both works were flagships for the study of architectural history of Old Cairo that remained descriptive more than analytical. Subsequent works that investigated specific examples or periods in a similar manner, include Jacques Revault and Bernard Maury's *Palais Et Maison Du Cairo: Du XIV au XVIII Siecle*, and, *Twentieth-Century Islamic Architecture in Cairo* by Tamer Sakr, and *The Doorways of Cairo Homes: 1872–1950* by Ahmed Abdel-Gawad. The majority of these contributions relied on documentation, such as images and drawings, rather than analysis, justification or determination of the forces behind the produced image or object. In this domain of research, the socio-spatial analysis of the Arab city by Jamil Akbar in his *Crisis in the Built Environment* and the sociological analysis of Nelly Hanna's study of seventeenth- and eighteenth-century houses in Cairo remained the most informative despite the reliance on archival records and deeds, with little attention to the social activities according to everyday patterns.

Despite the vast detailed and accurate survey of historical structures in Old Cairo, yet we find ourselves blind when it comes to understanding what these buildings stand for. What did they represent and why did they take such a particular form? The reading of architectural history requires, as Spiro Kostof advocates,[4] a *"life story"* that provides a much more important explanation of the object beyond our eyes. This book, in response, attempted to offer a comprehensive reading and analysis of architectural production based on social and anthropological justifications for previously surveyed objects and images. It offered an alternative method to read architectural history of the city different from the exclusive archaeological evidence and surveyed buildings. In fact, one of the central aspects of this work is the reading of buildings' life stories, complemented by real names, family structure, activities and archival records of the community within which these buildings operated.

I here challenged the preconception of the house as enclosed and isolated from the outside world: a notion advocated by western historians and orientalists. The evidence reported that the public sphere extended into houses, forming venues for collective gatherings that were spatially configured to respond to the local order of seniority. The presence of multiple venues for public gatherings around the central courtyard suggested that, for certain occasions and festivities, the ground floor merged with outdoor shared space to form one extended public sphere. The spatial configuration and decoration of al-mandharah had to reflect the wealth and power of the master to strangers, but not to the residents. Houses of the lower order, which lacked such emphasis, tend to offer enhanced image when a guest in received. This image could be a well-decorated corner, pieces of furniture or by enhanced arrangements of the space. The harah's public space, in fact, was a shared private sphere that was secured and protected by the whole community, while meals and private talks frequently took place in front of houses and shops.

INTERDISCIPLINARITY OF HOMES AND THEIR ENVIRONMENTS

On the side of the coin, sociological research in Cairo work on the ethos of social surveys and the tendency to explain a social phenomenon on the basis of changing culture and political situations. Anthropological work, on the other hand, relied on first-hand accounts, discovering the story through the words of human subjects along with visual observation.[5] Architectural research, however, used to align itself with the images (drawings or visual materials) of completed buildings. Recently, architectural work in Cairo has started to respond to the broader processes of community participation and integrate up-to-date social investigations and materials. It would be impossible to bring these diverse strands together without adopting an interdisciplinary strategy for gathering relevant, and potentially explanatory, fragments of information.

Understanding the way residents live and interact on daily basis was informative and influential in developing the notion of home as a set of social spheres. Their accounts of daily lives and participation in festivities and events allowed a privileged view from inside, reflecting on people's use and practice of their spaces. Interviews and images, based on the memories of senior residents, allowed for verification of spatial practices in private and public spaces of the early twentieth-century harah. Memories and accounts of daily life and activities from the 1940s complemented and brought to life Naguib Mahfous's narratives, as well as the social statistics and surveys of the same period by Mohamed El-Sioufi. The change in popular mood towards women was accompanied by a change in the social structure following the departure of wealthy merchants. Spatial organisation of houses around a central courtyard became an unnecessary luxury and was not affordable for the emerging low class community. It was replaced by the more affordable, compact and extrovert arrangement of multi-storey family houses on smaller plots of land.

As a result of inheritance and exchange laws, large houses were divided into smaller plots: with every portion later being developed as an independent unit/building.

This approach could be employed to coalesce the fragments from various disciplines concerned with human activities to complete the story of architecture. The history of architecture, I believe, is yet to be told, for the absence of stories that accompanied its production, use and spatial practice. The architecture of home, here, is a project that attempted to re-read architecture from a new lens that is not architectural, nor intellectual, but from the lens of ordinary people's everyday life, a process by which the building is consumed and reproduced. This allowed for exploration of other dimensions of the building processes and understanding how memory inform architecture and come to be shaped by it. Buildings are much complex construct than its physical characteristics or spatial orders; hence, the notions of social spheres are more capable of incorporating multi-dimensional information to bring the building to life, helping architects to know the social dynamics of their buildings.

CONTEMPORARY ARCHITECTURE AND URBAN DEVELOPMENT IN EGYPT

Architectural practice in Egypt tends to draw clear distinctions between the contemporary and the historical, present and past, and between the modern and the traditional. This could be related to the intellectual perception of the contradiction between tradition and modernity, mentioned earlier. This intellectual position influences architectural work on new buildings; which are expected to be modern in image and materials. Buildings in a historical context, however, have to look traditional and reflect the historical image and materials, even if only in the superficial manner which many architects term a "historical mask". Contemporary building laws, imposing specific standards on every quarter in the city, have become mandatory, regardless of how applicable they are to a particular context. Complete absence of any study, analysis or understanding of local context and the potential impact of proposed designs has had devastating effects in recent years, and many building permits were rescinded during construction.

While this project focused in principle on the architecture of Old Cairo, through investigation of people's memory, history and responses to architectural production in the long term, it redefined, in a wider sense, what architecture means to its users. It, implicitly, shed light on how architect-builders in Old Cairo were conscious of and responsive to many issues that are absent from the contemporary architect's agenda. This elaborates on the intellectual debate on the coherent identity of Cairene's architecture. A single decision could compromise the local integrity and homogeneity of the urban settings and urges that contextual analysis should become a compulsory aspect of planning applications. Architects are researchers, as denoted by Bryan Lawson and many other scholars,[6] yet the quality of their buildings should reflect adequate investigations, before being allowed to contribute to the existing and sensitive built fabric as that of Old Cairo. Distant contributions with universal values of creativity are no answer to these situations.

The tentative analysis of educational programmes in Cairo, gave partial but important answers to the dilemma of why architects exclude such significant areas of Old Cairo from their consideration. Despite a high historical and cultural profile, the hawari environment is not well integrated into current design and research tasks in professional training and courses of architecture. Mostly associated with western ideals of modern architecture, architectural culture looks at Old Cairo as distant history, but not a challenge that requires contemporary input. It is evident that incorporating the unique condition of the Old City into training programmes, would bring innovative ideas and encourage newly qualified architects to establish their practices in response to existing challenges and problems of their city.

MEMORY AND HISTORICAL CONTINUITY

The rational inductive approach used to investigate the homes of Old Cairo did not dismiss the phenomenon of the complex and ambiguous nature of these domestic environments. The influence of memory (individual and collective, humans and buildings), historical continuity, as well as the value of the fabric were essential factors that consolidated the stability of local socio-cultural norms, codes and practice over a long time. The centrality of memory in shaping everyday normative behaviour and guiding human-space association from late nineteenth to the end of the twentieth century was a discovery and enlightening at the same time. Considering the role of memory in connecting people to the built environment, I underlined the diverse interpretations of history that shape peoples way of thinking. While senior citizens used to value the historical continuity and inherent memory as established in buildings and humans, younger generations tend to protest against the constraints inherited from the past. The discourse of memory, thus, is prevalent in everyday life, communication and interaction.

As memory is of ambiguous nature that changes from one individual to another and from one community to another, it is perhaps debatable to display discrete findings and/or conclusions about its indeterminate influence or how it impacts on the mutual relationship between humans and old fabric. Nevertheless, interviews and recorded practices at occasions and events displayed multi-layered similarities of socio-spatial practice between the past and present. It is inevitable to declare the continuous presence of buildings' representatives of successive eras as influential in informing new generations of the meaning of *community-proper*, value system and traditions, as well as delivering a seminal education on the use of spaces. It is found that houses' models and the configuration of physical spaces recall certain socio-cultural systems such as women's position in society, tolerance about male-female interaction, or in certain cases dominant ideologies. These value systems are explicitly manifested in the presence of old houses, whose space organisation and order are demonstrable examples. While contemporary spaces remained the product of the present, social spheres and their temporal flexibility are guided by the memory of past experiences.

This, however, is applicable for most contemporary cities and urban environments. Not all quarters, neighbourhoods, and communities appear at once, or remain frozen in time. It is the very nature of urban experience and condition to be constructed out of a living repository of human experiences through successive waves of events, practices and buildings, which consequently construct inherent memories.[7] However, Contemporary architectural practice, in general, does not complement or promote such association with the past, when it comes to the interpretation of collective memory. Rather, there is an apparent preference for enhancing modernist versions in isolation from the past. The intrinsic differences between the house and the home were manifested in the experiences of the residents and values they attributed to the domestic environment. The memory embedded in buildings makes possible the connection between the present and the past, manifested in everyday practice and the use of particular social order, space organisation and sometimes, artefacts or decoration. Supported by memory and historical continuity, regardless of duration (a day, month, year, decade, or a century), occupied homes are valuable elements in the built fabric and the contemporary city. Homes are difficult elements to replace, while houses are consumable objects that can be disposed of without regret.[8] Architects aiming to make a valuable contribution to the built environment need to build on memory and historical continuity if the aim is building integrated environments, not isolated objects.

VIRTUES OF COMPLEXITY

The proposition of the socio-spatial practice as a discipline of architecture stems from its justification in the urban phenomenon of complexity in which a single building does not attend to a single need or have implications limited to its users. It is a protest against the singularity of the architect and his/her ignorance of the world around his/her work. Buildings are added elements to an already complex environment that impact on activities, events, behaviour of not only the users of the building, but anyone in its proximity. While the consequences might not be straightforward, the implications could be significant. When the dead end of a single harah was opened, though a single decision, commercial activities (by nature) found a room to protrude into an exclusively residential community, reshaping its social-structure and economic system. Shared public sphere, as a result, was polarised between day and night, and east and west in significant temporal and spatial divisions that compromised the deeply rooted practices.

   While this scholarly effort does not intend to promote certain actions, it is sensibly urgent to incorporate the investigation of local dynamics and potential implications a single building could have when a new design proposal is under consideration. I suggested the inclusion of a contextual study that addresses the implications and contextual adjustment of every building permission application. Such studies should make architects aware of the virtues of complexity inherent in their subject sites. Further, it highlights how proposed solutions could communicate meaning, influence rituals and contribute to the existing everyday life, through extensive participation and consultation processes. Architects need to understand that a building's effective area is not limited by its exterior physical boundaries nor surrounding pavement. Rather, it extends to non-visual domains of activities, and shared venues and social spheres in its locality.

* * * *

The built environments are definitely subject to successive waves of change with a continuous presence of old and new, then and now, past and present buildings. Such mutual existence makes a complex contribution to the city that extends my argument about social spheres and environment of home towards other types of building and communities, in which similar processes of human practices are making sensible use of space, enhancing private as well as public spheres. Practical consideration of social spheres during building design and planning, would liberate building programmes from their rigid and physical constraints, and allow architects to integrate buildings comfortably within the builtscape.

Architecture is much more than buildings and successful architecture could represent an effective enhancement of existing built fabric, a harmony of context and image, while elaborating on locally developed practices, meanings and memory. Practical implications of social spheres on the contemporary design theory could be seen in provision for the outdoor environment to extend inside buildings, and some private activities to move outdoors. Similarly, it would be useful for buildings to be planned for different functions during day and night, during weekday and weekends. Buildings that stand closed and unoccupied are dead statues and harsh on the surrounding environment. This book suggests that by virtue of their complexity, buildings could be made efficient, integrated and contextual. If architects consider multi-layered meanings of buildings, space organisation, and spectrum of social spheres as principal design tools, they could develop buildings as living organisms in the built scape with high potentials for an extended period of lifecycle.

The architecture of home in Cairo has been and will remain developing on the basis of the association between humans, space and socio-cultural assets. The transformation in the socio-spatial systems remained in line with local social and cultural values, traditions and norms, through the virtues of historical continuity materialised through old buildings and order of seniority. As old buildings attribute command on the built-scape, old members retain the power to lead, guide and assume control of local social codes, manners and interactions. This was the result of social and spatial cycles that reflect the continuous overlapping across generations of buildings and humans, represented in everyday practice. Through this process, the architecture of home emerges as a complex socio-spatial system of interconnected and flexible spaces that keep expanding, shrinking and merging with each other according to immediate needs.

It was not possible to understand or highlight the processes, complexity and historical continuity of buildings and people without the investigation of everyday practice of home. It is the everyday that reveals how people use their spaces, develop and transform them from private to public and from public to private on a daily basis. This architecture of home in fact challenges the study of buildings as physical spaces with fixed configurations. Social spheres, in this respect, are domains that allow the space to be liberated from its physical limitations, and allow private and public spaces to merge and dissolve in line with the occupants' needs. If architecture is to be sustainable, it must attend to the changing needs of people over long periods of time. The survival of the Cairene harah over the past centuries owes much to its particular practice of home and it can be expected that this will also enable it to endure for centuries to come.

NOTES

1   This is perceived as a consequence of a long belief in the dichotomy of tradition and modernity. This notion was adopted, according to Hania Sholkamy, by the majority Egyptian social scientists aligning themselves with encounters of American sociology, or perhaps, inherited from French-educated Egyptian elites of the late nineteenth century, who perceived tradition to oppose modernity. As such, to build a modern city, traditional quarters have to be abolished or ignored. See Sholkamy, *Why is Anthropology so Hard in Egypt?*, p. 141.

2   Said, *Rehabilitation with People's Participation*, p. 6.

3   Nasser Rabat provoked similar consensus on space-activity relationships in old Islamic cities in general. See Rabat, *The Culture of Building*. Stefano Bianca, in the meantime, referred to the *"matrix of behavioural archetypes, which, by necessity, generated correlated physical pattern"*. Bianca, *Urban Form in the Arab World: Past and Present,* p. 24.

4   Kostof and Castillo, *A History of Architecture: Settings and Rituals.*

5   See for example: Mona Abaza's *Changing Consumer Cultures of Modern Egypt*; Diane Singerman's *Avenues of Participation*; Homa Hoodfar's *Between Marriage and the Market*; and Farha Ghanna's *Remaking the Modern.*

6   Comments made during a research seminar in Sheffield School of Architecture during 2008. Bryan Lawson is a well-known author and writer in the field of architectural theory and practice; some of his books were used in this thesis.

7   Sandweiss, *Framing Urban Memory*, p. 26.

8   Schneider and Till, *Flexible Housing.*

References and Bibliographies

A. REFERENCES

Abada, G. (2004) Heterogeneity within Homogeneity: Fragmentation and the Possible Re-Coherence of Traditional Urban Forms in Cairo. *GBER*, 4(1), pp. 3–14.

Abaza, M. (2006) *Changing Consumer Cultures of Modern Egypt: Cairo Urban Reshaping*. Boston: Brill.

Abdelhalim, I.A. (1978) The Building Ceremony. Unpublished PhD thesis submitted to the University of California in Berkeley.

Abdelmonem, M.G. (2011) Understanding Everyday Homes of Urban Communities: The Case of Local Streets (Hawari) of Old Cairo. *Journal of Civil Engineering and Architecture*, 5(11), pp. 996–1010.

Abdelmonem, M.G. (2012) The Practice of Home in Old Cairo: Towards Socio-spatial Models of Sustainable Living. *Traditional Dwellings and Settlements Review*, 23(2), 2012, pp. 35–50.

Abdelmonem, M.G. (2012) Responsive Homes of Old Cairo: Learning from the Past, Feeding in the Future. *Hospitality and Society*, 2(3), 09.2012, pp. 251–71.

Abdelmonem, M.G. and Selim, G. (2012) Architecture, Memory and Historical Continuity in Old Cairo. *Journal of Architecture*, 17(2), pp. 163–89.

Abdel-Wahhab, H. (1957) *Takhtit al-Qahira wa Tanzimaha* (Planning and Organizing Cairo- Since Foundation) Arabic title. Cairo: Dar Al-Nashr LilGami'aat al-Masriyyah, pp. 23–4.

Abu-Loghud, J. (1971) *Cairo: 1001 Years of City Victorious*. Princeton: Princeton University Press.

Abu-Loghud, J. (1987) Islamic City: Historic Myth, Islamic Essence and Contemporary Relevance. *International Journal of Middle Eastern Studies*, 19(2), pp. 155–76.

Adham, K. (1997) The Building Border: A Hermeneutical Study of the Cultural Politics of Space. In *Egypt, The Case of El-Houd El-Marsoud Park in Cairo*. PhD Thesis. Texas: Texas A&M University.

Akbar, J.A. (1988) *Crisis in the Built Environment: The Case of the Muslim City*. Michigan: Brill Publishers.

Akbar, J.A. (1992) "The Architecture of the Earth in Islam", "*Imarat al-arḍ fi al-Islam: muqaranah al-shariah bi-anẓimat al-umran al-waḍiyah*". Arabic title. Beirut: Al-Resalah.

Akin, O. (2002) Case-based Instruction Strategies in Architecture. *Design Studies*, 23(4), pp. 407–31.

AKTC (2005) *Cairo: Urban Regeneration in the Darb al-Ahmar District, A Framework for Investment*. Aga Khan Trust for Culture: Historic Cities Support Programme. Rome: Artemide Edizioni.

Al-Aswany, A. (2004) *The Yacoubian Building*. Cairo: The American University in Cairo Press.

Alexander, C. (1979) *Notes on the Synthesis of Form*. London: Harvard University Press.

Alexander, C. (1979) *The Timeless Way of Building*. New York: Oxford University Press.

Alexander, C. (1985) *The Production of Houses*. Oxford: Oxford University Press.

Alexander, C., Ishikawa, S. and Silverstein, M. (1977) *A Pattern Language: Towns, Buildings, Construction*. New York: Oxford University Press.

Al-Guindi, F. (1999) *Veil: Modesty, Privacy and Resistance*. Oxford: Berg Publishers.

Al-Jabarti, A. (1754–1822) (1998) '*Ajāïb al-āthār fī al-tarājim wa-al-akhbār*. Cairo: Maṭba'at Dār al-Kutub al-Miṣrīyah.

Al-Jabarti, A. (1997) *Aja'ib al-Athar fi al-Tarajim wa al-Akhbar*. Cairo: Al-Hayah al-Ammah li-Dar al-Kutub (4 Volumes).

Al-Maqrizi, T. 1364–1442 (1959) *Al-Khitat Al-Maqriziyyah*. Safahat min Tarikh Misr Series. Cairo.

Al-Messiri, S. (1974) *The Role of the Futuwwa in the Social Structure of the City of Cairo*. Paper for the conference on Changing Forms of Patronage.

Al-Messiri-Nadim, N. (1979) The Concept of the Hara: A Historical and Sociological Study of al-Sukkariyya. In *Annale Islamogiques*, 15. Le Caire: Institut Francais d'archeologie Orientale, pp. 313–48.

Al-Muqaddasi (1897) *Ahsan al-Taqasim*. Calcutta: Baptist Mission Press.

Al-Sayyad, N. (1994) Cairo: Bayn al-Qasrayn: The Street between the Two Palaces. In Z. Celik, D. Favro and R. Ingersoll (eds), *Streets: Critical Perspectives on Public Space*. London: University of California Press.

Al-Sayyad, N. (1999) Virtual Cairo: An Urban Historian's View of Computer Simulation. *Leonardo*, 32(2), pp. 93–100.

Al-Sayyad, N. (2011) *Cairo: Histories of a City*. London: Harvard University Press.

Al-Sayyad, N. and Roy, A. (2006) Medieval Modernity: On Citizenship and Urbanism in a Global Era. *Space and Polity*, 10(1), pp. 1–20.

Al-Baghdadi, A. (1946) *Al-Ifadah wa Al-I'tibar, written 1204 A.D.* (Cairo, 1946).

Amin, Q. (1899) *Tahrir al-Mar'aa* (The Liberation of Women), Arabic Title. Cairo: The American University in Cairo Press (print 2000).

Amin, Q. (1900) *Al-Mar'ah al-jadīdah* (*New Woman*). Arabic Title. Cairo: The American University in Cairo Press (print 2000).

Arkoun, M. (2002) Spirituality and Architecture. In A. Petruccioli and K.K. Pirani, (eds), *Understanding Islamic Architecture*. London: Routledge Curzon, pp. 3–8.

Arnaud, Jean-Luc (1998) Cairo: Observations on the Modern City 1867–1907. *Le Caire: Mise en Place d'une ville modern 1867–1907, Des intérêts du prince aux sociétés privées*. French Title Arabic translation by H. Touson and F. El-Dahan (2002). Cairo: Higher Commission for Culture in association with French Centre for Culture and Cooperation in Cairo.

Asfour, K. (1990) Abdel Halim's Cairo Garden: An Attempt to Defrost History. *Mimar*, issue 36, pp. 72–7.

Aygen, Z. (2008) The Centennial of Post-Ottoman Identity: Contemporary Design Concepts in the Balkans and the Middle East. In P. Herrleand and E. Wegerhoff (eds), *Architecture and Identity*. Berlin: Lit Verlag, pp. 307–20.

Baer, G. (1982a) Decline and Disappearance of Guilds. In G. Baer (ed.), *Studies in the Social History of Modern Egypt*. Chicago: The University of Chicago Press.

Baer, G. (1982b) Fellah and Townsman in the Middle East. *Studies in Social History* Book Series Oxon: Frank Cass & Co.

Bahammam, O. (2006) The Role of Privacy in the Design of the Saudi Arabian Courtyard House. In B. Edwards (ed.), *Courtyard Housing: Past, Present and Future*. New York: Taylor & Francis Publishing, pp. 77–94.

Bahloul, J. (1996). *The Architecture of Memory: A Jewish-Muslim Household in Colonial Algeria, 1937–1962*. Cambridge: Cambridge University Press, p. 28.

Baker, C. (2002) *Rebuilding the House of Israel: Architectures of Gender in Jewish Antiquity*. Stanford: Stanford University press, pp. 113–14.

Barton, H. (2002) *Sustainable Communities: The Potential for Eco-neighbourhoods*. London: Earthscan.

Barton, H., Grant, M. and Guise, R. (2003) *Shaping Neighbourhoods: A Guide for Health, Sustainability and Vitality*. London: Spon Press.

Bauman, Z. (2000) *Liquid Modernity*. Cambridge: Polity Press.

Beattie, A. (1997) *Cairo: A Cultural History*. US: Oxford University Press.

Behrens-Abouseif, D. (1989) *Islamic Architecture in Cairo: An Introduction*. Fifth edition. Cairo: The American University in Cairo Press.

Beinin, J. (1981) Formation of the Egyptian Working Class. MERIP Reports, no. 94: *Origins of the Working Class, Class in the Middle East*. Washington: Middle East Research and Information Centre, pp. 14–23.

Beinin, J. and Lockman, Z. (1998) *Workers on the Nile: Nationalism, Communism, Islam, and the Egyptian Working Class, 1882–1954*. Cairo: The American University in Cairo Press.

Berke, D. and Harris, S. (1997) *Architecture of the Everyday*. New York: Princeton University Press.

Bianca, S. (2000) *Urban Form in the Arab World, Past and Present*. Zurich: vdf, p. 12.

Bickford, S. (2000) Constructing Inequality: City Spaces and the Architecture of Citizenship. *Political Theory*, 28(3), pp. 355–76. Sage Publication.

Birkeland, J. (ed.) (2002) *Design for Sustainability: A Sourcebook of Integrated Eco-logical Solutions*. London: Earthscan.

Blackmar, E. (1989) *Manhattan for Rent 1785–1850*. Ithaca: Cornell University Press.

Blackmar, E. (1989) The Social Meanings of Housing, 1800–1840. In B.M. Lane (ed.) (2007), *Housing and Dwelling: Perspectives on Modern Domestic Architecture*. New York: Routledge, Taylor & Francis, pp. 108–12.

Blundell Jones, P. (1987) *Social Construction of Space*. Spazio e societa, vol. 40. Gangemi editore, pp. 62–72.

Blundell Jones, P. (2005) Sixty-eight and After. In P. Blundell Jones, D. Petrescu, J. Till et al. (eds), *Architecture and Participation*. London: Spon, pp. 127–40.

Blundell Jones, P. (2007) *Peter Hübner: Building as a Social Process*. London: Edition Axel Menges.

Bourdieu, P. (1977) *Outline of a Theory of Practice*. English translation by Richard Nice. Cambridge: Cambridge University Press.

Bourdieu, P. (1990) *The Logic of Practice*. Translated by Richard Nice. Cambridge: Polity Press.

Bower, S.N. (1980) Territory in Urban Settings. In I. Altman, A. Rapoport and J.F. Wohlwill (eds), *Environment and Culture; Human Behaviour and Environment*, vol. 4. New York: Plenum Press, pp. 179–93.

Briggs, M.S. (1920) The Fatimid Architecture of Cairo (A.D. 969–1171). *The Burlington Magazine for Connoisseurs*, 37(211), pp. 190–95.

Butler, A.J. (1902) *The Arab Conquest of Egypt – And the Last Thirty Years of the Roman Dominion*. Oxford: Clarendon Press.

Campbell, R. and Comodromos, D. (2009) Urban Morphology + The Social Vernacular: A Speculative Skyscraper for Islamic Medieval Cairo. *Journal of Architectural Education*, 63(1), pp. 6–13.

Campo, J. (1991a) *The Other Sides of Paradise*. South Carolina: University of South Carolina Press.

Campo, J. (1991b) Orientalist Representations of Muslim Domestic Space in Egypt. *TDSR*, III(1), pp. 29–42.

Cannon, B.D. (1985) Nineteenth-Century Arabic Writings on Women and Society: The Interim Role of the Masonic Press in Cairo (al-Lataif, 1885–1895). *International Journal of Middle East Studies*, 17(4), pp. 463–84.

Carr, S., Francis, M., Rivlin, L.G. and Stone, A.M. (1992) *Public Space*. Environment & Behavior Series. Cambridge: Cambridge University Press.

Castell, M. (1977) *The Urban Question: A Marxist Approach*. London: Edward Arnold.

Castell, M. (2004) *The Power of Identity*. Oxford: Wiley-Blackwell.

Chandhoke, N. (2003) Transcending Categories: The Private, the Public and the Search for Home. In G. Mahajan and R. Helmut (eds), *The Public and the Private*. London: Sage Publication, pp. 181–204.

Clapham, D. (2005) *The Meaning of Housing: A Pathways Approach*. London: The Policy Press, p. 117.

Clerget, M. (1934) Cairo: Study of Urban Geography and Economic history, *Le Caire: Etude de geographie urbaine et d'histoire Economique*. Paris: Librarie Orientaliste Paul Geuthner.

Clot-Bey, A.B. (1840) *Apercu general sur L'Egypte*. Paris: Fortin, Libraires-Editeurs.

Cole, D. (2002) People, the State, and the Global in Cairo. *Anthropological Quarterly*, 75(4), pp. 793–800.

Cohen, A. (1998) *The Symbolic Construction of Community*. London: Routledge.

Cox, H.G. (1978) Sociology and the Meaning of History. Introduction to Sennett, R., *The Fall of Public Man*. London: Faber & Faber Publishing.

Crecelius, D. (1981) *The Roots of Modern Egypt: A Study of the Regimes of Ali Bey al-Kabr and Muhammad Bey abut al-Dhahab, 1760–1775*. Minneapolis, MN, USA: Bibliotheca Islamica.

Creswell, K.A.C. (1952) *The Muslim Architecture of Egypt. Volume 1, Ikhshids and Fatimids, A.D. 939–1171*. Oxford: Oxford University, re-issued (1978), New York: Hacker Arts Book.

Cromley, E.C. (1990) *Alone Together: A History of New York's Early Apartments*. New York: Cornell University Press.

Cruz, T. (2005) Tijuana Case Study Tactics of Invasion: Manufactured Sites. *Architectural Design*, 75(5), pp. 32–7.

Cuno, K. (1995) Joint Family Households and Rural Notables in 19th-century Egypt. *International. Journal of Middle East Studies*, 27(4), pp. 485–502.

De Certeau, M. (1984) *The Practice of Everyday Life,* vol. 1. London: University of California Press.

De Certeau, M., Giard, L. and Mayol, P. (1998) *The Practice of Everyday Life* (vol. 2). London: University of Minnesota Press.

De Nerval, G. (1851) *Voyage en Orient*. Paris: Charpentier.

Delaney, D. (2005) *Territory: A Short Introduction*. Oxford: Blackwell Publishing.

Denis, I. (2006) Cairo as Neoliberal Capital? From Walled City to Gated Communities. In Singerman, D. and Amar, P. (eds), *Cairo Cosmopolitan: Politics, Culture, and Urban Space in the New Globalized Middle East*. Cairo: The American University in Cairo Press, pp. 47–72.

Denoix, S. (2000) A Mamluk Institution for Urbanization: The Waqf. In D. Behrens-Abouseif (ed.), *The Cairo Heritage*. Cairo: The American University in Cairo Press.

Donohue, E.P. (2003) *A Heideggerian Reading of the Philosophy of Art of Susanne K. Langer with Special Reference to Architecture*. The proceeding of the first Hawaii International Conference on Arts and Humanities, 12–15 January 2003, Honolulu, Hawaii, USA.

Douglas, M. (1991) The Idea of Home: A Kind of a Space. *Social Research*, 58(1), pp. 287–308.

Douglas, M. (1996) *Natural Symbols: Exploration in Cosmology*. London: Routledge.

Dovey, K. (1999) *Framing Places: Mediating Power in Built Form*. London: Ashgate Publishing.

Dutton, T.A. (1989) Cities Cultures, and Resistance: Beyond Leon Krier and the Postmodern Condition. *Journal of Architectural Education*, 42(2), pp. 3–9.

Early, E. (1993) *Baladi Women of Cairo*. London: Lynne Rienner.

Editorial (2004) *Aims and Scopes: Home Cultures*. Oxford: Berg Publishers, vol. 1, no. 1.

El-Kholy, H.A. (2002) *Defiance and Compliance: Negotiating Gender in Low Income Cairo*. Oxford: Berghahn Publishing.

El-Rashidi, S. (2004) The History and Fate of al-Darb al-Ahmar. In S. Bianca and P. Jodidio (eds), *Cairo: Revitalising a Historic Metropolis*. Turin: Umberto Allemandi & C. for Aga Khan Trust for Culture, 55–65.

El-Sioufi, M. (1981) *A Fatimid Harah: Its Physical, Social and Economic Structure*. Cambridge, MA: Aga Khan Program for Islamic Architecture.

Eldridge, J. (1971) *Max Weber: The Interpretation of Social Reality*. London: Michael Joseph Publishers, p. 26.

Elsheshtawy, Y. (2006) Urban Transformation: Social Control at al-Rifa'l Mosque and Sultan Hasan Square. In D. Singerman and P. Amar (eds), *Cosmopolitan Cairo: Politics, Culture and Urban Space in the New Globalized Middle East*. Cairo: The American University Press, pp. 295–312.

Emara, M. (2008) *Qasim Amin: Complete Works*. Cairo: Dar al-Shorouq, pp. 319–417.

Fahmi, W.S. (2008) Global Tourism and the Urban Poor's Right to the City: Spatial Contestation within Cairo's Historical Districts. In P.M. Burns. and M. Novelli (eds), *Tourism Development: Growth, Myths, and Inequalities*. Oxfordshire, UK: CABI International, pp. 159–91.

Fargues, P. (2003) Family and Household in Mid-nineteenth-century Egypt. In B. Doumani (ed.), *Family History in the Middle East: Household, Property, and Gender*. New York: State University of New York Press, pp. 23–50.

Fathy, H. (1971) Urban Arab Architecture in the Middle East. A Lecture by Hassan Fathy to The Arab University in Beirut. In Y. El-Zeiny (2003), *Min Fikr Sheikh al-Mi'mari'een Hassan Fathy*. Cairo: The Higher Council of Culture, pp. 42–3.

Fathy, H. (1976) *Architecture for the Poor: An Experiment in Rural Egypt*. Chicago: The University of Chicago Press.

Feiler, G. (1992) Housing Policy in Egypt. *Middle Eastern Studies*, 28(2), pp. 295–312.

Fernandes, L. (2000) Istibdal: The Game of Exchange and its Impact on the Urbanization of Mamluk Cairo. In D. Behrens-Abouseif (ed.), *The Cairo Heritage*. Cairo: The American University in Cairo Press, pp. 203–22.

Festingr, L. (1972) Architecture and Group Membership. In R. Gutman (ed.), *People and Buildings*. London: Basic Books, pp. 120–34.

Flint, J. (2004) A Book Review Article for Low, S. (2003), *Behind the Gates: Life, Security and the Pursuit of Happiness in Fortress America*. In *Housing, Theory and Society*, 21(3), pp. 139–40.

Fourier, J.-B.-J. (1809–28) The Description of Egypt (*Préface Historique. Vol. I, of 'Le Description de l'Egypte'*). Paris: Imprimerie Royale. Part I.

Friedman, A.T. (1998) *Women and the Making of the Modern House: A Social and Architectural History*. New York: Harry N. Abrams Publishers.

Gabriel, R. (1998) The Failure of Pattern Languages. In L. Rising (ed.), *The Patterns Handbook*. Cambridge: Cambridge University Press, pp. 333–44.

Galal, K. (2006) Architectural Education as a Gate for "Glocal Architecture" in Egypt: A Proposed Approach. In J. Al-Qawasmi and G. Velasco (eds), *Changing Trends in Architectural Design Education*. Rabaat: The Centre for the Study of Architecture in the Arab Region (CSAAR), pp. 281–98.

Game, A. (2001) Belonging: Experience in Sacred Time and Space. In J. May and N.J. Thrift (eds), *TimeSpace: Geographies of Temporality*. London: Routledge, pp. 226–39.

Gauzin-Muller, D. and Favet, N. (2002) *Sustainable Architecture and Urbanism: Concepts, Technologies, Examples*. Basel: Birkhauser.

Gazda, E. and Haeckl, A. (eds) (1995) *Roman Art in the Private Sphere: New Perspectives on the Architecture and Decor of the Domus, Villa, and Insula*. Michigan: University of Michigan Press.

Gelernter, M. (1995) *Sources of Architectural Form: A Critical History of Western Design Theory*. Manchester: Manchester University Press, p. 266.

Ghannam, F. (2002). *Remaking the Modern: Space, Relocation, and the Politics of Identity in a Global Cairo*. California: The University of California Press.

Giddens, A. (1986) *The Constitution of Society: Outline of the Theory of Structuration*. London: Polity Press.

Giddens, A. (1991) *Modernity and Self Identity: Self and Society in the Late Modern Age*. Stanford: Stanford University Press.

Glover, W. (2004) A Feeling of Absence from Old England: The Colonial Bungalow. In K. Olwig and K. Hastrup (eds), *Siting Culture: The Shifting Anthropological Object*. London: Routledge, pp. 17–38.

Great Britain (2005) *Sustainable Communities: A Home for All*, by the Office of the Deputy Prime Minister.

Grindle, M.S. and Thomas, J.W. (1989) Policy Makers, Policy Choices and Policy Outcomes. The Political Economy of Reform in the Developing Countries. *Policy Sciences*, 22.

Groat, L. and Wang, D. (2002) *Architectural Research Methods*. New York: Wiley Publishers.

Guy, S. and Farmer, G. (2001). Reinterpreting Sustainable Architecture: The Place of Technology. *Journal of Architectural Education*, 54(3), pp. 140–48.

Habermas, J. (1991) *The Structural Transformation of the Public Sphere: An Inquiry Into a Category of Bourgeois Society*. Translated by Thomas Burger. London: MIT Press.

Hakim, B. (1986) *Arabic-Islamic Cities: Building and Planning Principles*, second edition. London: Kegan Paul International.

Hanna, N. (1984) Construction Work in Ottoman Cairo. *Supplement aux Annale Islamologiques*. Le Caire: Cahier N.4.

Hanna, N. (1991) Houses of Cairo in 17th & 18th Centuries. (*Habiter au Caire aux XVIIe et XVIIIe siècles*). Cairo: Institut Francais d'Archeologie Orientale.

Hanna, N. (1993) *Cairene Homes of the Seventeenth and Eighteenth Century: A Social and Architectural Study*, an Arabic translation by Halim Tosoun. Cairo: al-Araby Publishing, pp. 31–3.

Hank, I. (1984) Continuity and Change in Local Development Policies in Egypt: From Nasser to Sadat. *International Journal for Middle Eastern Studies*, 16(1), pp. 43–66.

Harries, K. (1975) *The Ethical Function of Architecture*. Massachusetts: MIT Press.

Harris, S. (1997) Everyday Architecture. In D. Berke and S. Harris, *Architecture of the Everyday*. Princeton: Princeton University Press.

Harvey, D. (1990) *The Condition of Postmodernity*. Oxford: Basil Blackwell.

Hastrup, K. and Olwig, K. (1997) *Siting Culture: The Shifting Anthropological Object*. London: Routledge.

Hatem, M. (1987) Class and Patriarchy as Competing Paradigms for the Study of Middle Eastern Women. *Comparative Studies in Society and History*, 29(4), pp. 811–18.

Hatem, M. (1987) Toward the Study of the Psychodynamics of Mothering and Gender in Egyptian Families. *International Journal of Middle East Studies*, 19(3), pp. 287–305.

Hertzberger, H. (2000) *Space and the Architect: Lessons in Architecture 2*. Rotterdam: 010 Publishers.

Hillenbrand, R. (1994) *Islamic Architecture: Form, Function and Meaning*. New York: Columbia University Press.

Holliss, F. (2012) *Space, Buildings and the Life Worlds of Home-based Workers: Towards better Design in 'Visualising the Landscape of Work and Labour'*. Sociological Research Online.

Hoodfar, H. (1996). Survival Strategies and the Political Economy of Low-income Households in Cairo. In H. Hoodfar and D. Singerman (eds), *Development, Change, and Gender in Cairo: A View from the Household*. Indiana: Indiana University Press.

Hoodfar, H. (1997) *Between Marriage and the Market: Intimate Politics and Survival in Cairo*. California: University of California Press.

Hoodfar, H. (1998) Women in Cairo's Invisible Economy. In R. Lobban and E. Fernea (eds), *Middle Eastern Women and the Invisible Economy*. Florida: University Press of Florida.

Hourani, A.H. and Stern, S.M. (eds) (1970) *The Islamic City: A Colloquium*. Oxford: Cassirer, pp. 9–24.

Hübner, P. (1989) The Joy of Building. An English translation in Blundell Jones, P. (2007), *Peter Hübner: Building as a Social Process*. London: Edition Axel Menges, p. 335.

Hübner, P. (1990) Shoes, Not Shoe Boxes. An English translation in Blundell Jones, P. (2007), *Peter Hübner: Building as a Social Process*. London: Edition Axel Menges, p. 337.

Hübner, P. (2007) Design Schools as Powerhouses. An English translation in Blundell Jones, P., *Peter Hübner: Building as a Social Process*. London: Edition Axel Menges, p. 341.

Hutcheon, L. (1986) *The Politics of Postmodernism: Parody and History*. Cultural Critique, No.5 Modernity and Modernism, Postmodernity and Postmodernism. Minnesota: University of Minnesota Press, pp. 179–207.

Hourani, A.A. (1962) *Arabic Thought in the Liberal Age, 1798–1939*. London: Oxford University Press.

Ibrahim, A. (2002) *Urban Harmony: A Dream yet to Come True*. Cairo: The Centre of Planning and Architectural Studies. AbdelBaki Ibrahim's Articles Collections, pp. 1–4. Available at: http:///www.cpas-egypt.com/Articles/Baki/ABR20026.htm (last accessed on 10th August 2009).

Ibn Manzur, Muḥammad ibn Mukarram, 1232–1311 or 12 (1955–1956) *Lisan al-Àrab*. 15 Volumes. Bayrut: Dar Ṣadir.

Ismail, S. (2000) The Popular Movement Dimensions of Contemporary Militant Islamic: Socio-Spatial Determinants in the Cairo Urban Setting. *Society for Comparative Study of Society and History*, p. 369.

Ismail, S. (2006) *Political Life in Cairo's New Quarters: Encountering the Everyday State*. Minnesota: University of Minnesota Press.

Jacob, J. (1962) *The Death and Life of Great American Cities*. London: Jonathan Cape.

Jairazbhoy, R.A. (1972) *An Outline of Islamic Architecture*. London: Asia Publishing House.

Jencks, C. (2005) *The Iconic Building*. London: Frances Lincoln.

Jenkins, R. (1992) *Pierre Bourdieu*. London: Routledge.

Jenkins, R. (2004) *Social Identity*, second edition. Key Ideas Series. London: Routledge.

Jomard, E. (1822) *Description Abrégée de la ville et de la citadelle du Kaire, Description de l'Egypte: Etat Moderne*. Tome II, Part II. Paris: L'imprimerie Royale, pp. 580–81.

Jreisat, J. (1997) *Politics without Process: Administering Development in the Arab World*. London: Lynne Rienner Publishers.

Junestrand, S. and Tollmar, K. (1998) The Dwelling as a Place for Work. *Lecture Notes in Computer Science*, vol. 1370, pp. 230–47.

Keddie, N. (1990) *Women in Middle Eastern History: Shifting Boundaries in Sex and Gender*. London: Yale University Press.

Keddie, N. (2006) *Women in the Middle East: Past and Present*. Princeton: Princeton University Press.

Kenzari, B. and El Sheshtawy, Y. (2003) The Ambiguous Veil: On Transparency, the Mashrabiy'ya, and Architecture. *Journal of Architectural Education*, 56(4), pp. 17–25.

Kern, S. (2003) *The Culture of Time and Space: 1880–1918*, second edition. USA: Harvard University Press.

Kemeny, J. (1981) *The Myth of Home-ownership: Private versus Public Choices in Housing Tenure*. London: Routledge.

Kemeny, J. (1992) *Housing and Social Theory*. London: Routledge.

Khaldun, I. (1958) *The Muqaddimah: An Introduction to History*.English Translation by F. Rosenthal. Princeton: Princeton University Press.

Khan, O. and Hannah, D. (2008) Performance/Architecture: An Interview with Bernard Tschumi. *Journal of Architectural Education*, 61(4), pp. 52–8.

Kholosy, M.M. (1997) *Hassan Fathi*. Beirut: Dar Qabess.

Kilian, T. (1998). Public and Private, Power and Space. In A. Light and J.M. Smith (eds), *Philosophy and Geography II: The Production of Public Space*. The Professional Geographer, 53(1), pp. 115–34.

Kostof, S. (1977) *The Architect: Chapters in the History of the Profession*. Oxford: Oxford University Press.

Kroll, L. (1986) *The Complexity of Architecture*. Translated by Peter Blundell Jones. London: BT Batsford.

Kubiak, W. (1987) *Al-Fustat: Its Foundation and Early Urban Development*. Cairo: The American University in Cairo Press.

Langer, S.K. (1953) *Feeling and Form: A Theory of Art*. New York: Charles Scribner's Sons.

Lane, B.M. (ed.) (2007) *Housing and Dwelling: Perspectives on Modern Domestic Architecture*. London: Routledge.

Lane, E.W. (1860) *An Account of the Manners and Customs of the Modern Egyptians*, fifth edition. London: John Murray.

Lane-Poole, S. (1845) *The Englishwoman in Egypt: Letters from Cairo*, two volumes. London: William Clowes and Sons.

Lane-Poole, S. (1902) *The Story of Cairo*. London: J.M. Dent and Co.

Lawrence, R. (1987) *Housing, Dwellings and Homes: Design Theory, Research and Practice*. London: John Wiley and Sons.

Lawson, B. (2006) *How Designers Think: The Design Process Demystified*. London: Elsevier.

Leavitt, J. and Saegert, S. (1989) *From Abandonment to Hope*. New York: Columbia University Press.

Lefebvre, H. (1991) *The Production of Space*. Translated by Donald Nicholson-Smith. Massachusetts: Blackwell Publishing, p. 68.

Lefebvre, H. (1997) The Production of Space (Extracts). In Neil Leach (ed.), *Rethinking Architecture*. London: Basil Blackwell Publishing.

Lethen, H. (1996) Between the Barrier and the Sieve: Finding the Border in the Modern Movement. *The Journal of Architecture*, 1(4), pp. 301–12.

Levi-Strauss, C. (1969) *The Elementary Structures of Kinship*. English translation by J.H. Bell, J.R. Sturmer and R. Needham. London: Eyre & Spottiswoode

Levine, N. (1997) *The Architecture of Frank Lloyd Wright*. Princeton: Princeton University Press.

Mahboub, S. (1935) Cairo: Some Notes on its History, Character and Town Plan. *The Journal of Town Planning Institute*, XXI(11), p. 289.

Mahgoub, Y. (2007) Design Studio Pedagogy: From Core to Capstone. In A. Salama and N. Wilkinson (eds), *Design Studio Pedagogy: Horizons for the Future*. Gateshead, UK: The Urban International Press, pp. 193–200.

Mahfouz, A. (2008) *Khabaya al-Qahira (Mysteries of Cairo)*. Cairo: Dar al-Shorouq.

Mahfouz, N. (2001) *Palace Walk*. Translated into English by William Maynard Hutchins and Olive E. Kenny. Cairo: The American University in Cairo Press.

Mallet, S. (2004) Understanding Homes: A Critical Review of the Literature. *The Sociological Reivew*, 52(1), pp. 62–89.

Massey, D. (2006) London Inside-Out. *Soundings*, (32), pp. 62–71.

McKee, A. (2005) *The Public Sphere: An Introduction*. Cambridge: Cambridge University Press.

McKenzie, S. (2004) *Social Sustainability: Towards Some Definitions*. Hawke Research Institute Working Paper Series, No. 27, University of South Australia, Magill, South Australia.

McKinlay, I. (2004) Social Work and Sustainable Development: An Exploratory Study. Unpublished PhD dissertation. Pretoria: University of Pretoria.

Miles, M. (2000) After the Public Realm: Spaces of Representation, Transition and Plurality. *International Journal of Art and Design Education*, 19(2), pp. 253–62.

Mitchell, T. (1988) *Colonizing Egypt*. Cambridge: Cambridge University Press.

Moghadam, M. (2003) *Modernizing Women: Gender and Social Change in the Middle East*. London: Lynne Rienner, p. 118.

Moore, S. and Chanock, M. (2000) *Law as Process: An Anthropological Approach*. Berlin: LIT Verlag Berlin-Hamburg-Münster, p. 39.

Morgan, D. (1996) *Family Connection: An Introduction to Family Studies*. Cambridge: Polity Press.

Mubarak, A. (1882) *Al-Khitat al-Tawfiqiyyah*, 20 volumes. Reproduced Cairo: Bulaq.

Myntti, C. (1999) *Paris along the Nile: Architecture in Cairo from the Belle Époque*. Cairo: The American University in Cairo Press.

Nabil, Y. (1994) Reconciliations and Continued Polarities in the Works and Theories of Halim and Bakri, Master's Thesis. Cambridge: Massachusetts Institute of Technology.

Nadim, N. (1975) The Relationship between the Sexes in a Harah of Cairo. Unpublished Doctoral dissertation. Bloomington: Indiana University.

Nadim, A. (2000) *Rehabilitation of al-Darb al-Asfar*. Project Report. Cairo: Supreme Council of Antiquities, Ministry of Culture, Egypt.

Nadim, A. (2002) Documentation, Restoration, Conservation and Development of Bayt Al-Suhaimy Area. In Government Document: *Historic Cairo*. Cairo: Supreme Council of Antiquities, Ministry of Culture, Egypt.

Nepomechie, M. (2004) Dwellings: The Vernacular House Worldwide, Book Review. *Journal of Architectural Education*, 58(1), pp. 72–3.

Nevett, L.C. (2001) *House and Society in the Ancient Greek World*. Cambridge: Cambridge University Press, p. 4.

Nippert-Eng, C.E. (1996) *Home and Work: Negotiating Boundaries through Everyday Life*. Chicago: University of Chicago Press.

Norberg-Schultz, C. (1971) *Existence, Space and Architecture*. London: Studio Vista.

Nydell, M. (2005) *Understanding Arabs: A Guide for Modern Times*. London: Intercultural Press.

Oliver, P. (2003) *Dwellings: The Vernacular House Worldwide*. London: Phaidon.

Olwig, K. (1997) Cultural Sites: Sustaining a Home in a Deterritorialized World. In K. Olwig and K. Hastrup (eds), *Siting Culture: The Shifting Anthropological Object*. London: Routledge, pp. 17–38.

Olwig, K and Hastrup, K. (eds) (2007) *Siting Culture: The Shifting Anthropological Object*. Oxon, UK: Routledge.

O'Kane, B. (2000) Domestic and Religious Architecture in Cairo: Mutual Influences. In A. Behrense (ed.), *The Cairo Heritage: Essays in Honor of Laila Ali Ibrahim*. Cairo: American University in Cairo Press, pp. 149–61.

Paddison, R. (2001) Communities in the City. In R. Paddison (ed.), *Handbook of Urban Studies*. London: Sage Publications, pp. 194–205.

Panayiota, I.P. (2007) Hassan Fathy Revisited: Postwar Discourses on Science, Development and Vernacular Architecture. *Journal of Architectural Education*, 60(3), pp. 28–39.

Pérez-Gómez, A. (1983) *Architecture and the Crisis of Modern Sciences*. London: MIT Press, p. x.

Perri 6 (1998) *The Future of Privacy*. London: Demo Publishing.

Petruccioli, A. (2006) The Courtyard House: Typological Variations. In B. Edwards (ed.), *Courtyard Housing: Past, Present & Future*. Oxon: Taylor & Francis Publishers, p. 9.

Phillips, S. (1860) *The Christian Home*. New York: Bill.

Pitts, A. (2003) *Planning and Design Strategies for Sustainability and Profit: Paradigmatic Sustainable Design on Building and Urban Scales*. Oxford: Architectural Press.

Pollard, L. (2000) The Family Politics of Colonizing and Liberating Egypt, 1882–1919, *Social Politics*, 7(1), pp. 47–79.

Pomeroy, S. (1991) *Women's History and Ancient History*. London: University of North Carolina Press, p. xiii.

Porteous, J.D. (1976) Home: The Territorial Core. *Geographical Review*, 66(4), pp. 383–90.

Prak, N.L. (1977) *The Visual Perception of the Built Environment*. Netherlands, Delft: Delft University Press.

Prochansky, H. (1974) Environmental Psychology and the Design Profession. In J.T. Lang (ed), *Designing for Human Behaviour: Architecture and Behavioural Sciences*. Stroudsburg, PA: Dowden, Hutchinson, and Ross, pp. 72–80.

Rabbat, N. (2002) *The Culture of Building and Building Culture* (Arabic Reference). Beirut: Riad El-Rayyes Books.

Rapoport, A. (1969) *House Form and Culture*. New Jersey: Prentice-Hall Inc.

Rapoport, A. (1981) Identity and Environment: A Cross-cultural Perspective. In J. Duncan (ed.), *Housing and Identity*. London: Francis & Taylor Publishing, p. 12.

Rapoport, A. (1982) *The Meaning of the Built Environment: A Nonverbal Communication Approach*, second edition. Arizona: The University of Arizona Press, p. 178.

Raymond, A. (1994a) Islamic City, Arab City: Orientalist Myths and Recent Views. *British Journal of Middle Eastern Studies*, 21(1), pp. 3–18.

Raymond, A. (1994b) *Cairo*. Arabic Translation by Latif Farag, Dar al-Fikr lildrasat, Cairo.

Raymond, A. (2000) *Cairo*. English Translation by Willard Wood. London: Harvard University Press.

Rendell, J. (2004) Architectural Research and Disciplinarity. *Architectural Research Quarterly (ARQ)*, 8(2), pp. 141–7.

Richardson, J.T.E. (1996) Introduction. In J.T.E. Richardson (ed.), *Handbook of Qualitative Research Methods for Psychology and the Social Sciences*. London: BPS Blackwell, pp. 3–10.

Riley, T. (1999) *The Un-private House*. New York: Museum of Modern Art.

Ross, H. (2002) Sustainability and Aboriginal Housing. In J. Birkeland (ed.), *Design for Sustainability: A Sourcebook of Integrated Eco-logical Solutions*. London: Earthscan, pp. 138–41.

Rossi, A. (1999) *The Architecture of the City*. London: MIT Press.

Ruiz, M.M. (2009) Orientalist and Revisionist Histories of 'Abd al-Rahman al-Jabarti. *Middle East Critique*, 18(3), pp. 261–84.

Safran, N. (1961) Egypt in Search of Political Community: An Analysis of the Intellectual and

Political Evolution of Egypt, 1804–1952. *Harvard Political Studies; Harvard Middle Eastern Studies*, no. 5. Harvard: Harvard University Press, p. 3.

Said, E.W. (1993) *Culture and Imperialism*. London: Knopf.

Said, E.W. (1995) *Orientalism*. London: Penguin Books.

Said, S.Z. (1999) Cairo: Rehabilitation with People Participation. *Alam Al Bena'a*, 216, pp. 6–9.

Salama, A. (1999) *Architecture Reintroduced: New Projects in Societies in Change*. Beirut: The Aga Khan Award for Architecture and American University of Beirut.

Salama, A. (2007) An Exploratory Investigation into the Impact of International Paradigmatic Trends on Arab Architecture. *GBER*, 6(1), pp. 31–43.

Sanders, P. (1994) *Ritual, Politics and the City in Fatimid Cairo*. New York: State University of New York Press.

Sandweiss, E. (2004) Framing Urban Memory: The Changing Role of History-Museums in the American City. In E. Bastea (ed.), *Memory and Architecture*. Mexico: The University of New Mexico Press, pp. 25–48.

Sannoff, H. (2007) Community-based Design Learning: Democracy and Collective Decision Making. In A. Salama and N. Wilkinson (eds), *Design Studio Pedagogy: Horizons for the Future*. Gateshead, UK: The Urban International Press, pp. 21–38.

Saunders, P. and Williams, P. (1988) The Constitution of the Home: Towards a Research Agenda. *Housing Studies*, 3(2), pp. 81–93.

Sayyid, H. (1987) The Development of the Cairene qa'a: Some Considerations. *Annales Islamogiques (AnIsl)*, 23. Le Caire: Institut Francais d'archeologie orientale.

Sayyid-Marsot, A. (1984) *Egypt in the Reign of Muhammad Ali*. Cambridge: Cambridge University Press.

Scheurman, J. (1983) *Research and Evaluation in the Human Services*. London: Free Press, Collier Macmillan.

Schneider, T. and Till, J. (2005) Flexible Housing: Opportunities and Limits. *Architectural Research Quarterly*, 9(2), pp. 157–66.

Schneider, T. and Till, J. (2007) *Flexible Housing*. London: Architectural Press.

Schneider, T. and Till, J. (2009) Beyond Discourse: Notes on Spatial Agency. *Footprint*, 4, *Agency in Architecture: Reframing Critically in Theory and Practice*. Spring Issue, pp. 97–111.

Sedky, A. (2001) The Living Past. *Al-Ahram Weekly Newspaper*, Issue 558. Cairo: Al-Ahram Publishing, p. 2.

Sedky, A. (2005) The Politics of Area Conservation in Cairo. *International Journal of Heritage Studies*, 11(2), pp. 113–30.

Sennett, R. (1977). *The Fall of Public Man*. London: Faber & Faber Publishing.

Shaarawi, H. (1986) *Harem Years: The Memoirs of an Egyptian Feminist*. Translated and Introduced by Margot Badran. New York: Feminist Press at the City University of New York.

Shehayyeb, D. (2002) *Tradition, Change, and Participatory Design: Community Participation in the Re-design of Tablita Market in Historical Cairo*. Al-Darb al-Ahmar Project paper. Cairo: Aga Khan Trust for Culture (AKTC).

Shehayeb, D. and Abdel-Hafiz, M. (ND) *Tradition, Change, and Participatory Design: Community Participation in the Re-design of Tablita Market in Historical Cairo*. A Project Report. Cairo: Aga Khan Trust for Culture and Near East Foundation.

Shidlo, G. (1990) *Housing Policy in Developing Countries*. London: Routledge.

Sibely, D. and Lowe, G. (1992) Domestic Control, Modes of Control, and Problem Behaviour. Geografiska Annaler. Series B, *Human Geography*, 74(3), pp. 189–98.

Silverman, D. (2006) *Interpreting Qualitative Data: Methods for Analysing Talk, Text and Interaction*, third edition. London: Sage Publications.

Siravo, F. (2004) Urban Rehabilitation and Community Development in al-Darb al-Ahmar. In S. Bianca and P. Jodidio (eds), *Cairo: Revitalising a Historic Metropolis*. Turin: Umberto Allemandi and C. for Aga Khan Trust for Culture, pp. 177–93.

Singerman, D. (1996) *Avenues of Participation: Family, Politics, and Networks in Urban Quarters of Cairo*. Princeton: Princeton University Press.

Singerman, D. (1998) Engaging Informality: Women, Work, and Politics in Cairo. In R. Lobban and E. Fernea (eds), *Middle Eastern Women and the Invisible Economy*. Gainesville: University of Florida Press, pp. 262–86.

Slavid, R. (2005) *Wood Architecture*. London: Laurence King Publishing.

Smith, N. (1993) Homeless/Global: Scaling Places. In J. Bird, B. Curtis, T. Putnam, G. Robertson and L. Tickner (eds), *Mapping the Futures: Local Cultures, Global Change*. London: Routledge, pp. 87–119.

Smith, P. (1973) *The Dynamics of Urbanism, The Built Environment*. London: The Anchor Press, p. 73.

Somerville, P. (1997) The Social Construction of Home. *Journal of Architectural and Planning Research*, 14(3), pp. 226–45.

Spector, T. (2001) *The Ethical Architect: The Dilemma of Contemporary Practice*. Princeton: Princeton University Press.

Staffa, S. (1977) *Conquest and Fusion: The Social Evolution of Cairo A.D. 642–1850*. Social, Economic and Political Studies of the Middle East Series, v.20. Leiden: Brill.

Stang, A. and Hawthorne, C. (2005) *The Green House: New Directions in Sustainable Architecture*. New York: Princeton Architectural Press.

Steel, J. (1997) *Sustainable Architecture: Principles, Paradigms and Case Studies*. London: McGraw-Hill.

Stieber, N. (2003) Architecture between Disciplines. *The Journal of the Society of Architectural Historians*, 26(2), pp. 176–7.

Storey, D. (2001) *Territory: The Claiming of Space*. Essex: Pearson Education Limited.

Sutton, K. and Fahmy, F. (2002) The Rehabilitation of Old Cairo. *Habitat International*, 26, pp. 73–93.

Taha, H. (2007) AlTafawut alTabaqi fi Masr, Mashrou'l infigar (Social Inequality in Egypt: A Project of Explosion). Journal Article (Arabic). *AlQuds al-Arabi Newspaper*, issue 5525, p. 12.

Taragan, H. (1999) Architecture in Fact and Fiction: The Case of the New Gourna Village in Upper Egypt. *Muqarnas*, 16, pp. 169–78.

Tietze, S. and Musson, G. (2005) Recasting the Home-Work Relationship: A Case of Mutual Adjustment? *Organization Studies*, 26(9), pp. 1331–52.

Till, J. (2005) The Negotiation of Hope. In P. Blundell Jones, D. Petrescu and J. Till (eds), *Architecture and Participation*. London: The Architectural Press, pp. 23–42.

Till, J. and Wiglesworth, S. (2001) The Future is Hairy. In J. Hill (ed.), *Architecture: The Subject is Matter*. Oxon: Routledge, pp. 11–28.

Tonkiss, F. (2005) *Space, the City and Social Theory: Social Relations and Urban Forms*. London: Polity Press, pp. 25–6.

Turner, V. (1969) *The Ritual Process: Structure and Anti-structure*. New York: Aldine de Gruyter.

Turner, V. (1980) Social Dramas and Stories about Them. *Critical Inquiry*, 7(1), pp. 141–68.

Turner, V. (1986) *The Anthropology of Performance*. New York: PJA Publications.

United Nation Centre for Human Settlements (Habitat) (1993) *Metropolitan Planning and Management in the Developing World: Spatial Decentralization Policy in Bombay and Cairo*. An UN-HABITAT Report.

Uraz, T.U. and Gulmez, N.U. (2007) Impact of Small Households on Housing Design. In D. Shehayeb (ed.), *Appropriate Home*. Cairo: HBRC-Housing & Building National Research Centre, pp. 68–80.

Urry, J. (1985) Chapter 1: Gregory, D. and Urry, J. (eds), *Social Relations and Spatial Structures*. Critical Human Geography Series. Basingstoke: Macmillan.

White, C. (2004) *The Idea of Home*. Normal, Ill.: Dalkey Archive Press.

Whitelaw, T.M. (1994) Order without Architecture: Functional, Social and Symbolic Dimensions in Hunter-Gatherer Settlement Organization. In M.P. Pearson and C. Richards (eds), *Architecture and Order: Approaches to Social Space*. London: Routledge, pp. 217–43.

Wikan, U. (1980) *Life among the Poor in Cairo*. London: Taylor & Francis.

Wigglesworth, S. and Till, J. (1998) *The Everyday and Architecture*. London: John Wiley & Sons Publishers.

Williams, C. (2006) Reconstructing Islamic Cairo: Forces at Work. In D. Singerman and P. Amar (eds), *Cairo Cosmopolitan: Politics, Culture and Urban Space in the New Globalized Middle East*. Cairo: The American University in Cairo Press, pp. 269–94.

Williamson, T.J., Radford, A. and Bennetts, H. (2003) *Understanding Sustainable Architecture*. London: Spon.

Wilson, G. (1945) *The Analysis of Social Change based on Observations in Central Africa*. Cambridge: Cambridge University Press, pp. 58–9.

Wilson, M.A. (1996) The Socialization of Architectural Preferences. *Journal of Environmental Psychology*, 16(1), pp. 33–44.

Woodland, W. and Hill, J. (2006) Water Management for Agriculture in Tunisia: Towards Environmentally Sustainable Development. In J. Hill, A. Terry and W. Woodland (eds), *Sustainable Development: National Aspiration, Local Implementation*. Aldershot: Ashgate, pp. 233–52.

B. REPORTS

GOPP (1991) Al-Darb al-Asfar Rehabilitation Project. Unpublished governmental report. Volume II. Cairo: General Organization for Physical Planning, p. 53.

National Organization for Urban Harmony (NOUH) website. Available at: http://www.urbanharmony.org/en/en_target.htm (last accessed on 1st June 2014).

UNDP and SCA (1997) Rehabilitation of Historic Cairo: Final Report. Unpublished report. Cairo: United Nation Development Program and Supreme Council for Antiquities.

UNDP and SCA (1997). *Rehabilitation of Historic Cairo*. Official report made by United Nations Development Programme (UNDP) in cooperation with Egyptian Supreme Council of Antiquities.

United Nations Report (2007) *World Urbanization Prospects*.

WCED (1990) *Our Common Future*. World Commission on Environmental and Development. Known as Brundtland Report.

C. DICTIONARIES

Cambridge Dictionary of American English. Cambridge Dictionary online. Available at: http://dictionary.cambridge.org/ (last accessed March 2013).

Ibn Mansur (1955) *Lisān al-Àrab (Arabic Dictionary)*. Beirut: Dar Sadir.

Oxford English Dictionary. Available at: www.oed.com (last accessed on 1st June 2014).

Simpson, J.A. and Weiner, E.S.C. (1989) *The Oxford English Dictionary*, second edition. Oxford: Clarendon Press.

D. NEWSPAPERS

D.1 Historical (Archives)

Al-lata'if magazine several issues, 1880s.

Al-Moqattam Newspaper, issues 3279, 80, 81 dated 8, 11, 15 January 1900. Cairo: Sarrouf and Co.

Al-Waqa'i El-Masriyyah, issue 72, dated 9 June 1886, p. 639.

Al-Waqa'i El-Masriyyah, issue 77, dated 7 July 1886, p. 639.

Al-Waqa'I al-Masriyyah, issue dated 28 September 1886.

D.2 Contemporary

Annual Magazine (1995–1997) Department of Architectural Engineering, Cairo University.

El-Aref, N. (2007) Thoroughfare. *Al-Ahram Weekly*, issue 865.

_____ (2000) Old Cairo's New Look. *Al-Ahram Weekly*, issue 478.

Unknown (2005) Crafting the Past. *Al-Ahram Weekly*, issue 754.

Unknown (2009) Decline is Not Inevitable. *Al-Ahram Weekly*, issue 969.

E. ARCHIVAL DOCUMENTS

E.1 Al-Salihiyyah al-Najmiyyah Court

Document no. 61, Record 537, dated Muharram 11th, 1218 *Hijri* (3 May 1803AD), p. 30. Housing property trade in al-Darb al-Asfar.

Document 97, Record 535, dated Muharram 17th 1207 *Hijri*, p. 56. Divorce statement document.

Document no. 102, Record 535, dated Muharram 18th, 1207 *Hijri*, p. 59. Property trade in *al-Gammliyya* quarter, Darb al-Mabyadha

Document no. 139, Record 537, dated Rabiee Thani 10, 1224 *Hijri*, p. 69. Housing property trade in al-Gammaliyya quarter.

Document 150, Record 537, dated Rajab 20, 1224 *Hijri*, p. 75. Housing property trade in al-Gammaliyya quarter.

Document no. 194, Record 537, dated Shawwal 17th 1225 Hijri (15 November 1810AD). Housing property trade in al-Darb al-Asfar.

Document no. 404, Record 535, dated Rajab 11, 1224 *Hijri*, p. 206. Housing property trade in al-Gammaliyya quarter.

Document no. 637, Record 534, dated Shawwal 1st, 1205 *Hijri* (3 June 1791AD), pp. 308–9. Housing property trade in al-Darb al-Asfar.

E.2 Al-Bab al-A'ali Court

Document no. 54, record 25 dated 1307H (1889AD). *Sigellat Masr Al-Shari'yyah* (Egypt Legal Records). Cairo: The National Centre for Archival Documents.

Document no. 64, record 24 dated 1307H (1889AD). *Sigellat Masr Al-Shari'yyah* (Egypt Legal Records). Cairo: The National Centre for Archival Documents.

Document no. 194, record 500 dated 16th Shaaban 1281H. Al-Bab al-A'ali Court Record. Cairo: The National Centre for Archival Documents.

Document no. 248, record 47, dated 1167 Hijri (1753AD), pp. 28–9.

E.3 Supreme Council of Antiquity Reports

Report of Monument no. 339, Bayt al-Suhaimy.

Report of Monument no. 471, Bayt Mustafa Ja'afar.

F. FORMAL REPORTS

F.1 Historical Reports and Decrees

Al-Waqa'I al-Munammarah, no. 64. In the decree, the numbers were written in black ink on a white background and framed in black.

Chief Architect Record: 5/2/1M (Architectural issued documents), no. 355 – Buildings, for the year 1877. Cairo: The National Centre for Archival Documents.

Decree dated 9 May 1889 (File 6/2/A, Public Works, Code: 0075-035989); 16 June 1895 (File 6/2/B, Public Works, Code: 0075-035990). Cairo: The National Centre for Archival Documents.

Decree dated 4 June 1896 (File 6/2/B, Public Works, Code: 0075-035991). Cairo: The National Centre for Archival Documents.

Decree dated 9 November 1908 (File 6/3/D, Public Works, Code: 0075-036183). Cairo: The National Centre for Archival Documents.

Decree dated 28 November 1910 (File 6/3/D, Public Works, Code: 0075-036184). Cairo: The National Centre for Archival Documents.

Decree for the Condition of Building Houses within Big Cities, 8 September 1883. Ministry of Public Works. Archival Document, file number 6/2/A Public Works (Archival Code: 0075-035972). Cairo: The National Centre for Archival Documents.

Decree for the Condition of Building Houses within Big Cities, 9 January 1899. Ministry of Public Works. Archival Document, file number 6/2/B Public Works (Archival Code: 0075036014). Cairo: The National Centre for Archival Documents.

Decree issued by the Government Cabinet Concerning the Conditions and the Structure of the *Tanzim* Department and its Responsibilities, 22 February 1882. Ministry of Public Works. Archival Document, file number 6/2/A Public Works (Archival Code: 0075-035967). Cairo: The National Centre for Archival Documents.

Decree on the Usage of Public Spaces, 23 April 1896. Ministry of Public Works. Archival Document, file number 6/2/A Public Works (Archival Code: 0075-036012). Cairo: The National Centre for Archival Documents.

Report for Restructuring of the Ministry of Public Works, 17 July 1881, Ministry of Public Works. Archival Document, file number 6/2/A Public Works (Archival Code: 0075-035966). Cairo: The National Centre for Archival Documents.

F.2 Contemporary Governmental Reports and Decrees

Cairo Governor's Decision 457 (1999).

Explanatory Report issued by Cairo Governor dated 3 September 2007, issued as an amendment to the Prime Minister decision no. 2003 for the year 2007, p. 206.

General Organization for Physical Planning GOPP (1991) *Al-Darb al-Asfar Rehabilitation Project* , vol. 2.

Minister of Culture Decision 250 (1990).

Population Census for the Year 1917, part two. Ministry of Finance. Cairo: Governmental Publishing Unit in Bulaq (1921).

Population Census for the Year 1937, part one, book 9. Ministry of Finance. Cairo: Governmental Publishing Unit in Bulaq (1940).

Prime Minister Decision no. 2,003 for the year 2007, pp. 203–4. This was the first to consider particular requirements for Old Cairo, even though it remained very general in its terms and conditions and lacked technical precision.

Rehabilitation of Historic Cairo Project (1997).

Rule no. 3 for the Year 1981 Concerning the Regulation of Urban Planning and its Implementation Procedures which were issued Under Decision no. 600 of the Minister of Urban Development and Housing for the Year 1982. Fifteenth edition, 2007, p. 14.

Temporary Building Code for Historic Cairo, June 2009.

UNDP Sponsored Projects of Historic Importance such as Gamalia Development Project (1994).

Index